METHODS IN MOLECULAR BIOLOGY

Series Editor
John M. Walker
School of Life and Medical Sciences
University of Hertfordshire
Hatfield, Hertfordshire, UK

For further volumes:
http://www.springer.com/series/7651

For over 35 years, biological scientists have come to rely on the research protocols and methodologies in the critically acclaimed *Methods in Molecular Biology* series. The series was the first to introduce the step-by-step protocols approach that has become the standard in all biomedical protocol publishing. Each protocol is provided in readily-reproducible step-by-step fashion, opening with an introductory overview, a list of the materials and reagents needed to complete the experiment, and followed by a detailed procedure that is supported with a helpful notes section offering tips and tricks of the trade as well as troubleshooting advice. These hallmark features were introduced by series editor Dr. John Walker and constitute the key ingredient in each and every volume of the *Methods in Molecular Biology* series. Tested and trusted, comprehensive and reliable, all protocols from the series are indexed in PubMed.

Pollen and Pollen Tube Analysis

Methods and Protocols

Edited by

Giampiero Cai

Department of Life Sciences, University of Siena, Siena, Italy

Chiara Piccini

National Research Council of Italy, Institute of BioEconomy (CNR-IBE), Sesto Fiorentino (FI), Italy

Veronica Conti

Department of Life Sciences, University of Siena, Siena, Italy

Sara Parri

Department of Life Sciences, University of Siena, Siena, Italy

Humana Press

Editors
Giampiero Cai
Department of Life Sciences
University of Siena
Siena, Italy

Chiara Piccini
National Research Council of Italy
Institute of BioEconomy (CNR-IBE)
Sesto Fiorentino (FI), Italy

Veronica Conti
Department of Life Sciences
University of Siena
Siena, Italy

Sara Parri
Department of Life Sciences
University of Siena
Siena, Italy

ISSN 1064-3745 ISSN 1940-6029 (electronic)
Methods in Molecular Biology
ISBN 978-1-0716-5307-4 ISBN 978-1-0716-5308-1 (eBook)
https://doi.org/10.1007/978-1-0716-5308-1

This Humana imprint is published by the registered company Springer Science+Business Media, LLC, part of Springer Nature.
The registered company address is: 1 New York Plaza, New York, NY 10004, U.S.A.

Preface

Pollen plays a fundamental role in the reproductive biology of seed plants, carrying the male sperm and enabling fertilization by developing into a pollen tube. In addition to its role in sexual reproduction, pollen has been pivotal in the evolution of terrestrial plants. Its emergence allowed plants to reproduce independently of water, enabling them to colonize diverse terrestrial habitats and driving the diversification of seed plants. This adaptation represented a significant evolutionary shift, facilitating long-distance dispersal and genetic exchange. In addition to its evolutionary significance, pollen plays a crucial role in maintaining genetic diversity, plant fitness, and ecosystem resilience. It sustains biodiversity and agricultural productivity while serving as a primary resource for pollinators, who are vital for ecosystem services. Furthermore, pollen grains are highly resilient and well-preserved in sediments, making them invaluable for paleoecological studies that reconstruct past vegetation, climate patterns, and evolution. Their biochemical composition and structural complexity also provide valuable insights into plant physiology, stress responses, and adaptation mechanisms.

At a molecular level, pollen development and pollen tube growth involve a tightly coordinated network of gene expression, signaling pathways, and structural components to ensure the successful delivery of sperm. When a pollen grain lands on a compatible stigma, it detects physicochemical signals via receptor-like kinases (RLKs) and lipid-binding proteins in its plasma membrane. These receptors then activate intracellular signaling cascades, which are mediated by calcium ions (Ca^{2+}), reactive oxygen species (ROS), and phosphoinositides. These signaling cascades regulate pollen hydration and germination, initiating the transition from a quiescent grain to an actively growing pollen tube. A crucial aspect of this process is the increased expression of genes that encode cell wall-modifying enzymes, such as pectin methylesterases. These enzymes remodel the pollen tube cell wall, ensuring its flexibility and rigidity. Maintaining appropriate cell wall architecture is essential, as the pollen tube must overcome mechanical resistance within the tissues of the stigma and style. The pollen tube exhibits polarized tip growth, which is driven by a dynamic actin cytoskeleton and regulated by Rho-like GTPases (ROPs). These ROPs control the trafficking of vesicles and their exocytosis at the growing tip. These vesicles deliver cell wall precursors, such as pectins, as well as cell wall synthesizing and modifying proteins to maintain the integrity and flexibility of the tube. Calcium gradients at the tip, which are maintained by Ca^{2+} channels and pumps, are essential for growth oscillations and directional guidance. Pollen tube development is genetically dependent on the expression of numerous male gametophyte-specific genes, including those involved in signal perception (e.g., LURE peptide receptors), cytoskeletal organization, and energy metabolism. Additionally, gamete fusion genes guarantee the delivery of sperm cells and fertilization. The interaction between pollen and pistil also involves species-specific signaling molecules. The pollen tube secretome, for example, might play a crucial role in fertilization by mediating communication between the male gametophyte and female tissues. Peptides are also secreted by synergid cells in the ovule and guide the pollen tube through the style toward the micropyle. This process relies on receptor complexes in the pollen tube membrane, as well as downstream signaling via MAPK cascades and second messengers.

Due to its intricate molecular and genetic mechanisms, the pollen tube is a powerful model for studying cell polarity, tip growth regulation, vesicle trafficking, and intercellular communication in plants. Beyond its value in basic research, pollen and pollen tube biology have significant implications for the health of ecosystems and environmental sustainability. Pollination is fundamental to terrestrial biodiversity as it enables sexual reproduction in seed plants, ensuring genetic diversity and resilience against environmental stress. This process supports the survival of plant species, which sustain food networks, stabilize soils, and contribute to carbon sequestration.

Understanding pollen viability, germination, and tube development, as well as their ecological roles, is essential in both fundamental and applied plant sciences. However, it is not enough to rely on traditional techniques alone to fully exploit this knowledge, given that methodological innovation in pollen biology is advancing rapidly, and new tools and protocols are emerging almost daily. Modern approaches, ranging from high-resolution imaging and live-cell tracking to omics technologies and computational modeling, are becoming increasingly sophisticated and more accessible to a broader scientific community. This rapid evolution creates both opportunities and challenges. Although researchers can now explore pollen biology with precision, the rapid pace of technological progress requires organization, evaluation, and integration of updated methods to ensure reproducibility and relevance. This book meets this need by bringing together the latest methodologies for studying pollen and pollen tubes. It covers structural, biochemical, and molecular approaches. This guide serves as a practical resource for researchers looking for up-to-date and state-of-the-art tools to advance the field of plant reproductive biology, by providing detailed protocols and explaining their significance.

Studying pollen viability and germination of pollen is crucial because these factors determine the reproductive success of seed plants, directly impacting fertilization efficiency, seed formation, and crop productivity. Furthermore, pollen viability and germination serve as sensitive indicators of plant health and environmental stress, offering insights for agriculture, biodiversity conservation, and ecological research. Chapter 1 introduces methods for assessing pollen and pollen tube viability. This is fundamental for determining reproductive success and evaluating stress tolerance in crops. Without accurate viability tests, breeding programs cannot reliably predict outcomes. Chapter 2 complements this by providing species-specific protocols for rice pollen viability and germination. These are essential for cereal breeding and ensuring high seed-set under variable environmental conditions.

Imaging and microscopy techniques are essential for studying pollen, as they enable the precise visualization of pollen grain structure, viability, and tube growth dynamics. High-resolution imaging provides insights into cellular organization, tip growth, and vesicle trafficking, while advanced microscopy methods enable the real-time observation of physiological processes. Chapter 3 introduces multispectral imaging flow cytometry as a method for the high-throughput characterization of pollen, enabling the rapid and quantitative analysis of morphological and internal features. This technique is particularly valuable for biodiversity studies and large-scale phenotyping. Chapter 4 describes the use of electron microscopy for the ultrastructural analysis of pollen grains and tubes. This technique, which is continuously renewed, provides information on the distribution of organelles and the architecture of the cell wall, both of which are critical for understanding growth mechanics. Chapter 5 explains the fluorescence and immunofluorescence techniques used to determine the spatial localization of proteins and cellular components in pollen and other plant cell types. Chapter 6 focuses on the use of automated fluorescence measurements to quantify dynamic processes such as vesicle trafficking and cytoskeletal rearrangements in defined regions of pollen tubes.

Analyzing the biochemical and molecular composition of pollen and pollen tubes is crucial for understanding the mechanisms underlying pollen viability, germination, and fertilization. Analysis reveals the roles played by proteins, lipids, carbohydrates, and signaling molecules in tube growth, guidance, and sperm delivery. Chapter 7 analyzes the sugars, lipids, and metabolites found in pollen tubes, which are essential for sustaining rapid tip growth and signaling. Metabolic profiling reveals how energy and structural precursors are allocated during elongation. Chapter 8 introduces FTIR and Raman spectroscopy as methods for the nondestructive chemical characterization of pollen grains. These methods are crucial for species identification and stress diagnostics in ecological and agricultural applications. Chapter 9 outlines the process of quantitative protein analysis using Western blotting. This technique allows researchers to study signaling pathways and structural proteins that regulate pollen tube growth and fertilization.

Studying the cellular dynamics and growth regulation of pollen and pollen tubes is crucial to understanding the mechanisms that drive polarized cell expansion, intracellular trafficking, and signal integration during fertilization. These processes are not only fundamental for successful reproduction, but also for broader aspects of plant development, stress adaptation, and evolutionary biology. Chapter 10 outlines methods for measuring reactive oxygen species, which play a crucial role in regulating tip growth and signaling. Imbalances in ROS can impair fertilization, making their quantification essential for understanding stress responses. Chapter 11 describes techniques for quantifying ion fluxes, such as calcium, that control tip growth and guidance. These measurements are essential for functional phenotyping and modeling pollen tube behavior. Chapter 12 examines the structural organization of the pollen tube tip, emphasizing the cytoskeletal and membrane structures that facilitate polarized growth, a prerequisite for successful fertilization. Chapter 13 focuses on measuring lipid–protein interactions. This process regulates two critical processes for maintaining growth rates: vesicle trafficking and cell wall expansion. Chapter 14 focuses on the composition and remodeling of the pollen tube cell wall. This determines the tube's mechanical properties and influences its ability to penetrate female tissues.

Studying pollen tube attraction and transformation, as well as the proteins released during this process, is essential for understanding the molecular dialogue between male and female tissues during fertilization. These processes involve signaling peptides, cell wall-modifying enzymes, and extracellular proteins that guide tube growth and enable sperm delivery. Chapter 15 explains how to measure self-incompatibility, which is a mechanism that prevents self-fertilization and promotes genetic diversity. Understanding this process is essential for developing effective breeding strategies and overcoming reproductive barriers. Chapter 16 describes in vitro and in vivo assays for pollen tube attraction in *Arabidopsis* and maize. These assays are essential for studying guidance mechanisms and the molecular dialogue between male and female tissues. Chapter 17 introduces the transient transformation of tobacco pollen tubes for the study of protein localization and interactions. This provides a powerful tool for functional genomics and live-cell imaging. Chapter 18 explores extracellular vesicles and the pollen tube secretome. These mediate communication with female tissues and influence reproductive success, offering new insights into intercellular signaling.

Studying the palynology and ecological applications of pollen and pollen tube analysis is important because it provides valuable insights into plant biodiversity, species distribution, and ecosystem dynamics. Pollen records can be used to reconstruct past vegetation and climate patterns, and current pollen studies can inform conservation strategies and pollination networks. Chapter 19 focuses on palynology, including reference to pollen collections and fresh preparations. These support taxonomy, paleoecology, and forensic studies. These resources are invaluable for identifying species and reconstructing the environment. Chapter 20 outlines the process of extracting pollen from mosses and sediments for palynological analysis. This analysis provides valuable insights into past vegetation and climate change, both of which are essential for comprehending long-term ecological dynamics.

Together, these chapters provide researchers specializing in pollen biology, plant reproduction, and related fields with practical protocols and conceptual insights. The methods described range from advanced imaging and biochemical analyses to ecological and evolutionary approaches. These tools enable researchers to explore pollen function at levels ranging from cellular to the ecosystem. By integrating structural, physiological, and molecular perspectives, this volume aims to support fundamental research and applied studies in agriculture, biodiversity, and plant breeding.

Siena, Italy

Sara Parri
Chiara Piccini
Veronica Conti
Giampiero Cai

Contents

Contributors

JEREMY ADLER • *Department of Medical Biochemistry and Cell Biology, Institute of Biomedicine, University of Gothenburg, Göteborg, Sweden*
LIDYANNE YURIKO SALEME AONA • *Universidade Federal do Recôncavo da Bahia (UFRB), Rua Rui Barbosa, s/n, Centro, 45300-000 Cruz das Almas, Bahia, Brazil*
STEFANIA BENNICI • *Department of Agriculture, Food and Environment, University of Catania, Catania, Italy*
DANIEL BOUYER • *Ecole Normale Supérieure de Lyon, Lyon Cedex 07, France*
LORENZO BRAGA • *Laboratorio di Palinologia e Paleobotanica, Dipartimento Scienze Vita, Università di Modena e Reggio Emilia, Modena, Italy*
MARIA BREYGINA • *Department of Plant Physiology, Biological Faculty, Lomonosov Moscow State University, Moscow, Russia*
GIAMPIERO CAI • *Department of Life Sciences, University of Siena, Siena, Italy*
ASLIHAN ÇETINBAŞ-GENÇ • *Department of Biology, Marmara University, Kadıköy, Istanbul, Türkiye*
JUNYI CHEN • *Key Laboratory of Pesticide and Chemical Biology of Ministry of Education, Hubei Key Laboratory of Genetic Regulation and Integrative Biology, School of Life Sciences, Central China Normal University, Wuhan, China*
ELEONORA CLÒ • *Laboratorio di Palinologia e Paleobotanica, Dipartimento Scienze Vita, Università di Modena e Reggio Emilia, Modena, Italy; National Biodiversity Future Center (NBFC), Palermo, Italy*
VERONICA CONTI • *Department of Life Sciences, University of Siena, Siena, Italy*
ERIN CULLEN • *American Society of Plant Biologists, Rockville, MD, USA*
DANIEL SANTA CRUZ DAMINELI • *Departamento de Bioquímica, Instituto de Química, Universidade de São Paulo, São Paulo, SP, Brazil*
EVERTON HILO DE SOUZA • *Fundação Getulio Vargas (FGV Conhecimento), Rua Jornalista Orlando Dantas, Rio de Janeiro, Rio de Janeiro, Brazil; Universidade Federal do Recôncavo da Bahia (UFRB), Rua Rui Barbosa, s/n, Centro, 45300-000 Cruz das Almas, Bahia, Brazil*
STEFANO DEL DUCA • *University of Bologna, Department of Biological, Geological, and Environmental Sciences, Bologna, Italy; University of Bologna, Interdepartmental Centre for Agri-Food Industrial Research, Cesena, Italy*
MARTINA DENAMI • *Laboratorio di Palinologia e Paleobotanica, Dipartimento Scienze Vita, Università di Modena e Reggio Emilia, Modena, Italy*
GAETANO DISTEFANO • *Department of Agriculture, Food and Environment, University of Catania, Catania, Italy*
THOMAS DRESSELHAUS • *Plant Cell Biology, Biochemistry, and Biotechnology, Institute of Plant Sciences, University of Regensburg, Regensburg, Germany*
SUSANNE DUNKER • *Helmholtz Centre for Environmental Research – UFZ, Leipzig, Germany; German Centre for Integrative Biodiversity Research – iDiv, Halle-Jena-Leipzig, Leipzig, Germany*
KEFENG FANG • *College of Landscape Architecture, Beijing University of Agriculture, Beijing, China*

JOSÉ A. FEIJÓ • *Cell Biology and Molecular Genetics Department, University of Maryland, College Park, MD, USA*
ASSUNTA FLORENZANO • *Laboratorio di Palinologia e Paleobotanica, Dipartimento Scienze Vita, Università di Modena e Reggio Emilia, Modena, Italy*
CAROLIN FRITZ • *Cell Biology, Department of Biology, Friedrich-Alexander-Universität Erlangen-Nürnberg, Erlangen, Germany*
ALESSANDRA GENTILE • *Department of Agriculture, Food and Environment, University of Catania, Catania, Italy*
DANIIL GORNOV • *Laboratorio di Palinologia e Paleobotanica, Dipartimento Scienze Vita, Università di Modena e Reggio Emilia, Modena, Italy*
THOMAS HORNICK • *Helmholtz Centre for Environmental Research – UFZ, Leipzig, Germany; German Centre for Integrative Biodiversity Research – iDiv, Halle-Jena-Leipzig, Leipzig, Germany*
JI-HYUN KIM • *Department of Life Science and Environmental Biochemistry, Life and Industry Convergence Research Institute, Pusan National University, Miryang, Republic of Korea*
YU-JIN KIM • *Department of Life Science and Environmental Biochemistry, Life and Industry Convergence Research Institute, Pusan National University, Miryang, Republic of Korea*
ACHIM KOHLER • *Faculty of Science and Technology, Norwegian University of Life Sciences, Ås, Norway*
BENEDIKT KOST • *Cell Biology, Department of Biology, Friedrich-Alexander-Universität Erlangen-Nürnberg, Erlangen, Germany*
AGATA LESZCZUK • *Institute of Agrophysics, Polish Academy of Sciences, Lublin, Poland*
MANEESH LINGWAN • *Donald Danforth Plant Science Center, St. Louis, MO, USA*
NEETA LOHANI • *Department of Biotechnology, Thapar Institute for Engineering and Technology, Patiala, Punjab, India*
OKSANA LUNEVA • *Department of Biophysics, Biological Faculty, Lomonosov Moscow State University, Moscow, Russia*
ANNA MARIA MERCURI • *Laboratorio di Palinologia e Paleobotanica, Dipartimento Scienze Vita, Università di Modena e Reggio Emilia, Modena, Italy; National Biodiversity Future Center (NBFC), Palermo, Italy*
ONDŘEJ NOVOTNÝ • *Institute of Experimental Botany of the Czech Academy of Sciences, Prague, Czech Republic; Department of Biochemistry and Microbiology, University of Chemistry and Technology, Prague, Czech Republic*
INGELA PARMRYD • *Department of Medical Biochemistry and Cell Biology, Institute of Biomedicine, University of Gothenburg, Göteborg, Sweden*
SARA PARRI • *Department of Life Sciences, University of Siena, Siena, Italy*
PŘEMYSL PEJCHAR • *Institute of Experimental Botany of the Czech Academy of Sciences, Prague, Czech Republic*
CHIARA PICCINI • *National Research Council of Italy, Institute of BioEconomy (CNR-IBE), Sesto Fiorentino (FI), Italy*
ANNA PODOBEDOVA • *Department of Plant Physiology, Biological Faculty, Lomonosov Moscow State University, Moscow, Russia*
MARIA TERESA PORTES • *Departamento de Botânica, Instituto de Biociências, Universidade de São Paulo, São Paulo, SP, Brazil*

Martin Potocký • *Institute of Experimental Botany of the Czech Academy of Sciences, Prague, Czech Republic; Department of Experimental Plant Biology, Faculty of Science, Charles University, Prague, Czech Republic*

Wantong Qu • *Key Laboratory of Pesticide and Chemical Biology of Ministry of Education, Hubei Key Laboratory of Genetic Regulation and Integrative Biology, School of Life Sciences, Central China Normal University, Wuhan, China*

Eleonora Rattighieri • *Laboratorio di Palinologia e Paleobotanica, Dipartimento Scienze Vita, Università di Modena e Reggio Emilia, Modena, Italy*

Cristina Ricucci • *Laboratorio di Palinologia e Paleobotanica, Dipartimento Scienze Vita, Università di Modena e Reggio Emilia, Modena, Italy*

Mônica Lanzoni Rossi • *Universidade de São Paulo (USP), Av. Centenário, Piracicaba, São Paulo, Brazil*

Patricia L. Seitz • *Plant Cell Biology, Biochemistry, and Biotechnology, Institute of Plant Sciences, University of Regensburg, Regensburg, Germany*

Eliška Škrabálková • *Institute of Experimental Botany of the Czech Academy of Sciences, Prague, Czech Republic; Department of Experimental Plant Biology, Faculty of Science, Charles University, Prague, Czech Republic*

Taliane Leila Soares • *Embrapa Mandioca e Fruticultura, Rua Embrapa, Cruz das Almas, Bahia, Brazil*

Fernanda Vidigal Duarte Souza • *Embrapa Mandioca e Fruticultura, Rua Embrapa, Cruz das Almas, Bahia, Brazil*

Chiara Suanno • *University of Bologna, Department of Biological, Geological, and Environmental Sciences, Bologna, Italy*

Yifang Sui • *Plant Cell Biology, Biochemistry, and Biotechnology, Institute of Plant Sciences, University of Regensburg, Regensburg, Germany*

Ewa Szczuka • *Department of Cell Biology, Maria Curie-Skłodowska University, Lublin, Poland*

Paola Torri • *Laboratorio di Palinologia e Paleobotanica, Dipartimento Scienze Vita, Università di Modena e Reggio Emilia, Modena, Italy*

Elisabetta A. M. Verderio • *University of Bologna, Department of Biological, Geological, and Environmental Sciences, Bologna, Italy; Nottingham Trent University, Interdisciplinary Biomedical Research Centre, School of Science and Technology, Nottingham, UK*

Franziska Walther • *Helmholtz Centre for Environmental Research – UFZ, Leipzig, Germany; German Centre for Integrative Biodiversity Research – iDiv, Halle-Jena-Leipzig, Leipzig, Germany*

Jessica Zappa • *Laboratorio di Palinologia e Paleobotanica, Dipartimento Scienze Vita, Università di Modena e Reggio Emilia, Modena, Italy*

Weiwei Zhang • *College of Plant Science and Technology, Beijing University of Agriculture, Beijing, China*

Liang-Zi Zhou • *State Key Laboratory of Wheat Improvement, College of Life Sciences, Shandong Agricultural University, Tai'an, China*

Boris Zimmermann • *Faculty of Science and Technology, Norwegian University of Life Sciences, Ås, Norway*

Chapter 1

Assessment of Pollen Viability and Germination Through Histochemical, In Vivo and In Vitro Methods

Everton Hilo de Souza, Taliane Leila Soares, Mônica Lanzoni Rossi, Lidyanne Yuriko Saleme Aona, and Fernanda Vidigal Duarte Souza

Abstract

Pollen viability is a key parameter for understanding the functioning of plant reproductive systems and for supporting strategies aimed at genetic resource conservation and plant breeding. Accurate assessments of this attribute are essential for identifying genotypes with a higher reproductive rate, monitoring prezygotic barriers, testing compatibility between individuals, and planning controlled crosses, especially in cases of floral asynchrony or spatial isolation. Accordingly, this chapter compiles the main methods used for this purpose, providing detailed and up-to-date protocols. It covers histochemical tests using different stains (TTC, FDA, modified Alexander stain, and acetocarmine), in vitro pollen germination assays using specific culture media (BK, BM, RM, and PEG), and in vivo analysis of pollen tube growth through fluorescence microscopy with aniline blue staining, as well as viability tests based on fruit and seed set. Each methodology is presented with corresponding materials, procedures, and explanatory notes. The techniques described are broadly applicable and can be adapted to a wide range of plant species and research objectives.

Key words Histochemical staining, In vitro pollen germination, Fluorescence microscopy, Pollen tube growth, Reproductive biology, Conservation, Plant breeding

1 Introduction

Pollen grain viability is crucial for studies of floral and reproductive biology, with direct applications in plant breeding programs, genetic resource conservation, and species management strategies, particularly for agronomically important plant species [1–5]. An accurate assessment of pollen viability allows researchers to infer a species' fertilization capacity, contributing to the elucidation of reproductive mechanisms for the success of artificial pollination [4–6]. Furthermore, it is an essential tool for selecting genotypes with higher reproductive fitness and is thus highly valuable for germplasm banks and hybridization programs [6–8].

Giampiero Cai et al. (eds.), *Pollen and Pollen Tube Analysis: Methods and Protocols*, Methods in Molecular Biology, vol. 3042,
https://doi.org/10.1007/978-1-0716-5308-1_1,

A wide range of methodologies has been developed to assess pollen viability from different plant species. Each has its own advantages, limitations, and specific applications. Histochemical tests are widely used in the determination of pollen viability because of their easy operation and short, time-consuming process. Several dyes are used to measure viability, such as 2,3,5-triphenyltetrazolium chloride (TTC), Alexander's stain, fluorescein diacetate (FDA), and acetocarmine, which offer rapid and relatively low-cost evaluations based on cell integrity, the presence of viable cytoplasmic content, or metabolic activity [2, 4–8]. However, studies have shown that these tests may overestimate viability compared to more direct approaches, such as in vitro pollen germination [4–8].

In vitro pollen germination provides a more direct estimate of germinative potential by simulating stigma conditions and enabling observation of pollen tube growth in culture medium, making it a more accurate method [4–12].

In vivo fertilization testing via fluorescence microscopy following aniline blue staining allows for the direct visualization of pollen grain germination on the stigma and the subsequent growth of pollen tubes through the pistil [13–16]. This technique is also employed in studies of reproductive compatibility, prezygotic barriers, and pollen performance under different conditions [13–16], providing comprehensive information about pollen viability and fertilization potential [7, 13–16].

Pollen viability analysis is also critical for ex situ conservation strategies, such as pollen cryopreservation, which enables long-term storage of male gametes without significant loss of viability [3, 17–21]. When associated with pollen viability assessments, cryopreservation allows for the establishment of pollen cryobanks that can be used in breeding programs, facilitating crosses between individuals that flower asynchronously or are geographically distant [3, 20].

Considering the diversity of available methods for assessing pollen viability, the choice of the most appropriate technique depends on the study's objective, the species, and the available financial resources. Integrating different approaches, such as histochemical tests, in vitro germination, and in vivo fertilization, can offer a more comprehensive understanding of pollen viability to support breeding and conservation programs as well as studies on floral and reproductive biology [1–21].

This chapter aims to present and thoroughly describe the main methods and protocols used to evaluate pollen viability, including histochemical tests, in vitro germination assays, and in vivo fertilization analysis via fluorescence microscopy. To this end, it provides a comprehensive and up-to-date methodological reference applicable to both basic research and practical applications in genetic resource conservation and breeding program development.

These methodologies can be adapted according to the specific goals of the study, the biological characteristics of the plant species, and the available resources. Factors such as pollen morphology, physiological development stage, longevity, desiccation sensitivity, and stigma receptivity may influence the choice and effectiveness of each technique. Likewise, logistical aspects, including laboratory infrastructure, reagent availability, and experimental scale, must also be considered. The flexible nature of these protocols allows for their adjustment and optimization across a wide range of experimental conditions, ensuring reliable and reproducible results in both basic and applied research contexts.

2 Materials

Prepare all solutions using ultrapure water and analytical grade reagents. Diligently follow all waste disposal regulations when disposing of waste materials.

2.1 Histochemical Tests

1. 1% TTC sucrose Solution [22]: 0.01 g/mL 2,3,5-Triphenyltetrazolium chloride, 0.05 g/mL sucrose, distilled water (*see* **Note 1**). Prepare the TTC-sucrose solution by dissolving 1 g TTC and 5 g sucrose in 100 mL of distilled water. Stir using a magnetic stirrer until fully dissolved. Adjust pH to 7.0 if needed (*see* **Note 2**). Use it fresh or store it at 4 °C protected from light for no longer than 48 h.
2. FDA Solution [23]: Prepare the stock solution by adding 1 mg fluorescein diacetate in 1 mL of acetone and store it in amber vials at −20 °C (*see* **Note 3**). Prepare the working solution just before use by adding 1–2 μL of stock to 1 mL of 0.1 M phosphate buffer, pH 7.4. Mix gently and protect from light. Use immediately (*see* **Note 4**).
3. Modified Alexander Stain Solution [24, 25]: 10 mL 95% ethanol, 1 mL glacial acetic acid, 25 mL glycerol, 25 mg brilliant green, 10 mg acid fuchsin, 5 mL distilled water (*see* **Notes 5** and **6**). First, dissolve brilliant green and acid fuchsin in 95% ethanol in a beaker. Then add glacial acetic acid, glycerol, and distilled water, mixing thoroughly either with a magnetic stirrer or by vortexing. Store the solution in a tightly sealed amber bottle at room temperature and protect it from light. Mix well before each use.
4. Acetocarmine Solution [26]: 0.01 g/mL carmine, glacial acetic acid. In a fume hood, place 1 g of carmine powder in a beaker and add 100 mL of glacial acetic acid. Gently heat the mixture (without boiling) while stirring continuously until the dye is completely dissolved (*see* **Note 7**). Filter the solution through filter paper to remove undissolved particles. Store the filtered solution in an amber glass bottle at room temperature (*see* **Note 8**).

5. 0.1 M Phosphate Buffer, pH 7.4 [23]: 0.614 g $Na_2HPO_4{\cdot}2H_2O$, 0.392 g $NaH_2PO_4{\cdot}H_2O$, distilled water, up to 100 mL total volume. Stir the solution with a magnetic stirrer until fully dissolved. Use the buffer fresh, or store in tightly closed vials at 4 °C for up to 1 week. For longer storage, freeze in aliquots at −20 °C (*see* **Note 9**).
6. Glass microscope slides.
7. Cover slips.
8. Fine-tip paintbrush.
9. Pipettes and micropipette tips.
10. Filter paper.
11. pH meter.
12. Precision scale.
13. Beaker.
14. Erlenmeyer flask.
15. Spatulas.
16. Magnetic stirrer.
17. Tweezers.
18. Fume hood (acetocarmine).
19. Light microscope.
20. Fluorescence microscope (for the FDA method).
21. Incubator or heating plate (for TTC and Alexander).

2.2 In Vitro Germination

1. BK culture medium [9]: 100 mg/L H_3BO_3, 300 mg/L $Ca(NO_3)_2{\cdot}4H_2O$, 200 mg/L $MgSO_4{\cdot}7H_2O$, 100 mg KNO_3, 15% sucrose, 0.8% agar as solidifier, distilled water, pH to 6.5 (*see* **Note 10**).
2. BM culture medium [11]: 100 mg/L H_3BO_3, 15% sucrose, 0.8% agar as solidifier, distilled water, pH 6.5 (*see* **Note 11**).
3. RM culture medium [27]: 10 mg/L H_3BO_3, 362 mg/L $CaCl_2{\cdot}6H_2O$, 0.01% KNO_3, 60–130 mg/L Tris, 30% sucrose, 0.8% agar as solidifier, distilled water, pH 6.5 (*see* **Note 12**).
4. PEG culture medium [28]: 0.1 to 1.1 M PEG 400, 100 mg/L $H_3BO_3KNO_3$, 15% sucrose, 0.8% agar as solidifier, distilled water, pH 6.5 (*see* **Note 13**).
5. Precision scale.
6. Beaker.
7. Erlenmeyer flask.
8. Magnetic stirrer.
9. Spatulas.
10. Petri dishes.

11. Small tweezers.
12. Autoclave (*see* **Note 14**).
13. BOD incubator.
14. Laminar flow chamber.
15. Stereoscopic microscope.
16. 0.05% Toluidine blue (*see* **Note 15**).
17. Glass funnel.
18. Filter paper.
19. Amber flask.

2.3 In Vivo Pollen Tube Growth Assessment via Fluorescence Microscopy

1. Dissecting tools (forceps and scalpel).
2. Histological microscope slides and coverslips.
3. Autoclave.
4. Refrigerator (−5 °C).
5. Light-protected containers.
6. Fluorescence microscope with UV filter (*see* **Note 16**).
7. Fixative Solution: Carnoy 3:1. Mix 75 mL absolute ethanol and 25 mL glacial acetic acid (*see* **Note 17**).
8. Tissue Softening Solution: 10% Sodium Sulfite. Dissolve 10 g Na_2SO_3 in 100 mL distilled water (*see* **Note 18**).
9. Aniline Blue Staining Solution: 0.01% in 0.1 M K_3PO_4 buffer. Dissolve 2.1 g K_3PO_4 in 100 mL distilled water, then add 0.01 g aniline blue [15, 29] (*see* **Note 19**).

2.4 In Vivo Viability Testing and Fruit and Seed Set Assessment

1. Tweezers and a scalpel.
2. Fine paintbrush or toothpicks (for pollen transfer).
3. Fine mesh bags (e.g., voile) or paper envelopes.
4. Labels.
5. Tags.
6. Strings.
7. Permanent markers (for identifying pollination treatments).
8. Petri dishes.
9. Growth chamber or greenhouse (*see* **Note 20**).

3 Methods

3.1 Histochemical Staining with the TTC Sucrose Solution

1. Collect fresh pollen using tweezers (Fig. 1a) or a fine-tip paintbrush.
2. Place the pollen grains on a clean microscope slide.

3. Add 1–2 drops of TTC sucrose solution (Fig. 1b).
4. Incubate at 30–37 °C for 1–4 h in a humid chamber (*see* **Notes 21** and **22**).
5. Observe under a light microscope (Fig. 1c).
6. Viable pollen grains will appear red (Fig. 1d), indicating dehydrogenase activity and metabolic viability.

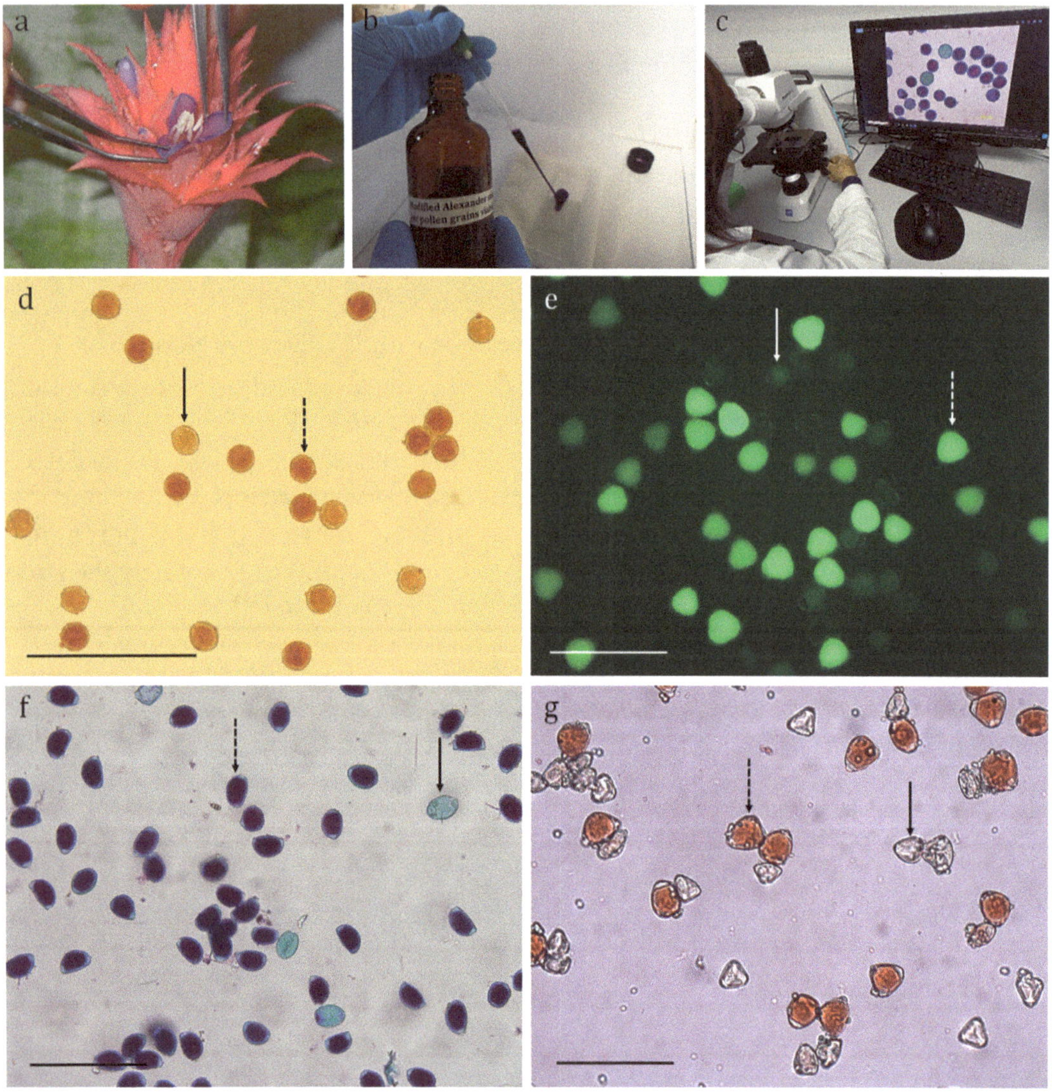

Fig. 1 Pollen viability assessed by histochemical tests. (**a**) Pollen grain collection from a bromeliad flower. (**b**) Deposition of pollen grains and modified Alexander stain solution onto a microscope slide. (**c**) Observation of pollen viability under a light microscope equipped with an image capture system. (**d**) Banana pollen grains stained with TTC sucrose solution. (**e**) Eucalyptus pollen grains stained with FDA solution. (**f**) Bromeliad pollen grains stained with modified Alexander solution. (**g**) Eucalyptus pollen grains stained with acetocarmine solution. Non-viable pollen (full arrow); viable pollen (dotted arrow). Bars: (**d**, **e**) 200 μM, (**f**, **g**) 100 μM

3.2 Histochemical Staining with the FDA Solution

1. Place fresh pollen on a clean slide.
2. Add 1–2 drops of freshly prepared FDA working solution.
3. Incubate in the dark at room temperature for 5–15 min.
4. Observe under a fluorescence microscope.
5. Viable pollen fluoresces bright green; non-viable pollen shows no fluorescence (Fig. 1e).

3.3 Histochemical Staining with Modified Alexander Stain Solution

1. Place pollen on a microscope slide.
2. Add 1–2 drops of modified Alexander stain.
3. Heat gently (do not boil) to enhance staining (*see* **Note 23**).
4. Cover with a coverslip.
5. Observe under a light microscope.
6. Viable pollen stains red/purple; non-viable pollen appears blue-green (Fig. 1f).

3.4 Histochemical Staining with Acetocarmine Solution

1. Place fresh pollen grains on a microscope slide.
2. Add 1–2 drops of acetocarmine solution.
3. Heat gently if needed to improve stain penetration (avoid boiling).
4. Cover with a coverslip.
5. Observe under a light microscope.
6. Viable pollen grains will stain deep red, indicating intact cytoplasm; aborted or empty grains remain lightly stained or colorless (Fig. 1g).

3.5 In Vitro Germination of the Pollen Grains

1. With the aid of tweezers and a fine-tip paintbrush, distribute the pollen grains uniformly on the surface of the culture medium in sterile Petri dishes (Fig. 2a, *see* **Note 24**).
2. Keep the Petri dishes in the dark inside a BOD incubator set at 27 ± 1 °C (*see* **Note 25, 26**).
3. Germination should be evaluated within 8–24 h after inoculation (*see* **Note 27**).
4. Spread droplets of Toluidine Blue Solution homogeneously in each dish to improve contrast (Fig. 2b, *see* **Note 28**).
5. Germination can be evaluated by direct counting or using photomicrographs taken with the aid of a stereomicroscope (Fig. 2c). All pollen grains within each field of view (or in the entire dish, when feasible) should be counted. A grain is considered germinated when the pollen tube is equal to or longer than the diameter of the pollen grain (Fig. 2d, e, *see* **Note 29**).

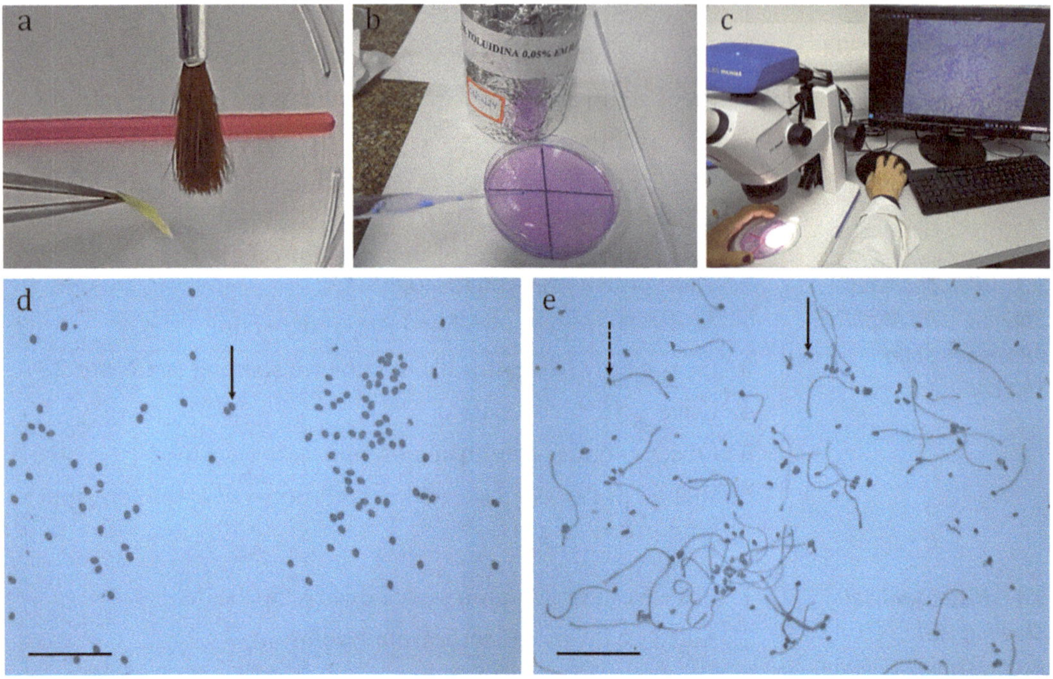

Fig. 2 In vitro pollen germination procedure. (**a**) Distribution of Bromeliad pollen grains on BK culture medium. (**b**) Staining of pollen grains and pollen tubes with toluidine blue to enhance contrast for stereomicroscopic observation. (**c**) Image capture under a stereomicroscope of *Eucaliptus* in vitro germinated pollen grains after 24 h of incubation. (**d**) Image of non-viable bromeliad pollen grains. e) Image of bromeliad pollen grains with high viability and well-developed pollen tubes. Non-viable pollen (full arrow); viable pollen (dotted arrow). Bars: 0.5 mm

3.6 In Vivo Viability Testing and Fruit and Seed Set Assessment

1. Controlled pollinations should be performed according to the study objectives. For cross-pollinations (xenogamy), recipient flowers (female parental) from different plants of the same species, preferably from distinct populations (*see* **Note 30**), must be emasculated the day before anthesis (pre-anthesis) to prevent self-fertilization. In self-pollinations, flowers should not be emasculated.
2. Protect the flowers with voile bags to prevent contamination by pollen grains from other species (Fig. 3a, *see* **Note 31**).
3. Collect fresh pollen grains from donor plants on the day of pollination. Using tweezers or a fine brush, carefully distribute the fresh pollen grains onto the surface of the stigmas (Fig. 3b).
4. Protect the flowers again with voile bags to prevent contamination by undesired pollen.
5. Identify and label the pollinated flowers (Fig. 3c, *see* **Note 32**).
6. After fruit maturation, evaluate the fruit set and count the number of seeds per fruit (Fig. 3d, *see* **Note 33**).

Fig. 3 In vivo viability test and assessment of fruit and seed set. (**a**) Bromeliad inflorescence covered with a voile bag to prevent uncontrolled pollination. (**b**) Cross-pollination procedure between flowers of the same species. (**c**) Inflorescence with pollinated and properly labeled flowers. (**d**) Fruits formed with viable seeds after pollination, demonstrating the efficiency and pollen viability

3.7 In Vivo Pollen Tube Growth Assessment Using Aniline Blue Staining and Fluorescence Microscopy

1. Collect samples of pollinated flowers at specific time intervals after pollination (e.g., 6, 12, 24 h) to analyze pollen tube growth. Sampling times should be adjusted according to the species studied and the research objectives. Samples must be immediately fixed for subsequent processing (*see* **Note 34**).
2. Collect pistils (including stigma, style, ovary). If the pistil is large, it is necessary to dissect the structures separately to facilitate handling (*see* **Note 35**).
3. Fix in Carnoy's solution for 1–24 h at 4 °C. For long storage, transfer to 70% ethanol (Fig. 4a, *see* **Note 36**).
4. Transfer pistils to 10% sodium sulfite solution (*see* **Note 37**).
5. Autoclave samples at 120 °C for 20 min (time depends on tissue hardness) (*see* **Note 38**).
6. Immerse pistils in 0.01% aniline blue solution for 4 h or overnight at −5 °C in the dark (*see* **Note 39**).
7. Mount on microscope slides, cover with a coverslip, and gently squash tissues (Fig. 4b, *see* **Note 40**).
8. Observe under a fluorescence microscope using a UV filter (Fig. 4c, *see* **Notes 41–43**).

4 Notes

1. The presence of sucrose enhances osmotic balance and promotes enzymatic activity in pollen grains, increasing staining efficiency. Incubation time may require adjustment depending on the species and pollen condition.

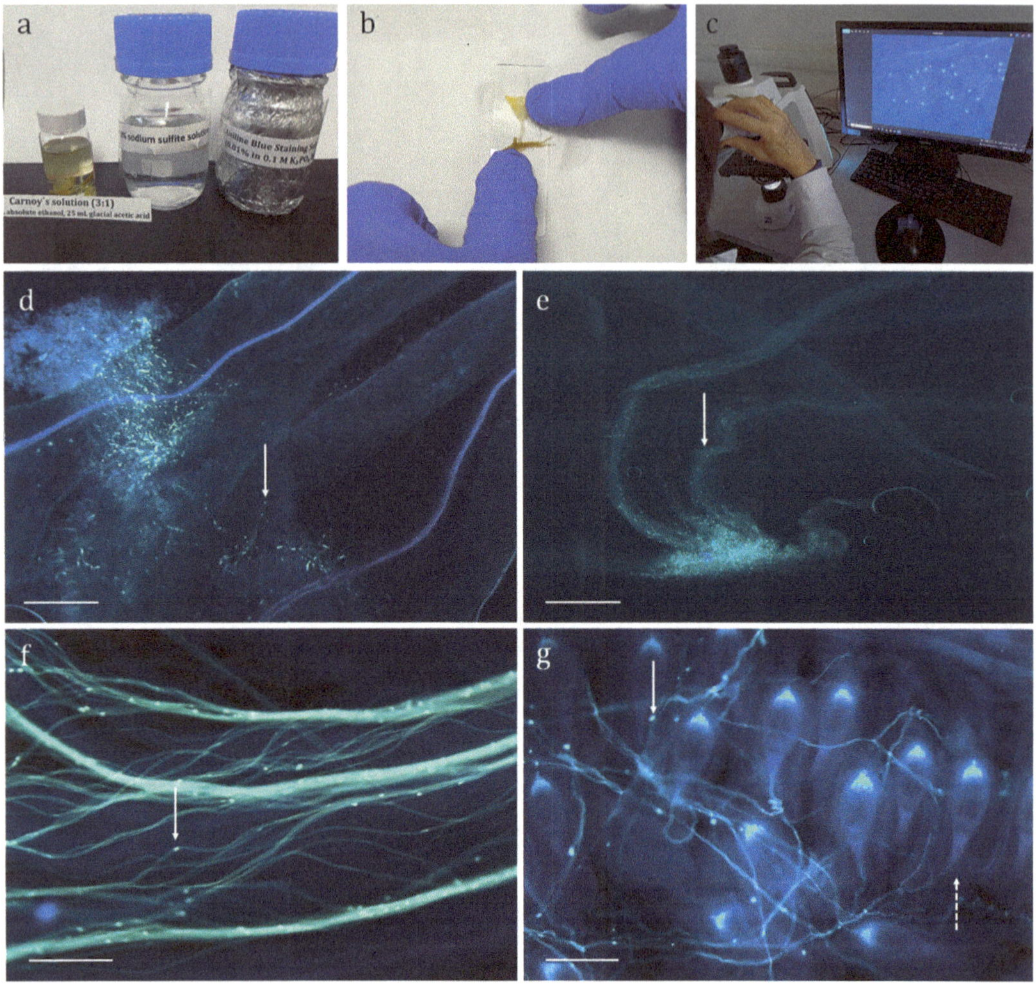

Fig. 4 In vivo fertilization protocol via fluorescence microscopy, with pollen tubes stained with aniline blue and visualized under UV light. (**a**) Solutions used for sample fixation and processing: Carnoy's solution, sodium sulfite solution, and aniline blue solution. (**b**) Slide preparation by squashing for analysis under fluorescence microscopy. (**c**) Slide observation under a fluorescence microscope with a UV filter, equipped with an image capture system. (**d**) Pollen grains with low viability on the stigma surface of a bromeliad. (**e**) Coiled pollen tubes in the style, indicating the presence of a reproductive barrier in the bromeliad. (**f**) Pollen tubes in the apical third of the style, showing sequential callose deposition along the tubes. (**g**) Image showing ovules being penetrated by pollen tubes; the dotted arrow indicates the micropyle, and the solid arrow highlights callose plugs in the pollen tubes. Bars: (**d**, **e**) 500 μM, (**f**, **g**) 200 μM

2. TTC solution is pH-sensitive; adjusting the solution to pH 7.0 enhances mitochondrial dehydrogenase activity and improves staining contrast.
3. The stock solution of the FDA should be stored in tightly sealed amber vials at −20 °C to protect it from light and hydrolysis. When properly stored, the stock remains stable for several weeks. Avoid repeated freeze–thaw cycles by aliquoting

the solution. Incubate the microscope slides in the dark to prevent FDA photodegradation and fluorescence quenching. The incubation time (5–15 min) may need adjustment depending on species and pollen viability. Avoid prolonged exposure to light after staining to preserve fluorescence intensity during observation.

4. FDA is light-sensitive and hydrolyzes rapidly in aqueous solution. Always prepare the working solution fresh and use it within 10–15 min. Keep all solutions protected from direct light.
5. Due to the addition of solvents and viscous components such as glycerol, the final volume of the solution may slightly exceed the sum of its individual components. Expect approximately 66 mL of final volume from this formulation.
6. This modified Alexander stain method avoids the use of toxic substances present in the original formulation [25], such as phenol and formic acid, replacing them with safer alternatives. This modification reduces health and environmental risks while maintaining effective differentiation of pollen viability [24].
7. Heating must be gentle and controlled to dissolve the dye effectively without boiling, which can degrade the carmine and reduce staining effectiveness.
8. Acetocarmine solution is acidic and photosensitive; store it in a tightly sealed amber bottle at room temperature, protected from light and air to prevent degradation and preserve staining quality [26].
9. Phosphate buffer should be freshly prepared or stored in tightly sealed vials at−20 °C in aliquots to avoid contamination and degradation. Avoid repeated freeze–thaw cycles, as they may alter pH and ionic balance. Prior to use, allow the buffer to reach room temperature and verify pH if necessary.
10. BK culture medium is one of the most widely used formulations for in vitro pollen germination across a broad range of plant species [9]. However, the optimal concentrations of sucrose, agar, and the final pH may vary depending on the species studied. Adjustments should be made empirically to optimize pollen tube growth and viability.
11. BM culture medium is a simplified formulation initially developed for Bromeliaceae [11]. Like other media, it may require adjustments in sucrose concentration, agar content, and pH according to the specific requirements of the species or cultivar under study.
12. RM culture medium is designed for more demanding or recalcitrant pollen types, offering a high sucrose concentration (30%) and buffering capacity via Tris, which helps stabilize

pH. The inclusion of $CaCl_2 \cdot 6H_2O$ supports pollen tube elongation [27]. Due to its composition, RM medium may better mimic the in vivo environment of certain species.

13. PEG-based media are commonly used to simulate osmotic stress conditions or to control water potential during pollen tube growth studies [28]. The osmotic potential can be finely tuned by adjusting PEG 400 concentrations (from 0.1 to 1.1 M) [28]. As with other media, pH and other component concentrations may require optimization for specific applications.
14. Do not exceed 121 °C and 20 min during autoclaving to avoid caramelization of the medium, particularly due to its high sucrose concentration.
15. Toluidine blue is not used to induce or evaluate pollen germination, but may be applied solely to enhance contrast in microscopy images. It should be filtered through filter paper before use and applied carefully to the sample. After 2–4 min, staining may begin to fade, and image acquisition should be done promptly. Alternative contrast stains may be used depending on availability and imaging goals.
16. Use a fluorescence microscope equipped with a UV filter set (excitation: 365–400 nm; emission: 420–460 nm), which is suitable for detecting the fluorescence of aniline blue bound to callose in pollen tubes. Ensure the microscope is properly calibrated and fluorescence intensity is not saturated during imaging.
17. Carnoy's fixative solution should preferably be prepared fresh just before use to ensure optimal fixation quality. If necessary, it may be stored for short periods in a tightly sealed container at 4 °C, protected from light and evaporation.
18. Sodium sulfite must be completely dissolved in distilled water using a magnetic stirrer to prepare the 10% tissue-softening solution [29]. Incomplete dissolution can result in uneven softening and may affect tissue transparency during fluorescence microscopy.
19. Aniline blue solution should be allowed to stand at room temperature in the dark for 1–2 h before use to ensure complete dissolution and stability. Store the solution in amber bottles at −5 °C and protect it from light to maintain fluorescence properties [29].
20. The use of a growth chamber or greenhouse is optional and should be considered based on the species studied. Some species may require controlled environmental conditions for proper development, while others can tolerate greater variation and may be grown in open or less controlled environments.

21. The incubation temperature significantly affects the reduction rate of TTC to formazan. Temperatures between 30 °C and 37 °C are optimal for mitochondrial dehydrogenase activity in most species. However, preliminary tests are recommended to determine the ideal incubation conditions for each species, as excessive temperature may denature enzymes and reduce staining specificity.
22. Incubation time should be optimized based on pollen species and viability. Short incubation (1–2 h) may suffice for metabolically active pollen, while recalcitrant or partially viable grains may require longer periods (up to 4 h). Over-incubation can lead to background staining or false positives due to non-specific formazan precipitation.
23. Heat the slide gently to enhance staining intensity, taking care not to boil the solution, as excessive heat can cause tissue damage or alter stain specificity.
24. Take care not to damage the medium. Clumping of grains should be avoided, as it compromises counting accuracy and hinders observation.
25. If a BOD incubator is not available, a growth chamber with controlled temperature and humidity may be used. Light or dark incubation should be selected based on the species' requirements. When germination needs to occur in the dark, Petri dishes should be wrapped in aluminum foil to prevent light exposure, which may interfere with pollen tube development.
26. The incubation temperature must be adjusted according to the physiological requirements of the species under study. Temperate species may require lower temperatures compared to tropical-adapted species.
27. Germination is typically observable between 8 and 24 h after inoculation. If the aim is to measure pollen tube length, observations should preferably be conducted at 24 h. After this period, microbial contamination may occur, especially since the pollen grains are not disinfected, and the medium is rich in sucrose, favoring microbial growth.
28. Toluidine blue is used to enhance contrast, facilitating microscopic visualization and image capture. The staining solution should be freshly prepared, filtered through filter paper, and stored at room temperature in a tightly sealed container.
29. The length of pollen tubes can be measured using an ocular micrometer attached to a stereomicroscope/microscope or through image analysis software, such as ImageJ [30], which allows precise measurements from images captured under microscopy.

30. Emasculation should be performed carefully to remove all anthers before pollen dehiscence without damaging the pistil or other floral structures. Timing is critical; emasculation 1 day before anthesis ensures prevention of self-pollen release. When possible, use flowers from multiple populations to account for genetic variability.
31. Voile bags are lightweight, breathable mesh bags that prevent unwanted pollen contamination while allowing air circulation. Ensure bags are securely fastened to avoid accidental pollen ingress. Replace bags if damaged during the experiment.
32. After pollination, each flower should be clearly labeled using waterproof tags or fine-tipped permanent markers on attached labels. Include information such as date, type of pollination (e.g., self, cross, or treatment group), and identifier codes. Accurate labeling is essential for sample tracking, data reliability, and avoiding confusion during subsequent analyses of fruit set, seed formation, or pollen tube growth.
33. Fruit set and seed count are important indicators of reproductive success following controlled pollination. Fruit maturation time varies among species and should be monitored closely. Seeds should be counted manually or with image analysis software for accuracy, considering potential seed abortion or incomplete development.
34. The optimal sampling intervals for collecting pollinated flower samples vary depending on the species' reproductive biology and the specific goals of the study, such as early pollen tube growth, tube elongation, or fertilization timing. Researchers should tailor collection times accordingly to capture relevant developmental stages.
35. If pistils are too large to be processed or mounted properly, they should be dissected into manageable segments (e.g., separating stigma, style, and ovary) to ensure adequate stain penetration and ease of handling without damaging critical reproductive structures.
36. Carnoy's fixative preserves both cellular structure and pollen tube integrity. Fixation should be performed at 4 °C for at least 1 h and up to 24 h. For long-term storage, transfer samples to 70% ethanol and keep them refrigerated and protected from light.
37. Sodium sulfite (10%) acts as a softening agent and helps prevent tissue browning due to oxidation, improving the uptake of aniline blue and enhancing fluorescence signal clarity. Samples should be fully submerged and handled carefully to prevent mechanical damage.

38. Autoclaving assists with tissue clearing and softening by breaking down cell walls. Standard parameters are 120 °C for 20 min. However, duration may need adjustment depending on tissue consistency; less time for soft tissues and more for lignified or thick tissues.
39. Aniline blue binds to callose deposits in pollen tubes and fluoresces under UV light. Incubation in the dark prevents dye degradation. Overnight staining at−5 °C enhances penetration and contrast, especially in thicker or more lignified styles.
40. When mounting the samples, apply gentle but consistent pressure to squash the pistils under the coverslip without rupturing tissues or damaging pollen tubes. Avoid excessive force that could distort or obscure structures.
41. Observations should be performed under a fluorescence microscope equipped with a UV filter (Fig. 4c). Adjust brightness, exposure time, and contrast settings to optimize visualization of the fluorescent pollen tubes.
42. Callose deposits are typically distributed along the pollen tube, forming fluorescent structures visible under fluorescence microscopy (Fig. 4d–g). Interpreting this pattern is essential for assessing pollen viability, germination, and the extent of pollen tube growth toward the ovule micropyle (Fig. 4g), as well as for detecting potential incompatibility barriers (Fig. 4d, e). This technique allows for the estimation of pollen viability and the evaluation of pollen tube development or arrest within the pistil.
43. Pollen viability assessments may also be influenced by the stigma receptivity, which varies among species and developmental stages. Inadequate stigma receptivity can hinder pollen hydration, germination, and tube growth, potentially leading to underestimated viability in in vivo assays.

Acknowledgments

The authors acknowledge Fundação Getulio Vargas (FGV), Embrapa Mandioca e Fruticultura, Universidade de São Paulo (USP), Universidade Federal do Recôncavo da Bahia (UFRB), Conselho Nacional de Desenvolvimento Científico e Tecnológico (CNPq) (TLS - PDS102715/2024-6, FVDS - 307128/2021-0, and LYSA - 314502/2023-8), Coordenação de Aperfeiçoamento de Pessoal de Nível Superior (CAPES), and Fundação de Amparo à Pesquisa do Estado da Bahia (FAPESB).

References

1. Althiab-Almasaud R, Teyssier E, Chervin C, Johnson MA, Mollet JC (2024) Pollen viability, longevity, and function in angiosperms: key drivers and prospects for improvement. Plant Reprod 37:273–293. https://doi.org/10.1007/s00497-023-00484-5
2. Mota SS, Faro IAM, Soares TL, Almeida PS, Souza FVD, Aona LYS, Costa MAPC, Rossi ML, Souza EH (2024) Pollen morphology and viability of *Lymania* (Bromeliaceae) species with ornamental potential. Sci Hortic 328: 112890. https://doi.org/10.1016/j.scienta.2024.112890
3. Silva RL, Souza EH, Vieira LJ, Pelacani CR, Souza FVD (2017) Cryopreservation of pollen of wild pineapple accessions. Sci Hortic 219: 326–334. https://doi.org/10.1016/j.scienta.2017.03.022
4. Silva TA, Nascimento GQS, Almeida PS, Cavalcante BP, Souza FVD, Aona LYS, Costa MAPC, Souza EH (2024) Pollen morphology, viability, and stigma receptivity of *Hohenbergia* (Bromeliaceae) species occurring in the Atlantic Forest biome with potential for landscaping and cut flowers. Genet Res Crop Evol 72: 2007–2029. https://doi.org/10.1007/s10722-024-02077-1
5. Nascimento GQS, Silva TA, Almeida PS, Rossi ML, Aona LYS, Souza FVD, Costa MAPC, Souza EH (2025) Pollen and stigma morphology, pollen viability and stigma receptivity of *Wittmackia* species (Bromeliaceae) by light, fluorescence and scanning electron microscopy. Microsc Res Tech 88:1–20. https://doi.org/10.1002/jemt.24795
6. Souza EH, Souza FVD, Rossi ML, Packer RM, Cruz-Barros MAV, Martinelli AP (2017) Pollen morphology and viability in Bromeliaceae. Ann Braz Acad Sci 89:3067–3082. https://doi.org/10.1590/0001-3765201720170450
7. Sousa MBE, Andrade LRB, Souza EH, Alves AAC, Oliveira EJ (2021) Reproductive barriers in cassava: factors and implications for genetic improvement. PLoS One 16:e0260576. https://doi.org/10.1371/journal.pone.0260576
8. Souza SO, Oliveira RS, Aona LYS, Souza FVD, Soares TL, Souza EH (2021) Pollen morphology and viability of *Tillandsia* L. (Bromeliaceae) species by light microscopy and scanning electron microscopy. Microsc Res Tech 84:441–459. https://doi.org/10.1002/jemt.23601
9. Brewbaker JL, Kwack BH (1963) The essential role of calcium ion in pollen germination and pollen tube growth. Am J Bot 50:859–865
10. Li M, Jiang F, Huang L, Wang H, Song W, Zhang X, Zhang Y, Niu L (2023) Optimization of in vitro germination, viability tests and storage of *Paeonia ostii* pollen. Plants 12(13): 2460. https://doi.org/10.3390/plants12132460
11. Parton E, Vervaeke I, Delen R, Vandenbussche B, Deroose R, De Proft M (2002) Viability and storage of bromeliad pollen. Euphytica 125:155–161. https://doi.org/10.1023/A:1015884019619
12. Soares TL, Silva SO, Costa MAPC, Santos-Serejo JA, Souza AS, Lino LSM, Souza EH, Jesus ON (2008) In vitro germination and viability of pollen grains of banana diploids. Crop Breed Appl Biotechnol 8:111–118. https://doi.org/10.12702/1984-7033.v08n02a03
13. Mota SS, Faro IAM, Souza FVD, Aona LYS, Costa MAPC, Souza EH (2023) Reproductive systems and hybridization of *Lymania* species (Bromeliaceae) endemic to Northeast Brazil threatened with extinction. Sci Hortic 322: 112447. https://doi.org/10.1016/j.scienta.2023.112447
14. Soares TL, Jesus ON, Souza EH, Oliveira EJ (2015) Reproductive biology and pollen-pistil interactions in *Passiflora* species with ornamental potential. Sci Hortic 197:339–349. https://doi.org/10.1016/j.scienta.2015.09.045
15. Souza EH, Versieux LM, Souza FVD, Rossi ML, Costa MAPC, Martinelli AP (2017) Interspecific and intergeneric hybridization in Bromeliaceae and their relationships to breeding systems. Sci Hortic 221:1–20. https://doi.org/10.1016/j.scienta.2017.04.027
16. Souza GKB, Souza FVD, Aona LYS, Costa MAPC, Souza EH (2025) Reproductive systems and hybridization for ornamental use of *Wittmackia* (Bromeliaceae) endemic to Brazil and occurring in the Atlantic Forest fragments. Euphytica 221:1–21. https://doi.org/10.1007/s10681-025-03495-7
17. Calić D, Milojević J, Belić M, Miletić R, Zdravković-Kirać S (2021) Impact of storage temperature on pollen viability and germinability of for Serbian autochthon apple cultivars. Front Plant Sci 12:709231. https://doi.org/10.3389/fpls.2021.709231
18. Galgóci M, Kormuťák A, Klobučník M, Gömöry D, Lukáčik I, Boleček P (2024) Pollen viability, dispersion ability and longevity during deep-freezer storage in Juniperus taxa native to Slovakia. Sci Rep 14:3706. https://doi.org/10.1038/s41598-024-53152-7

19. Souza EH, Souza FVD, Rossi ML, Brancalleão N, Ledo CAS, Martinelli AP (2015) Viability, storage and ultrastructure analysis of *Aechmea bicolor* (Bromeliaceae) pollen grains, an endemic species to the Atlantic Forest. Euphytica 204:13–28. https://doi.org/10.1007/s10681-014-1273-3
20. Tai PYP, Miller JD (2002) In vivo viability assay of sugarcane pollen stored at ultra-low temperature following preservation treatments. J Am Soc Sugarcane Technol 22:135–136
21. Vendrame WA, Carvalho VS, Dias JMM, Maguire I (2008) Pollination of dendrobium hybrids using cryopreserved pollen. HortScience 43:264–267
22. Norton JD (1966) Testing of pollen viability with tetrazolium salts. Proc Am Soc Hortic Sci 89:132–134
23. Heslop-Harrison J, Heslop-Harrison Y (1970) Evaluation of pollen viability by enzymatically induced fluorescence; intracellular hydrolysis of fluorescein diacetate. Stain Technol 45:115–120. https://doi.org/10.3109/10520297009085351
24. Peterson R, Slovin JP, Chen C (2010) A simplified method for differential staining of aborted and non-aborted pollen grains. Int J Plant Biol 1:e13. https://doi.org/10.4081/pb.2010.e13
25. Alexander MP (1969) Differential staining of aborted and nonaborted pollen. Stain Technol 44:117–122. https://doi.org/10.3109/10520296909063335
26. Shivanna KR, Rangaswamy NS (1992) Pollen biology: a laboratory manual. Springer-Verlag, Berlin, p 119. https://doi.org/10.1007/978-3-642-58035-0
27. Roberts IN, Gaude TC, Harrod G, Dickinson HG (1983) Pollen-stigma interactions in *Brassica oleracea*: a new pollen germination medium and its use in elucidating the mechanism of self-incompatibility. Theor Appl Genet 65:231–238. https://doi.org/10.1007/BF00308074
28. Zhang HQ, Croes AF (1982) A new medium for pollen germination *in vitro*. Acta Bot Neerlandica 31:113–119. https://doi.org/10.1111/j.1438-8677.1982.tb01597.x
29. Martin FW (1959) Staining and observing pollen tubes in the style by means of fluorescence. Stain Technol 34:125–128. https://doi.org/10.3109/10520295909114663
30. Schneider CA, Rasband WS, Eliceiri KW (2012) NIH image to ImageJ: 25 years of image analysis. Nat Methods 9:671–675. https://doi.org/10.1038/nmeth.2089

Chapter 2

Evaluating Pollen Viability and Tube Growth in Rice (*Oryza sativa*)

Ji-Hyun Kim and Yu-Jin Kim

Abstract

Pollen maturity, viability, and pollen tube growth are essential indicators of male fertility in rice (*Oryza sativa*). This chapter presents standardized protocols to evaluate these traits using a range of staining methods, including iodine–potassium iodide (I_2–KI), DAPI, calcofluor white, auramine O, LM19/LM20 antibodies, and FM4-64. Pollen germination is assessed through in vitro assays on sucrose-based germination medium optimized for rice, while in vivo pollen tube growth is visualized with aniline blue staining. Given the rapid and short-lived nature of rice pollen under laboratory conditions, practical tips are included to enhance reproducibility. Together, these methods provide a reliable toolkit for the functional analysis of rice pollen in developmental, genetic, and breeding research.

Key words Rice, Pollen maturity, Pollen viability, Pollen germination, In vitro germination assay, In vivo pollen tube growth, Pectin localization, Plant reproduction

1 Introduction

In flowering plants, the male gametophyte, or pollen grain, plays a central role in fertilization and reproductive success. The functional competence of pollen—its maturity, viability, and capacity to germinate and produce a pollen tube—is a critical determinant of seed set and yield, particularly in self-pollinated cereal crops such as rice (*Oryza sativa*). Accurate assessment of pollen quality is therefore essential for studies in reproductive biology, abiotic stress responses, transgenic line screening, and hybrid breeding programs involving cytoplasmic male sterility (CMS) or environment-sensitive genic male sterility (EGMS) [1].

In temperate climates such as Korea, rice is typically transplanted in May and flowers in August (Fig. 1). Anthesis occurs between 10:00 and 14:00, depending on environmental factors such as temperature, humidity, and photoperiod. Pollen grains mature in the days before flowering and are short-lived once

Giampiero Cai et al. (eds.), *Pollen and Pollen Tube Analysis: Methods and Protocols*, Methods in Molecular Biology, vol. 3042,
https://doi.org/10.1007/978-1-0716-5308-1_2,

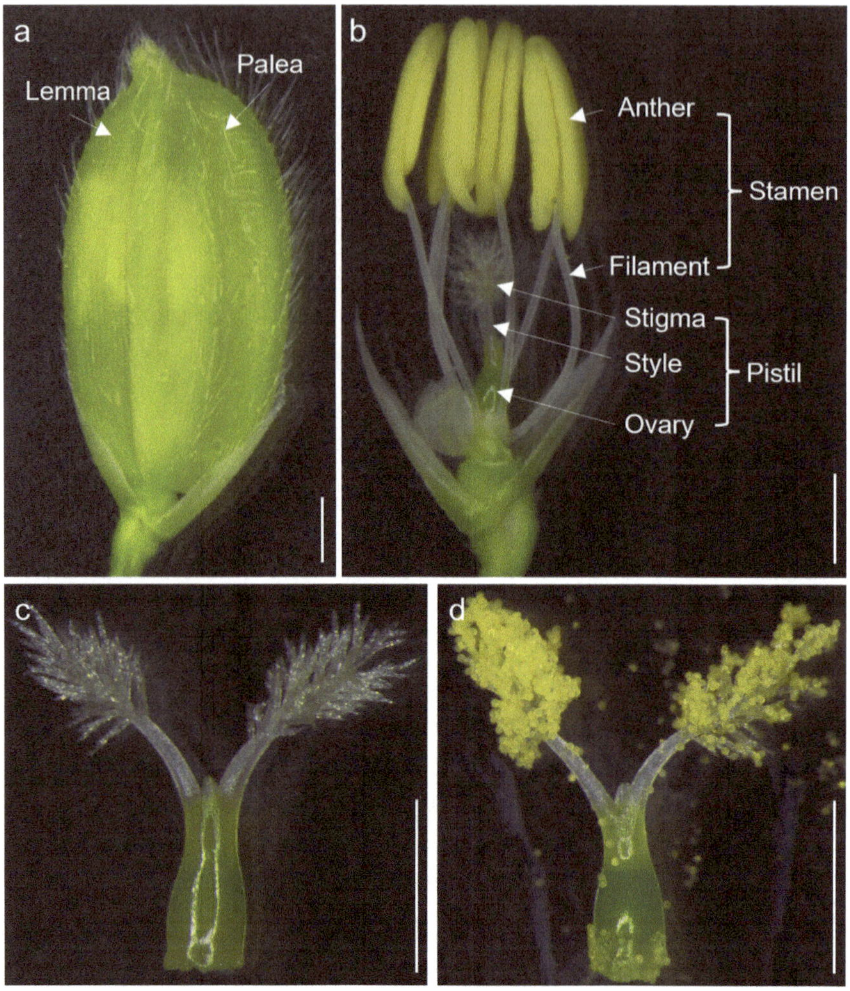

Fig. 1 Floral structure of rice (*Oryza sativa*). (**a**) A rice spikelet enclosed by the lemma and palea. (**b**) Floral structure of a rice floret after removal of the lemma and palea. The six stamens each consist of an anther and a filament, and the central pistil comprises two feathery stigmas, styles, and an ovary. (**c**) Pistil prior to pollination. (**d**) Pistil after pollination, with numerous pollen grains attached to the stigmatic surface. Scale bars = 1 mm

released. Under in vitro conditions, hydration rapidly induces germination, often resulting in bursting within 15 min. These time-sensitive dynamics make rice pollen handling particularly challenging, requiring precise timing and practiced technique.

To enable accurate and reproducible analysis of rice pollen function, this chapter presents a set of validated and standardized methods. Pollen maturity is evaluated using I_2–KI staining [2], nuclear organization is visualized with DAPI [3], and pollen wall architecture—including exine and intine layers—is examined with calcofluor white and auramine O [2]. Cell wall remodeling during tube growth is further assessed by analyzing pectin distribution with LM19 and LM20 antibodies [4]. Plasma membrane dynamics during germination and tube elongation are examined using FM4-64 staining [2].

The chapter also describes both in vitro and in vivo germination assays. In vitro assays employ a sucrose-based pollen germination medium (PGM) optimized for rice to quantify germination rate and tube elongation [5]. In vivo pollen tube progression is visualized by staining with aniline blue after pollination [5]. Importantly, this chapter is intended not only as a collection of protocols but also as a source of practical tips and troubleshooting advice based on extensive experience with rice pollen. These insights help mitigate technical variability, timing sensitivity, and environmental influences. Together, the methods and recommendations presented here provide a robust and adaptable framework for studying rice pollen in developmental, physiological, and agronomic contexts.

2 Materials

Unless otherwise stated, all reagents should be prepared using sterilized ultrapure water (resistivity ≥18 MΩ·cm at 25 °C), obtained by sterilizing triple-distilled water of Milli-Q–equivalent quality. All solutions should be prepared with analytical-grade or higher-purity reagents to minimize contamination. Care should be taken to prevent microbial or chemical contamination during reagent preparation and handling. Glassware, plasticware, and tools should be sterilized before use. Solutions requiring sterilization should be either autoclaved or filter-sterilized (0.22 μm), as specified in the relevant protocol sections. No sodium azide or antimicrobial agents should be added to reagents to avoid cytotoxic effects on pollen or pollen tubes. All waste materials, including unused stains and germination media, must be disposed of according to institutional biosafety and chemical waste regulations.

2.1 Plant Material and Growth Conditions

Rice plants (*Oryza sativa* L. cv. Dongjin) were used in this protocol. Seeds were dehulled to remove the lemma and palea, then surface-sterilized with 50% (v/v) commercial bleach solution (approximately ~2–3% sodium hypochlorite) for 20 min. Seeds were rinsed once with sterile distilled water, briefly immersed in 70% ethanol, and washed at least three times with sterile distilled water before sowing on Murashige and Skoog (MS0) medium. Seeds were germinated under long-day conditions (16 h light/ 8 h dark) at 28 °C for approximately 10 days before transplantation into soil for genotyping via leaf DNA extraction and sequencing.

Soil-grown seedlings were cultivated in a growth chamber or greenhouse under long-day conditions at 28 °C (16 h light)/22 °C (8 h dark) for an additional 25–30 days prior to transplantation into the paddy field. For chamber-based experiments, rice plants were grown under long-day conditions for 6–8 weeks to induce tillering, followed by a switch to short-day conditions (12 h light/12 h dark) to promote flowering.

2.2 Basic Cytological Assessment of Pollen: Maturity (I_2–KI), Nuclear Organization (DAPI), Exin/Intine Structure (Calcofluor White, Auramine O), and Plasma Membrane (FM4-64)

1. Microcentrifuge tubes (1.5 mL, 2 mL).
2. Humid box (plastic container with moist paper towels).
3. Fine forceps.
4. Microscope slides (76 × 26 mm) and cover slips (18 × 18 mm, 24 × 40 mm).
5. Vacuum pump (optional for improved pollen release).
6. I_2–KI solution: 2% potassium iodide, 0.2% iodine in distilled water. Store at 4 °C.
7. DAPI: 5 μg/mL in PBS (pH 7.0). Store at 4 °C protected from light.
8. Calcofluor white: 0.1% solution in distilled water. Store at −20 °C protected from light.
9. Auramine O: 0.1% stock solution in Tris–HCl buffer (pH 7.5). Store at 4 °C protected from light. Prepare a 0.001% working solution in distilled water before use.
10. FM4-64: 1 μg/mL stock solution in DMSO. Store at −20 °C, protected from light. Dilute to a 3 ng/mL working solution in liquid PGM prior to application.
11. Pollen germination medium (PGM): Add 7 mL of sterile distilled water to a 15 mL conical tube. Dissolve 2 g of sucrose and 1 g of PEG 4000 in the water (*see* **Note 1**). Sequentially, add the following stock solutions, mixing after each addition to avoid unwanted reactions: 100 μL of $Ca(NO_3)_2{\cdot}4H_2O$ stock (70 mg/mL), 100 μL of boric acid stock (4 mg/mL), and 334 μL of vitamin B1 stock (0.3 mg/mL). Adjust the final volume to 10 mL with sterile distilled water. Allow the medium to cool to room temperature before use (*see* **Note 2**).
12. Solid PGA (agar-solidified PGM): add 5 mL of sterile distilled water into a 15 mL conical tube. Dissolve 2 g of sucrose and 1 g of PEG 4000 in water to prepare 2×PGM (*see* **Note 3**). Sequentially, add the following stock solutions, mixing after each addition to prevent unwanted reactions: 100 μL of Ca $(NO_3)_2{\cdot}4H_2O$ stock (70 mg/mL), 100 μL of boric acid stock (4 mg/mL), and 334 μL of vitamin B1 stock (0.3 mg/mL). Keep the 2×PGM warm in an oven at 60 °C for up to 1 h during preparation. In a 50 mL beaker, dissolve 65 mg of agar in 7 mL of sterile distilled water by heating in a microwave for 1.5 min, until the agar is completely dissolved. As the heating cycle ends, prepare a Styrofoam container filled with hot water and place both the beaker with molten agar and the 2×PGM

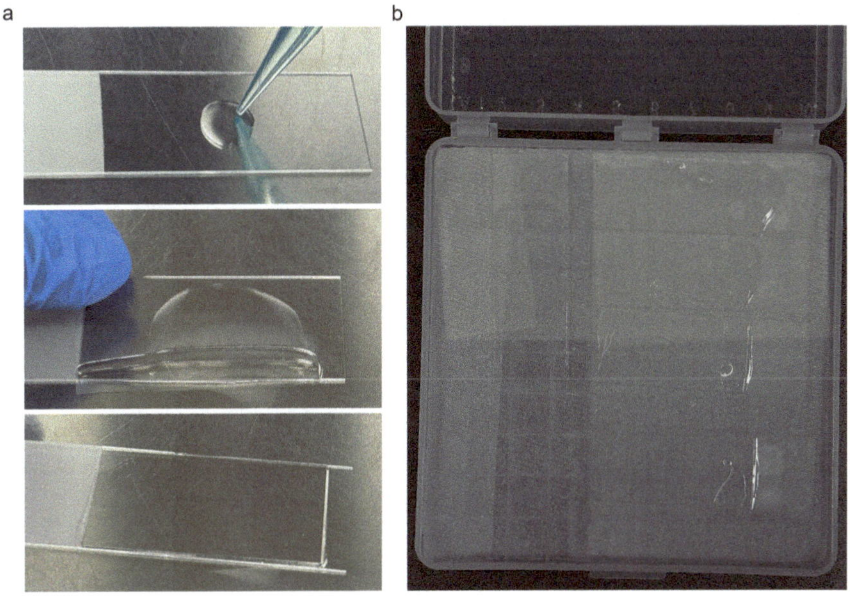

Fig. 2 Preparation and storage of PGA. (**a**) Dispensing molten PGA onto a slide glass and overlaying with a cover glass. Approximately 435 μL of PGA is applied per slide; the cover glass is placed carefully to avoid air bubbles. This step must be performed quickly to prevent premature solidification and ensure even distribution. (**b**) Storage of prepared PGA slides in a humid box to prevent dehydration. Slides are placed on moist paper towels to maintain humidity and can be stored at room temperature for several hours before pollen application

tube inside to maintain temperature during transfer to the clean bench. In the clean bench, gently mix 5 mL of warm 2×PGM with molten agar, avoiding bubble formation. Immediately dispense 435 μL of the mixture onto a microscope slide and overlay with a 24 × 40 mm cover slip, minimizing air bubbles (*see* **Note 4**; Fig. 2a). Allow the medium to solidify for approximately 15 min at room temperature. Carefully remove the cover slip with sterile forceps. Store prepared slides in a humid box until use (Fig. 2b).

2.3 Pectin Localization in Pollen Tubes

1. Liquid PGM.
2. 37% formaldehyde solution.
3. Rice PEM buffer: 100 mM PIPES, 10 mM EGTA, 5 mM $MgSO_4$, 0.3 M mannitol, and 2% glycerol, adjusted to pH 6.9.
4. Phosphate-buffered saline (PBS, pH 7.0).
5. 50% (v/v) glycerol.
6. LM19 monoclonal antibody.
7. LM20 monoclonal antibody.
8. FITC-conjugated IgG secondary antibody.
9. Distilled water.

2.4 Aniline Blue Reagents for In Vivo Pollen Tube Staining

1. Pollinated rice pistils (approximately 2 h after anthesis).
2. Carnoy's solution (ethanol:acetic acid = 3:1).
3. 0.05% aniline blue staining solution.
4. Distilled water.
5. 2 mL microcentrifuge tubes.
6. Fine forceps.
7. Micropipettes and sterile tips.
8. Microscope slides and cover slips.
9. UV fluorescence microscope.
10. Dark storage box or aluminum foil.

3 Methods

3.1 Growing the Plant Material

1. Dehull rice seeds using a hand seed dehusker (Fig. 3) to remove the lemma and palea (*see* **Note 5**).
2. Transfer dehulled seeds to a 1.5 mL microcentrifuge tube, add 1 mL of 50% (v/v) commercial bleach solution (approximately ~2–3% sodium hypochlorite), and incubate for 20 min with occasional inversion (*see* **Note 6**).
3. Remove the bleach solution and rinse the seeds once with sterile distilled water.
4. Immediately immerse the seeds in 70% ethanol for a few seconds and promptly remove to avoid reduced germination (*see* **Note 7**).
5. Wash the seeds three times with sterile distilled water.
6. Using sterile forceps, place sterilized seeds evenly on MS0 germination medium.
7. Transfer plates to a growth chamber maintained at 28 °C under long-day conditions (16 h light/8 h dark). Germinate and grow seedlings for approximately 10 days.
8. Transfer seedlings to soil and grow in a greenhouse or growth chamber at 28 °C (light phase) and 22 °C (dark phase) under long-day conditions for 25–30 days to establish before field transfer.
9. For greenhouse or growth chamber-based experiments, continue growing seedlings for 6–8 weeks under long-day conditions to promote tillering, maintaining relative humidity above 70% (*see* **Note 8**).
10. For greenhouse or growth chamber-based experiments, switch to short-day conditions (12 h light/12 h dark) to induce

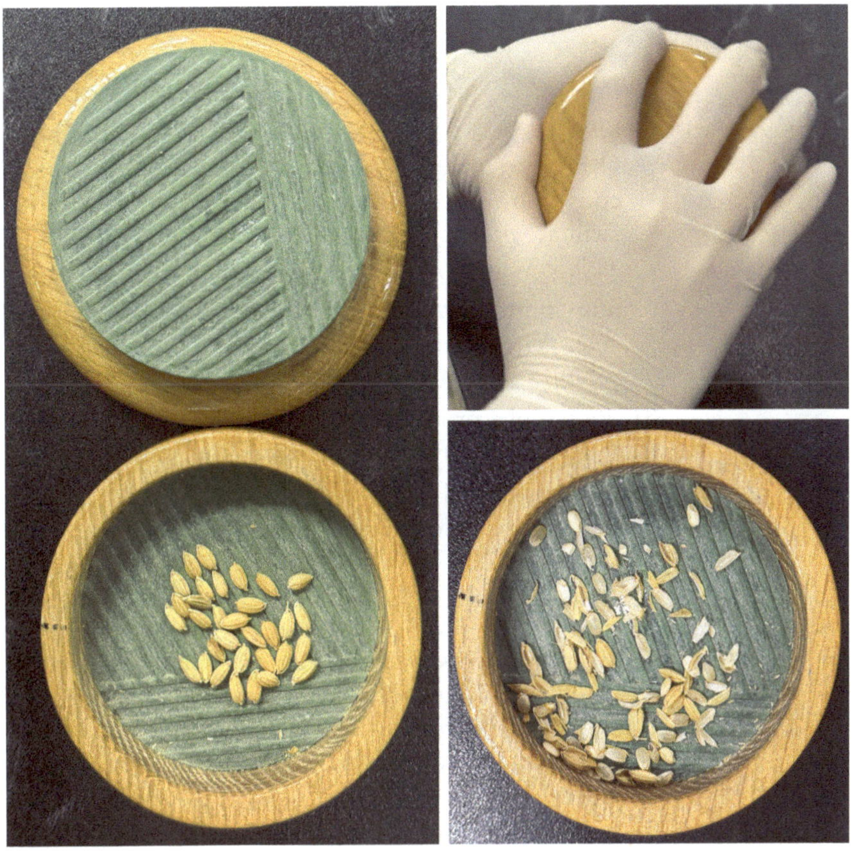

Fig. 3 Dehulling of rice seeds using a hand-operated seed dehusker. Manual dehulling of rice seeds removes the lemma and palea from dry seeds prior to surface sterilization. This step improves seed-to-medium contact and reduces contamination risk during in vitro culture

flowering and adjust relative humidity to above 80% for proper anther development.

3.2 Cytological Staining of Rice Pollen

This section describes four complementary cytological staining methods for assessing developmental stage, nuclear configuration, and cell wall integrity. Depending on the staining reagent, these protocols enable evaluation of pollen maturity (I_2–KI), nuclear organization (DAPI), and exine/intine structure (calcofluor white, auramine O).

1. Collect rice spikelets (*see* **Note 9**). To prevent drying, place them in a humid box or a 2 mL tube.
2. For DAPI staining, gently open the lemma and palea with forceps to expose the anthers. Fix spikelets in Carnoy's solution and apply vacuum infiltration for 30 min to enhance penetration (*see* **Note 10**).

3. Remove Carnoy's solution and replace it with PBS (*see* **Note 11**).
4. Prepare 1.5 mL tubes containing 50 μL of each staining solution (I_2–KI, DAPI, calcofluor white, or auramine O).
5. Using fine forceps, carefully detach mature anthers. Trim both ends with a sterile syringe blade, then cut through the center to facilitate pollen release.
6. Transfer the cut anthers into the staining solution. Apply vacuum infiltration for 10 min to release pollen grains; if unavailable, gently squash the anthers directly in ~50 μL of staining solution on a microscope slide.
7. For samples processed by vacuum infiltration, briefly centrifuge at 1000 × g for 20 s to sediment the pollen.
8. For centrifuged samples, cut the end of a pipette tip and gently transfer pollen to a slide. Immediately place a cover glass on top and observe under a microscope. Since staining solutions dry quickly, mount and cover immediately before observation.

3.3 In Vitro Pollen Germination

1. Monitor anthesis time under the specific growth conditions (paddy field, greenhouse, or growth chamber). Although anthesis timing varies with light, temperature, and humidity, rice generally flowers at a consistent time once acclimated (*see* **Note 12**).
2. Immediately after anthesis, gently tap the panicle over the PGM to release pollen. For reproducibility, tap each spikelet no more than 10 times. Repeated tapping may dislodge non-viable peripheral pollen grains and lower germination rates (*see* **Note 13**).
3. Select panicles 30–60 min before flowering (second day after heading). Insert panicles into a 500 mL plastic bottle and cover with a transparent plastic bag to prevent premature drying (Fig. 4) (*see* **Note 14**).
4. Transfer the covered bottle to a growth chamber at 28 °C under light conditions to induce flowering. Maintain high humidity inside the bag. Do not remove the bag prematurely, as dry air can cause anthers to desiccate before pollen collection (*see* **Note 15**).
5. At flowering, gently remove the bag and allow humidity to equilibrate for 20–30 s before collecting pollen.
6. In a wind-protected location, gently tap spikelets over the PGM. If using solidified PGA, apply approximately 50 μL of liquid PGM on top before placing the cover glass (*see* **Note 16**).

Fig. 4 Preparation of panicles for controlled anthesis under indoor conditions. Panicles are sampled 30–60 min before flowering and placed in a water-filled container (e.g., 500 mL PET bottle) covered with a plastic bag to prevent premature drying of anthers. This setup enables flowering under controlled indoor conditions, allowing reliable pollen collection for in vitro experiments. Black arrows indicate flowers that have undergone anthesis inside the plastic bag

7. Pollen on PGM begins germinating within 1 min. Observe under a microscope immediately after mounting. In vitro germination typically completes within approximately 15 min.
8. For quantitative analysis of germination rates or pollen tube length, incubate pollen in liquid PGM for 10 min, then add 4% paraformaldehyde to fix and prevent tube bursting (*see* **Note 17**).
9. After fixation, tilt the 24-well plate to concentrate pollen at the bottom. Cut the tip of a 200 μL pipette tip and gently transfer germinated pollen to a slide (Fig. 5; *see* **Note 18**).

3.4 Plasma Membrane Staining Using FM4-64

1. Follow **steps 1–6** of Subheading 3.3, then prepare liquid PGM supplemented with FM4-64 at a final concentration of 3 ng/mL.

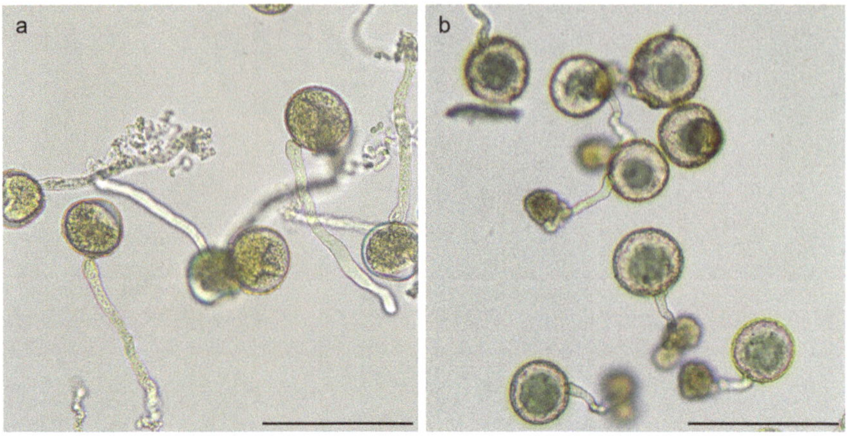

Fig. 5 Pollen germination in liquid PGM and on PGA. (**a**) Pollen germination in liquid PGM. Liquid medium allows uniform reagent access, facilitating chemical treatment experiments. However, pollen tube rupture may release contents into the surrounding solution, potentially affecting observations. (**b**) Pollen germination on solidified PGA. Solid medium provides a stable environment for imaging, making it more suitable for microscopic observation, though precise chemical concentration control is more challenging. Scale bars = 100 μm

2. Apply 50 μL of FM4-64-containing liquid PGM to the sample and place a cover glass on top.
3. Observe fluorescence under a CLSM using an excitation laser line at 561 nm and detect emission at 594/46 nm (*see* **Note 19**).

3.5 Pectin Localization in Pollen Tubes

This section describes an immunofluorescence-based method to visualize pectin distribution in rice pollen tubes using two monoclonal antibodies, LM19 and LM20, which distinguish different degrees of homogalacturonan methylesterification. LM19 recognizes low-methylesterified pectin, whereas LM20 recognizes highly methylesterified regions [5].

1. Set a 15 min timer and collect pollen into 1 mL of liquid PGM at approximately 2 min intervals for a total of 15 min (*see* **Note 20**).
2. At the end of 15 min, add 100 μL of 37% formaldehyde to achieve a final concentration of approximately 4% formaldehyde.
3. Incubate the fixed pollen at room temperature for 30 min.
4. Aliquot approximately 500 μL fixed pollen into 2 mL microcentrifuge tubes.
5. Centrifuge at 5000 × g for 2 min and carefully discard the supernatant.
6. Prepare PEM buffer containing 3.7% (v/v) formaldehyde (prepared by mixing PEM buffer and 37% formaldehyde

at a 9:1 ratio, v/v) immediately before use. Add 2 mL of the solution to each tube.

7. Incubate on a shaker at room temperature for 30 min.
8. Centrifuge at 5000 × g for 2 min and remove the supernatant.
9. Wash the pollen pellet three times with PBS (pH 7.0), centrifuging at 4000 × g for 2 min each.
10. Prepare antibody working solutions by diluting LM19 or LM20 in PBS (4 μL antibody + 200 μL PBS per tube).
11. Resuspend each pellet in 200 μL of the respective antibody solution and incubate at room temperature for 2 h.
12. Wash the samples three times with PBS (pH 7.0), centrifuging at 4000 × g for 2 min per wash.
13. Prepare FITC-conjugated secondary antibody solution (1:200 dilution; 1 μL antibody + 200 μL PBS).
14. Add 200 μL of the secondary antibody solution to each tube and incubate in the dark for 2 h.
15. Wash the samples five times with PBS to remove unbound secondary antibodies.
16. Resuspend the final pellet in 50 μL of 50% (v/v) glycerol.
17. Mount the sample on a microscope slide and cover with a cover glass.
18. Observe FITC under a CLSM using an excitation laser line at 488 nm and detect emission at 525/50 nm.

3.6 In Vivo Pollen Tube Visualization Using Aniline Blue

1. At ~2 h after anthesis, collect rice pistils from pollinated florets. Using fine forceps, carefully isolate the pistil, excluding lodicules and filaments (*see* **Note 21**).
2. Immerse excised pistils in Carnoy's solution and incubate overnight (16–24 h) at 4 °C in a sealed 2 mL microcentrifuge tube.
3. Remove Carnoy's solution and wash the pistils five times with distilled water (1 mL per wash).
4. After the final wash, remove the supernatant completely and add 100 μL of 1 M NaOH. Incubate pistils at room temperature for 6 h to soften tissues. After NaOH treatment, pistils typically turn yellow (Fig. 6a, b).
5. Wash pistils twice with distilled water (200 μL per wash). At this stage, pistils become partially transparent and extremely fragile, requiring gentle handling.
6. Stain pistils with 80 μL of 0.05% aniline blue solution for 1 h in the dark.
7. After staining, transfer carpels onto a slide with the aniline blue solution used during staining. Cover with a cover glass and

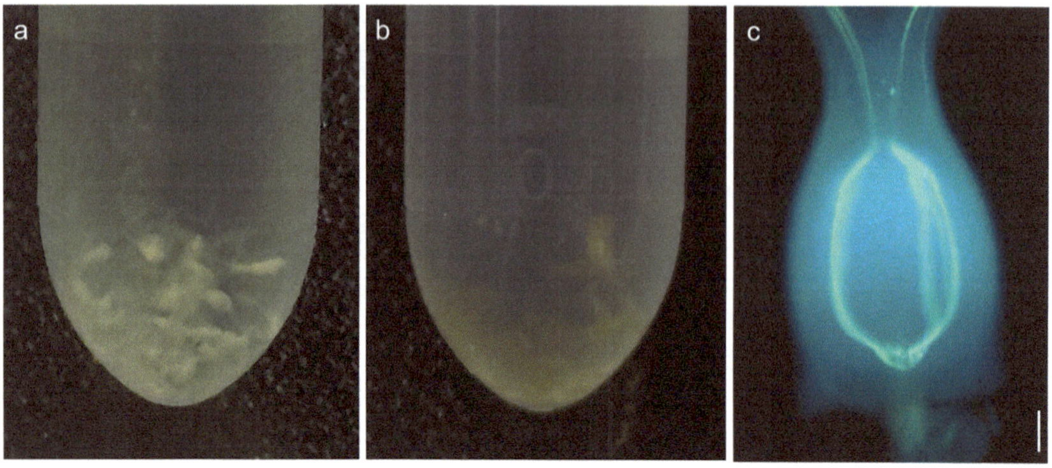

Fig. 6 In vivo staining of rice pistils for pollen tube visualization. (**a**) Pistils fixed in Carnoy's solution, with chlorophyll pigments leached out, appear white. (**b**) Pistils treated with 1 M NaOH become softened and semi-transparent; tissues turn yellow and are fragile, requiring careful handling. (**c**) Pollen tube growth visualized by aniline blue staining under UV fluorescence microscopy. Pollen tubes exhibit bright fluorescence and grow through the style toward the micropyle. Scale bar = 100 μm

observe pollen tubes under a UV fluorescence microscope (Fig. 6c; *see* **Note 22**).

4 Notes

1. Some medium ingredients do not dissolve readily at room temperature. Incubate the tube at 60–65 °C in a dry oven for at least 30 min until fully dissolved. Incomplete dissolution may cause crystal formation, reducing pollen germination efficiency. To save time, dissolve ingredients the night before use, but handle under sterile conditions to prevent contamination. Liquid PGM is easier to prepare and generally offers higher reproducibility than solid PGM. Its quality is less affected by environmental conditions such as temperature and humidity. For improved consistency and reliability, most experiments should therefore be conducted using liquid PGM.
2. At room temperatures above 35 °C, rice pollen often germinates and bursts simultaneously, producing very short tubes. Perform assays in an air-conditioned environment, ideally at ~24 °C. At lower temperatures (20~24 °C), germination is slower, but bursting is delayed, which can benefit imaging or treatments. These thresholds were established under summer conditions in southern Korea; researchers should adjust conditions based on their own environments.

3. PGA is essential for visualizing subcellular localization of organelles or fluorescent proteins within pollen grains or pollen tubes. High temperature and humidity can interfere with solidification; use air conditioning or other measures to control conditions. Both PGM and PGA should be prepared fresh on the day of use for best results. This step is especially critical for preparing a solid PGM. If sucrose and PEG 4000 are not fully dissolved, granules may form during agar solidification, reducing germination rates. Incubate the mixture at 60 °C for at least 1 h to ensure complete dissolution. For best results, dissolve sucrose and PEG 4000 the day before use.
4. To avoid premature solidification, prepare slides individually by dispensing and covering one at a time, rather than all at once.
5. Sowing seeds without removing the lemma and palea increases the risk of microbial contamination on germination media.
6. To further reduce contamination, seeds can be immersed for 1 h in a commercial seed treatment solution containing 6% copper hydroxide and 8% ipconazole at the manufacturer's recommended concentration, prior to sterilization.
7. Prolonged exposure of seeds to 70% ethanol (>20 min) significantly reduces germination rates. Seeds should therefore be disinfected briefly (a few seconds) and immediately transferred to sterile water.
8. Continuous short-day conditions from the seedling stage promote earlier flowering but reduce tiller number and spikelet production, limiting usable flowers per plant. For rapid experiments (e.g., fluorescent protein localization) rather than quantitative analyses of yield or pollen development, plants may be grown under short-day conditions from the start.
9. For mature pollen-based assays, anthers should be collected from spikelets at anthesis or 1 day prior. Spikelets with filaments extending beyond half the lemma height are considered at the appropriate stage. Anthers from lower-positioned or immature spikelets should be avoided due to poor pollen quality.
10. Although DAPI can stain unfixed pollen, fixation in Carnoy's solution is highly recommended to enhance nuclear morphology preservation and dye penetration. Fixed spikelets can be stored in Carnoy's solution at 4 °C for up to 1 year, protected from light and tightly sealed.
11. If Carnoy's solution is not removed prior to DAPI staining, poor mixing of solutions can significantly reduce staining efficiency.
12. Record the time of anthesis for at least 3 days prior to pollen sampling to predict flowering windows and improve reproducibility. Flowering in a single panicle typically occurs over

~3 days, peaking on the second day. For quantitative analyses (e.g., pollen germination rate), panicles from the second day after heading are recommended.

13. For in vitro germination assays, pollen must be collected immediately upon spikelet opening, as rice pollen is highly sensitive to desiccation. Viability declines sharply within 5–10 min of dry air exposure, and dehydrated pollen does not reliably rehydrate or germinate.

14. Under field conditions with temperatures above 35 °C, premature pollen rupture may occur. If the weather is excessively hot or overcast, use the following alternative method. Under typical summer conditions in Korea, anthesis occurs between 10:00 and 12:00 on sunny days, though flowering may be delayed or suppressed under high humidity or rain. On rainy days, anther extrusion may fail, leading to internal pollen shedding. However, brief sunshine before 14:00 can still trigger flowering.

15. Once a panicle is exposed to dry air by removing the plastic bag, the anthers can only be used once for pollen collection. Each panicle should therefore be covered individually. If anthesis occurs inside the bag, pollen can typically be collected for up to 1 h as long as the bag remains sealed.

16. For subcellular localization using transgenic lines expressing fluorescent proteins, solid PGM is preferred because pollen grains move excessively in liquid medium, complicating imaging. For quantitative assays such as pollen germination rates, liquid PGM is recommended.

17. If the ambient temperature is below 23 °C (e.g., in air-conditioned rooms), extend incubation to 15 min to allow sufficient germination before fixation.

18. To minimize spreading, use a slightly smaller cover glass, or omit it if pollen is concentrated enough for direct imaging.

19. If the microscopy room is excessively warm, pollen may germinate and burst too quickly, resulting in poor FM4-64 staining. Maintain room temperature below 23 °C using air conditioning. For time-course imaging (>30 min), treatment with RALF17 or RALF19 peptides can delay pollen tube rupture [2, 6]. This approach should only be used after confirming that the protein of interest is unaffected by RALF signaling.

20. Significant sample loss may occur during washing steps. Collect as much pollen as possible initially to ensure sufficient material for downstream analysis.

21. Sampling is optimal between 2–4 h after anthesis. After 6 h, pollen tubes may no longer be clearly visible even after staining. Strict adherence to timing ensures successful observation.

22. If same-day observation is not possible, stained pistils can be stored overnight in 0.05% aniline blue. However, quality declines with prolonged storage. Do not store samples in glycerol, which causes the stain to wash out.

Acknowledgments

This research was supported by Brain Pool program (RS-2025-25455550, Pusan National University) and grants (RS-2024-00411930 to JHK, RS-2023-00217064, RS-2024-00344197 to YJK) from the National Research Foundation (NRF) and Ministry of Education, Science and Technology.

References

1. Moon S et al (2025) Advancements in hybrid rice production: improvements in male sterility and synthetic apomixis for sustainable agriculture. Plant Biotechnol J 23:2330–2345
2. Kim EJ et al (2023) Rice pollen-specific OsRALF17 and OsRALF19 are essential for pollen tube growth. J Integr Plant Biol 65:2218–2236
3. Kim EJ et al (2025) Global identification of key genes for pollen germination in rice through high-throughput screening and gene editing. J Integr Plant Biol 67:1665–1684
4. Kim YJ et al (2021) OsMTD2-mediated reactive oxygen species (ROS) balance is essential for intact pollen-tube elongation in rice. Plant J 107:1131–1147
5. Kim YJ et al (2020) Physiological importance of pectin modifying genes during rice pollen development. Int J Mol Sci 21:4840
6. Kim JH et al (2025) Pollen tube-expressed RUPO forms a complex with OsMTD2 and OsRALF17 and OsRALF19 peptides in rice (Oryza sativa). J Plant Physiol 305:154421

Chapter 3

Analyzing Pollen with Multispectral Imaging Flow Cytometry (MIFC)

Thomas Hornick, Franziska Walther, and Susanne Dunker

Abstract

Multispectral Imaging Flow Cytometry (MIFC) using the Imagestream X MkII instrument was currently introduced for pollen analysis of recent and fossil pollen. The method allows processing up to 2000 objects/s by simultaneously taking up to 12 microscopic images per object. Based on the images, multispectral patterns, scatter information, and morphological parameters of the pollen can be easily and accurately accessed in high throughput. This enables a powerful way of upscaling pollen analysis and pollen phenotyping to sample amounts, rarely feasible to be processed by classical microscopy, while at the same time providing accurate pollen trait measurements based on hundreds to thousands of processed pollen images. In addition, pollen images can be used for automated pollen identification via machine learning techniques such as Convolutional Neural Networks (CNNs).

Key words Pollen fluorescence, Pollen morphology, Multispectral imaging flow cytometry, Pollen phenotyping, Machine learning

1 Introduction

Pollen analysis by manual microscopy for identification, phenotyping, and quantification of pollen is time-consuming and requires an expert processing it. In order to speed up these analyses, new approaches were proposed recently using multispectral imaging flow cytometry (MIFC) [1–4]. This method, originally applied in medical applications, can be used for the quick and extensive analysis of pollen from different origins, including anthers, corbiculae, aerosol pollen traps, fossil or honey samples [5]. Pollen in a fluid suspension is transferred to the cytometer cuvette and aligned by hydrodynamic focusing with a faster-moving sheath fluid around the core stream containing the pollen. The pollen is physically separated due to the narrowing diameter of the cuvette. Ideally, isolated single pollen can then be imaged by an illuminating light source (brightfield LEDs or a laser). The Imagestream X MkII

Giampiero Cai et al. (eds.), *Pollen and Pollen Tube Analysis: Methods and Protocols*, Methods in Molecular Biology, vol. 3042,
https://doi.org/10.1007/978-1-0716-5308-1_3,

instrument allows for up to 2000 pollen/s to be processed, and from every particle, pollen images can be taken (max. 12 different brightfield and spectral images per pollen). The resulting images allow for a detailed analysis of pollen size [3], for quantifying pollen trait variability (e.g., size, shape, or fluorescence intensity based on image features) [3, 4], for pollen phenotyping [6], and for AI-based recognition of pollen using pollen traits or pollen images [1, 2]. The sample preparation is similar to manual microscopic analyses, but in comparison, the analysis can be much more automated. For example, pollen suspensions can be measured automatically using an autosampler in 96-well plates for automated sample handling (Fig. 1). Standardized data analysis protocols facilitate precise pollen trait estimation [3] and allow for reuse and long-term data archiving of pollen images and image features, enabling verification, re-analysis, and facilitating opportunities for future methodological innovations using the generated datasets.

2 Materials

Protocols for pollen sampling, extraction, MIFC measurement, and data analysis are based on [1, 3, 4].

2.1 Pollen Sample from Anthers

1. Collect mature anthers or, in the case of small inflorescences, the entire inflorescence or parts of it directly into a 1.5 mL or 2 mL sample tube with clean forceps. Alternatively, it is also possible to collect the anthers or, in the case of large flowers, in paper bags or Falcon tubes (*see* **Notes 1–5**).
2. If possible, measure all samples immediately. Alternatively, freeze samples at −20 °C until further analysis.

2.2 Pollen Samples from Corbiculae Pollen

1. Collect corbiculae pollen either directly from pollinators or alternatively by using pollen collection traps in front of honeybee or bumblebee hives (*see* **Note 6**) and transfer these into a 1.5 mL or 2 mL sample tube with clean forceps.
2. If possible, measure all samples immediately. Alternatively, freeze samples at −20 °C until further analysis.

2.3 Pollen Samples from Honey

1. Weigh 10 g of the honey sample into a 50 mL centrifuge tube (LABSOLUTE®, TH Geyer, Germany) and process it in accordance with DIN 10760.
2. Briefly, add twenty milliliters of distilled water to dissolve the honey, mix, and centrifuge it afterward for 10 min at 1000 *g* (Heraeus Megafuge 16R centrifuge, ThermoFisher Scientific, Osterode, Germany).

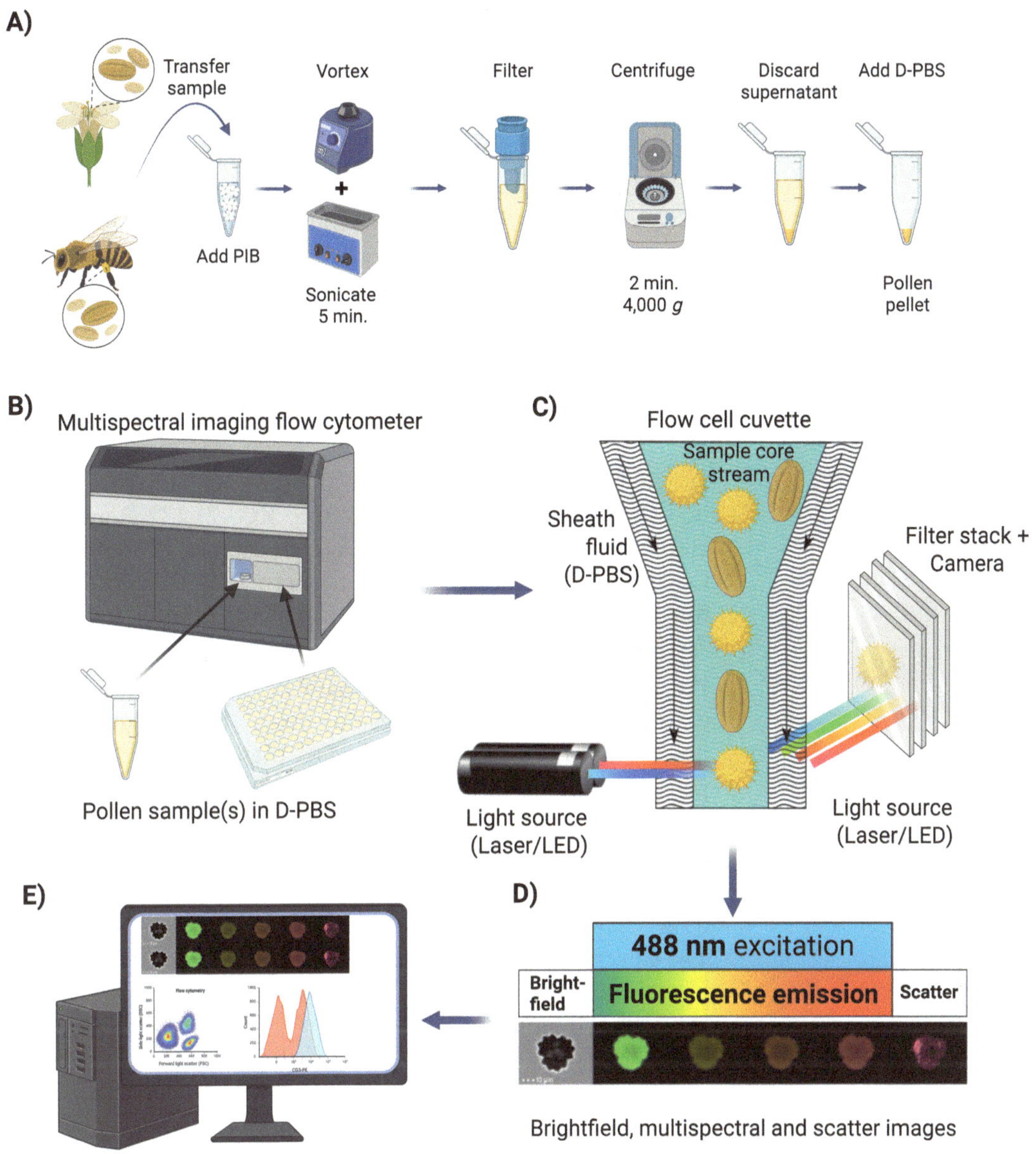

Fig. 1 Sample preparation and pollen measurements via multispectral imaging flow cytometry (MIFC). (**a**) Schematic illustration of major steps for pollen sample processing for MIFC. (**b**) Samples can be measured manually from tubes or automatically from microtiter plates. (**c**) Pollen gets aligned in the flow cell cuvette by hydrodynamic focusing, physically separated by a narrowing diameter of the cuvette. Pollen is imaged by a filter stack and a CCD camera system. (**d**) Recorded brightfield, fluorescence, and scatter images allow for (**e**) detailed phenotyping and trait analysis. (Created in BioRender. Dunker, S. (2025) https:/BioRender.com/gm4 xop3)

3. Discard the supernatant, and add another 20 mL of distilled water to fully dissolve the sugar crystals in the honey.

4. Centrifuge the solution again for 5 min at 1000 *g*, and decant the supernatant.
5. Add 1 mL of 1X D-PBS (Dulbecco's phosphate-buffered saline, without calcium and magnesium; Biowest, Nuaillé, France) (*see* Subheading 2.2) to the centrifuge tube and mix it thoroughly.
6. Pipette this sample into a 1.5 mL sample tube by filtering it through a 100 μm filter (CellTrics, Sysmex, Norderstedt, Germany), and centrifuge it for 2 min at 4000 *g*.
7. Remove the supernatant and add 70 μL of 1X D-PBS.

2.4 Fluid Reagents for Pollen Sample Processing and MIFC Measurements

Prepare the following solutions for pollen sample processing and MIFC measurement:

1. Pollen isolation buffer (PIB): 100 mM KH_2PO_4 (e.g., cat. no. 1648, Th. Geyer GmbH & Co. KG, 71272 Renningen, Germany), pH 7.5; 1 mM EDTA (e.g., cat. no. 131669, AppliChem GmbH, 64291 Darmstadt, Germany), 0.1% (v/v); Triton X-100 (e.g., cat. no. 3051.3, Carl Roth GmbH & Co. KG, 76185 Karlsruhe, Germany) (*see* **Note 7**).
2. Prepare 1X Dulbecco's Phosphate-Buffered Saline (D-PBS) w/o Calcium w/o Magnesium from 10× solution (cat. no. X0515, Biowest, 49340 Nuaillé, France).
3. Flow cytometry reagents: Sheath fluid—Dulbecco's Phosphate Buffered Saline (D-PBS) 1× w/o Calcium w/o Magnesium prepared from 10× solution (cat. no. X0515, Biowest, 49340 Nuaillé, France) (*see* **Note 8**); Sterilizer: Sodium hypochlorite 0.6% (e.g., prepare dilution from cat. no. 1305.1000; Th. Geyer GmbH & Co. KG, Renningen, Germany); Debubbler: 70% 2-Propanol (HPLC grade); Speed Bead® Image Stream®X System Calibration Reagent (cat. no. 400041, Cytek Bioscience, Inc., Seattle WA 98119, USA); Cleanser: FlowClean Cleaning Agent (cat. no. C48093; Beckman Coulter Ireland Inc., Lismeehan, Ireland); Rinse Reagent–0.2 μm filtered, deionized water.

3 Methods

3.1 Processing of Anther and Corbiculae Pollen Samples for MIFC Measurement

1. Use the prepared pollen sample from anther or corbiculae samples for the subsequent processing.
2. Add 200–700 μL of PIB to fully cover the sample.
3. Place the sample in an ultrasonic bath for 5 min to separate pollen and avoid aggregated pollen from being retained during the following filtration steps.
4. Vortex the sample thoroughly.

5. Filter the sample through a cell-strainer with an appropriate filter size depending on the expected pollen size into 1.5 mL sample tubes with appropriate filters (e.g., CellTrics® Filter: 30, 50, 100 µm (cat. no. 04-0042-2316, 04-0042-2313, 04-0042-2313), Sysmex Deutschland GmbH, Norderstedt, Germany; Filcon® Filter: 70 µm (cat. no. 12170-67), Süd-Laborbedarf GmbH, Gauting, Germany) (*see* **Note 9**).
6. Centrifuge the filtrate at 4000 *g* for 2 min and carefully discard the supernatant.
7. Resuspend the pollen pellet in 50–100 µL of D-PBS and vortex to homogenize.
8. Measure the pollen sample using the instrument-specific INSPIRE-software. If immediate measurement is not possible, store the sample at −20 °C until analysis.

3.2 Multispectral Imaging Flow Cytometry

1. Check that reagent containers (Sheath, Sterilizer, Debubbler, Cleanser, Rinse, and Speed Bead® Calibration Reagent) are filled and the waste container is emptied. Run the instrument-specific ASSIST test for flushing, cleaning, and calibration of the ImageStreamX MkII system daily before the start of the measurement. In case of not proper functioning of fluidics or air bubbles and/or blocking particles in the tubing system, the ASSIST test will result in an error. Check the manufacturer's instructions for troubleshooting.
2. Use the following settings for the ImageStreamX MkII instrument measurements:
   - 488 nm laser intensity: 5 mW.
   - 561 nm laser intensity: 20 mW + neutral density filter (ND 1.0).
   - 785 nm laser intensity: 0.1 mW.
   - For ImageStreamX MkII use channel 01 (Ch01) and channel 09 (Ch09) as brightfield channels, fluorescence channels (Ch02, Ch03, Ch04, Ch05, Ch07, Ch08, Ch10, Ch11, Ch12) as well as channel 06 (Ch06) as scatter channel (see Fig. 2).
   - Magnification: Use the 40× objective for pollen <60 µm (pixel resolution at "Low" speed 0.5 × 0.5 µm^2), 20× objective for pollen >60 µm (pixel resolution at "low" speed 1 × 1 µm^2). The upper size limit depends on the used cell-strainer filter size during sample preparation. We recommend using a maximum filter size of 100 µm to prevent clogging of fluidics. However, elongated pollen with a large size in the longest direction that can pass the filter can still be measured (approx. 120 µm).

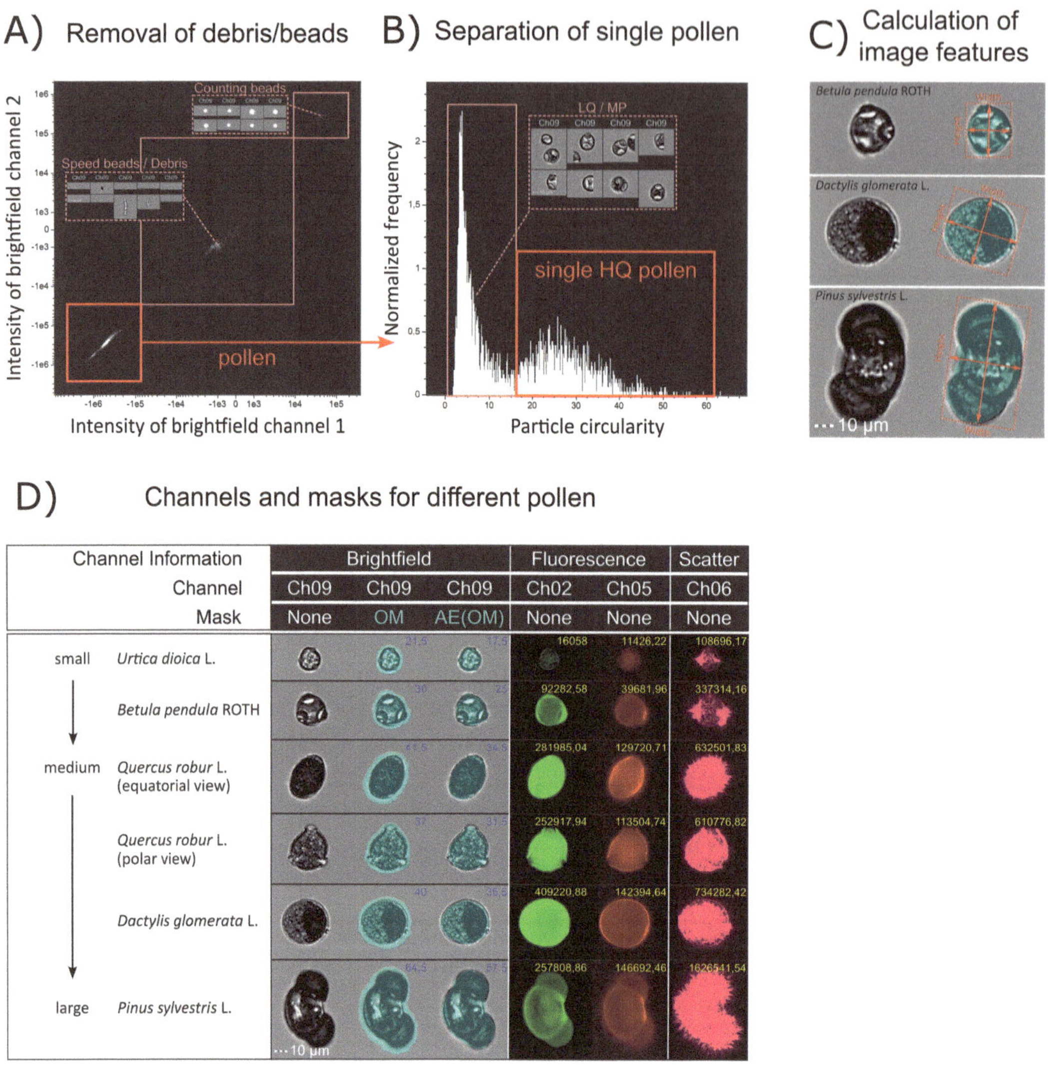

Fig. 2 Workflow for image processing in IDEAS cytometry software. (**a**) Pollen (*Corylus avellana* L.) is distinguished from beads and debris using histograms or biplots of brightfield intensities from Cameras 1 and 2. (**b**) Based on a density plot of particle circularity, high-quality (HQ) pollen, defined as intact single grains without debris, are presorted from low-quality (LQ) or multiple pollen grains; HQ pollen were manually verified, with automation possible via outlier detection. (**c**) From single HQ pollen, size features were computed using an AdaptiveErode object mask (AE(OM)) on brightfield channel 09 (Ch09). As an example, size estimation for spherical (*Betula pendula* ROTH & *Dactylis glomerata* L.) and elongated (*Pinus sylvestris* L.) pollen is shown. (**d**) The AE(OM) mask reliably captures morphological features across pollen size classes, yielding more accurate measurements than default (M09) or other masks. Small numbers on the top right of the brightfield images denote the calculated "Length" feature for the two mask types (OM object mask, AE(OM) Adaptive erode mask on an object mask) as well as fluorescence and scatter intensities (default mask)

- Flow speed: Set the flow speed to "Low," otherwise image quality is not sufficient to derive accurate pollen size estimates or other features.
- Acquired particle number: Collect data of at least 5000 particles, depending on sample concentration and data demand, or restrict the time to a maximum of 10 min (*see* **Notes 10–12**).

3.3 Data Analysis

1. Analyze the collected images using the IDEAS-software. For pollen trait analysis, high-quality (HQ) single pollen images need to be discriminated from images containing multiple pollen, beads, or non-pollen particles. First, remove debris images based on a bivariate plot of brightfield image intensities from both CCD-cameras (*see* **Notes 13** and **14**) (Fig. 2). Discriminate brightfield images with single pollen and multiple pollen by using the circularity feature that measures the degree of particle shape deviation from a circle (see manual of IDEAS: selecting "Help" → "Contents...") (Fig. 2a/b).
2. Once you have defined the population of interest, export the images and/or image features as fcs-file/txt-file for user specific tasks (e.g., training of a convolutional neural network (CNN) or trait analysis using image features) (IDEAS: selecting "Tools" → "Export tif. Images..." or "Export Feature Values...").
3. For pollen trait analyses extract median values per measurement to reveal robust average values and account for non-gaussian trait distributions (IDEAS: 1) selecting "Reports" → "Define Statistics Report" → "Add Statistic Report Columns" by defining "Statistics" (→ Median) the "Selected population" (→ HQ) and "Features" (→ Length feature) → "Add Statistics"; 2) selecting "Reports" → "Generate Statistics Report").

4 Notes

1. When pollen samples are collected directly from plants in the field, the plant should be identified at the species level using identification books or apps (e.g., Flora Incognita). It is advisable to take photos and, if possible, record GPS data and additional metadata (e.g., plant species, location, time of sampling, number of flowers or anthers, and name of the collector).
2. Metadata documentation is very important for a FAIR data management practice and a sustainable data life cycle. It is recommended to store all relevant information on sample collection and processing, instrument details, measurement, and

data analysis templates, purpose of the experiment according to the MIFlowCyt standard [7, 8].

3. The pollen count per anther and flower varies greatly between plant species. Generally, wind-pollinated species produce more pollen than insect-pollinated ones; however, even within pollination syndromes, the amount of pollen produced can vary largely between species. Additionally, the efficiency of pollen extraction can depend on the accessibility of pollen in the flower (closed (e.g., Laminaceae, Fabaceae) vs. open (Asteraceae) flower types). Therefore, no precise information on the amount of anthers or flowers can be given that should be used for sample processing, but as a rough guideline, we found that 3 ripe anthers or at least 20 flowers for species with rather low pollen amounts have generally been sufficient for our sample processing. Paper bags and Falcon tubes might be advantageous for sampling larger quantities of pollen. For most wind-pollinated species, we recommend sampling whole inflorescences (e.g., catkins) directly into paper bags and gently drying them at no more than 30 °C in a ventilated drying chamber with forced airflow to prevent mold growth [3]. For insect-pollinated species, larger quantities of flowers might be sampled and processed in Falcon tubes. In this case, the protocol for further sample processing needs to be adjusted to suitable centrifuges and volumes of used liquids. Otherwise, follow the recommended protocol based on sample processing in 1.5 or 2 mL Eppendorf tubes.
4. To obtain as much mature pollen as possible, it is important to estimate the correct ripeness of the pollen, ideally based on phenological observations. Mature pollen can be recognized by clearly visible pollen grains on opening anthers. If pollen is collected from plants that are not yet in full bloom, a larger proportion of the pollen will still be undeveloped. Collecting pollen at two time points, e.g., in the morning from recently opened flowers and early afternoon, could limit this issue and reduce the contamination with heterospecific pollen.
5. To minimize contamination between samples from different plant species, hands and forceps should be cleaned between samples (e.g., using wet wipes).
6. Individual corbiculae samples of honeybees and bumblebees contain sufficient material for one measurement or do not even need to be diluted. Depending on the research questions, it might be advisable to homogenize samples of corbiculae pollen and take a subsample from the homogenate.
7. PIB is prepared according to [9], who used the buffer to resuspend pollen for microscopic analysis, with the modification of using KH_2PO_4 instead of Na_2HPO_4. EDTA weakens

cell–cell or cell–surface adhesion, and the non-ionic surfactant Triton X-100 lowers the surface tension of hydrophobic pollen surfaces to better disperse pollen in the buffer. KH_2PO_4 helps to maintain a stable pH in the near-neutral range. EDTA is available in various chemical forms that vary in solubility and the degree of neutralization. For this protocol, we used Disodium EDTA, which is highly soluble in water and less alkaline than other forms such as Tetrasodium EDTA.

8. In order to avoid optical effects caused by differences in the refractive index between the sheath fluid and the sample stream, use the same buffer (1X D-PBS) as the sheath fluid and the liquid to resuspend pollen for MIFC measurement (sample stream).
9. Pouring samples directly into the cell-strain filters may leave some pollen adhering to the walls of the tubes and filters. Rinse the original sampling tubes with at least 200 μL of PIB and re-filter to minimize pollen loss.
10. A minimum sample volume of 50 μL should be prepared, as smaller volumes could interfere with the fluidics system. When using manual mode for measurements, it is recommended to vortex the sample sufficiently before measurement, while avoiding air bubbles in the sample, which can interfere with stable laminar flow.
11. Measurements with images of 5000 acquired particles usually result in a storage space of about 100–500 Mb (.rif-file). Conversion of files into.cif- and.daf-files during data analysis will increase disk space by a factor of about 2.5.
12. A threshold on size should be used during image collection in order to reduce data storage of non-pollen images (i.e., small debris particles). But if the size threshold is set too high, a systematic exclusion of potentially relevant pollen might happen. For single-species samples, the threshold can be more restricted in comparison to mixed species samples or environmental samples with an unknown pollen composition.
13. The numbering/naming of image channels depends on the user-specific instrument configurations. In case of cytometers that are only equipped with one CCD-camera also a histogram of one brightfield image intensity can be used for separation of debris and beads from pollen-like particles instead. Also, the gating strategy described here has proven to be optimal for us. Nonetheless, differentiation of various particle classes can also be achieved based on other features (e.g., using the elongatedness instead of circularity feature in case of more elongated pollen types).
14. The image analysis workflow (Fig. 2) depends, for example, on the purity of the sample or the pollen type and is established to

achieve the best possible determination of image features, which is why great care was taken to discriminate single HQ pollen images. Manual inspection involves removing other remaining particles, hetero-specific pollen, cropped pollen, and pollen with debris attachments, which allows for unbiased estimation of pollen size features. For other purposes, such as training of CNNs, however, it might be beneficial to also include other images in the dataset to increase the observed variability across pollen images (e.g., images with multiple pollen, pollen with dirt attachments, etc.).

Acknowledgments

Funding for this research was provided by the German Centre for Integrative Biodiversity Research (iDiv) Halle–Jena–Leipzig (DFG Research Centre FZT 118, iCyt–Support Unit 346001057-01), PolDiv Flexpool project (34600830-13), by the Federal Ministry of Food and Agriculture based on a resolution of the German Bundestag as part of the Federal Program for Organic Farming and Other Forms of Sustainable Agriculture via the project Nutri-Bee (2819NA102), the Carl-Zeiss-Foundation funded Phenology-based pollen prediction and EEG-based allergic response assessment using AI (PollenNet) project (Grant Number: P2022-08-006) and the Helmholtz Association.

References

1. Dunker S, Motivans E, Rakosy D, Boho D, Mäder P, Hornick T, Knight TM (2021) Pollen analysis using multispectral imaging flow cytometry and deep learning. New Phytol 229(1): 593–606. https://doi.org/10.1111/nph.16882
2. Barnes CM, Power AL, Barber DG, Tennant RK, Jones RT, Lee GR, Hatton J, Elliott A, Zaragoza-Castells J, Haley SM, Summers HD, Doan M, Carpenter AE, Rees P, Love J (2023) Deductive automated pollen classification in environmental samples via exploratory deep learning and imaging flow cytometry. New Phytol 240(3):1305–1326. https://doi.org/10.1111/nph.19186
3. Hornick T, Harpole WS, Dunker S (2025) High-throughput assessment of anemophilous pollen size and variability using imaging cytometry. New Phytol 246(4):1875–1888. https://doi.org/10.1111/nph.70070
4. Walther F, Hofmann M, Rakosy D, Plos C, Deilmann TJ, Lenk A, Römermann C, Harpole WS, Hornick T, Dunker S (2025) Multispectral imaging flow cytometry for spatio-temporal pollen trait variation measurements of insect-pollinated plants. Cytometry A 107(5): 293–308. https://doi.org/10.1002/cyto.a.24932
5. Dunker S, Boyd M, Durka W, Erler S, Harpole WS, Henning S, Herzschuh U, Hornick T, Knight T, Lips S, Mäder P, Švara EM, Mozarowski S, Rakosy D, Römermann C, Schmitt-Jansen M, Stoof-Leichsenring K, Stratmann F, Treudler R, Virtanen R, Wendt-Potthoff K, Wilhelm C (2022) The potential of multispectral imaging flow cytometry for environmental monitoring. Cytometry A 101(9): 782–799. https://doi.org/10.1002/cyto.a.24658
6. Waesch C, Gao Y, Koch N, Gaede N, Hornick T, Dusny C, Fuchs J, Börner A, Himmelbach A, Mascher M, Pillen K, Dunker S, Dreissig S (2025) Pollen and anther morphological variation in rye was shaped by domestication. BMC Plant Biol 25(1):389. https://doi.org/10.1186/s12870-025-06416-x
7. Spidlen J, Breuer K, Brinkman R (2012) Preparing a Minimum Information about a Flow

Cytometry Experiment (MIFlowCyt) compliant manuscript using the International Society for Advancement of Cytometry (ISAC) FCS file repository (FlowRepository.org). Curr Protoc Cytom Chapter 10(1), Unit 10.18. https://doi.org/10.1002/0471142956.cy1018s61

8. Lee JA, Spidlen J, Boyce K, Cai J, Crosbie N, Dalphin M, Furlong J, Gasparetto M, Goldberg M, Goralczyk EM, Hyun B, Jansen K, Kollmann T, Kong M, Leif R, McWeeney S, Moloshok TD, Moore W, Nolan G, Nolan J, Nikolich-Zugich J, Parrish D, Purcell B, Qian Y, Selvaraj B, Smith C, Tchuvatkina O, Wertheimer A, Wilkinson P, Wilson C, Wood J, Zigon R, Scheuermann RH, Brinkman RR (2008) MIFlowCyt: the minimum information about a flow cytometry experiment. Cytometry A 73(10):926–930. https://doi.org/10.1002/cyto.a.20623

9. Aloisi I, Cai G, Tumiatti V, Minarini A, Del Duca S (2015) Natural polyamines and synthetic analogs modify the growth and the morphology of Pyrus communis pollen tubes affecting ROS levels and causing cell death. Plant Sci 239:92–105. https://doi.org/10.1016/j.plantsci.2015.07.008

Chapter 4

Imaging Pollen and Pollen Tubes by Scanning and Transmission Electron Microscopy

Weiwei Zhang and Kefeng Fang

Abstract

Pollen and pollen tubes are important structures for sexual reproduction in plants. The study of these structures is of great significance for understanding biological processes such as plant fertilization mechanisms, cell polarity growth, cell wall synthesis, and signal transduction. With the advancement of microscopic imaging technology, researchers have gained a deeper understanding of the ultrastructure of pollen and pollen tubes. This article summarizes the research methods for the ultrastructure of pollen and pollen tubes, including scanning electron microscope (SEM) and transmission electron microscope (TEM).

Key words Pollen, Pollen tube, Ultrastructure, Scanning electron microscope, Transmission electron microscope

1 Introduction

Scanning electron microscopy (SEM) is irreplaceable in plant reproductive biology owing to its high resolution, large depth of field, and three-dimensional imaging capabilities [1, 2]. It exhibits distinct advantages, particularly in the investigation of pollen morphology and pollen tube growth mechanisms, serving as a robust tool for analyzing the surface ultrastructure of pollen and pollen tubes. Key observable features include pollen surface ornamentation, germination furrows/pores, polarity and symmetry, as well as the textural characteristics of pollen tube tips and walls [3–5]. Beyond structural analysis, SEM also plays a pivotal role in plant taxonomy, evolutionary biology research, assessment of plant responses to environmental stress and plant physiological states, and qualitative/semi-quantitative analysis of surface elements [6–8].

Transmission electron microscopy (TEM) possesses ultra-high resolution and exceptional magnification capacity, which compensate for the limitations of light microscopy in resolving

Giampiero Cai et al. (eds.), *Pollen and Pollen Tube Analysis: Methods and Protocols*, Methods in Molecular Biology, vol. 3042, https://doi.org/10.1007/978-1-0716-5308-1_4,

organelle-level details. As such, it is an indispensable tool for investigating the ultrastructure of pollen and pollen tubes. TEM has widespread application in key research areas, including pollen wall architecture (exine, intine, and germination furrow/pore morphology) [9–11]; cellular mechanisms underlying pollen tube polar growth (polarized organelle distribution, cytoskeletal dynamics, and exocytotic events) [12–14]; organelle morphology and function (e.g., vegetative nuclei and sperm cells, sperm cell transport and double fertilization regulation, mitochondria, Golgi apparatus, endoplasmic reticulum, and nuclei) [12, 15, 16]; cellular chemistry and immunolocalization studies [12]; stress-induced structural abnormalities [14] and programmed cell death (PCD) in pollen/pollen tubes [17].

Although transmission electron microscopy (TEM) has inherent limitations—such as labor-intensive sample preparation and the inability to capture dynamic processes—the ultra-high-resolution structural details it provides are irreplaceable by other techniques. This ensures TEM remains a core technology in plant reproductive biology and palynology research. The following are the detailed methods of the SEM and TEM sample preparation for pollen and pollen tubes.

2 Materials

1. Pollen grains: collect mature anthers from the target plant and place them under ambient light to facilitate pollen dehiscence, followed by the collection of the released pollen. For larger pollen grains, direct use (for observation, without prior culture, if applicable when mature pollen tubes are already present) is feasible.
2. Pollen tubes: liquid culture is the preferred method for pollen tube induction, as it enables convenient harvesting.

2.1 Reagents and Consumables for Scanning Electron Microscopic Observation

1. 2.5% glutaraldehyde fixative (adjusted to pH 7.4).
2. 0.1 M phosphate-buffered saline (PBS) buffer.
3. Osmium tetroxide (OsO_4): (*see* **Note 1**).
4. Absolute ethanol (100%).
5. Deionized water.
6. Tert-Butanol.
7. Gold target (for sputtering).
8. Toothpicks.
9. Conductive adhesive tapes.
10. SEM sample stubs.

2.2 Reagents and Consumables for Transmission Electron Microscopy

1. Phosphate-buffered saline (PBS) buffer (pH 7.2): 0.1 M solution in water.
2. Glutaraldehyde fixative (for primary fixation): 2.5% solution in PBS buffer.
3. Osmium tetroxide.
4. Absolute ethanol (100%, for dehydration).
5. Deionized water.
6. Acetone (for resin substitution).
7. Spurr resin (for embedding): Prepare the resin mixture in a fume hood using the following proportions: ERL-4221: 10 g, DER-736: 5 g, NSA (nonenyl succinic anhydride): 26 g, DMAE (dimethylaminoethanol, catalyst): 0.4 g. Mix components thoroughly until homogeneous. Degas under vacuum for about 40 min to eliminate air bubbles (critical for embedding quality). Store the degassed resin at −20 °C to prevent premature polymerization.
8. Uranyl acetate staining solution (for contrasting membrane structures).
9. Lead citrate staining solution (for enhancing cytoplasmic and organelle contrast).
10. Formvar−/carbon-coated copper grids (100 mesh, for section mounting).
11. 1 mL syringes (for liquid transfer).
12. Sulfuric paper (weighing paper, for labeling).
13. Pencils (for non-erasable labeling).
14. Adhesive tapes (e.g., parafilm or waterproof tape, for water trough assembly).

3 Methods

3.1 Pollen and Pollen Tube Sampling and Observation by Scanning Electron Microscopy

1. Pre-Fixation Rinsing (Exposing Surface Features): To preserve the samples in a near-physiological state, surface impurities must be removed prior to fixation. Transfer the specimens to 0.1 M PBS buffer and perform gentle rinsing three times, with each rinse lasting 20 min. Ensure all manipulations are conducted gently, rapidly, and precisely to prevent physical damage to pollen or pollen tubes.
2. Primary Fixation (Pre-Fixation): Immediately immerse the rinsed samples in 2.5% glutaraldehyde fixative (prepared in 0.1 M PBS, pH 7.4). Apply vacuum to the fixation system for 2 min to eliminate air bubbles trapped on the sample surface (critical for uniform fixative penetration). Subsequently, incubate the samples at room temperature for 2–4 h, with

intermittent shaking during the incubation period to ensure consistent fixation.

3. Post-Fixation Rinsing: Rinse the fixed samples with 0.1 M PBS buffer (pH 7.4) 8 times, with each rinse duration of 10 min (*see* **Note 2**).
4. Secondary Fixation: Transfer the rinsed samples to a 1% osmium tetroxide solution and incubate at room temperature for 1–2 h. Do not exceed this duration, as prolonged OsO_4 exposure may cause over-fixation and make samples brittle.
5. Rinsing Post-Secondary Fixation: Rinse the samples with 0.1 M PBS buffer 8 times, with each rinse lasting 10 min (*see* **Note 3**).
6. Gradient Dehydration: Perform dehydration at room temperature using ethanol as the dehydrating agent, following a sequential concentration gradient to minimize sample shrinkage. Incubate the samples in each ethanol concentration for 15 min: 10%→30%→50%→70%→80%→90%→95%→100%→100% (*see* **Note 4**).
7. Tert-Butanol Substitution: After dehydration, transfer the samples to a tert-butanol solution and incubate for 30 min or overnight at room temperature (*see* **Note 5**).
8. Freeze-Drying: Pre-cool the cold trap of the freeze-dryer to the preset low temperature and stabilize the temperature before sample loading. Transfer the tert-butanol-treated pollen/pollen tubes into 2 mL centrifuge tubes and secure the tubes onto a cardboard support. Place the cardboard with centrifuge tubes into the freeze-dryer's cold trap, seal the chamber, and initiate vacuum drying. Maintain the freeze-drying process for 4–6 h to ensure complete removal of tert-butanol.
9. Sample Mounting and Gold Sputtering: Using double-sided conductive adhesive tape, mount the freeze-dried samples onto SEM sample stubs, ensuring the samples are firmly attached and oriented for optimal imaging. Place the mounted stubs into an ion sputtering coater and deposit a thin gold layer onto the sample surface for 60 s (*see* **Note 6**).
10. SEM Observation: Transfer the gold-coated samples to the SEM chamber, evacuate the chamber to the required vacuum level, and perform imaging at appropriate accelerating voltages (typically 5–20 kV, depending on sample size and desired resolution). The normal and abnormal pollen of the strawberry are shown in Fig. 1a and b, we can find that normal pollen is full and round, and abnormal pollen is sunken. By contrast, the normal and abnormal pollen of Chinese chestnut had no

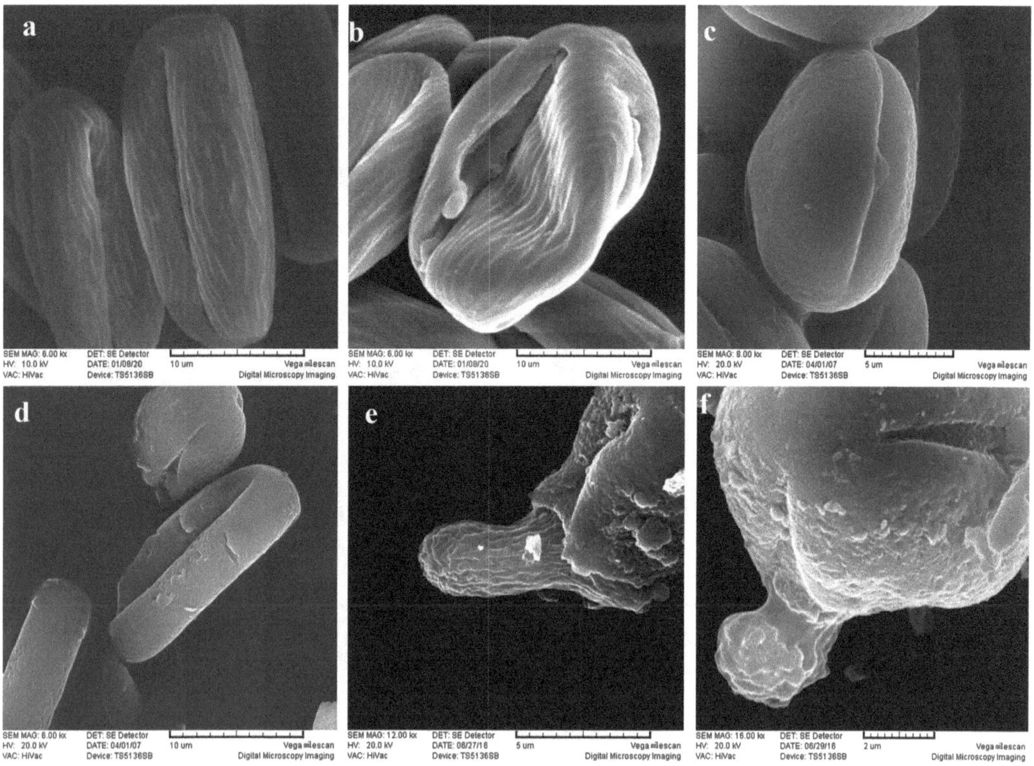

Fig. 1 Observation of pollen and pollen tube using SEM. (**a**) Normal pollen of the strawberry is full and round, with smooth strip-shaped patterns on the surface. (**b**) Abnormal pollen of the strawberry exhibits altered surface patterning due to concavity. (**c** and **d**), Normal and abnormal pollen of the Chinese chestnut. (**e** and **f**), Morphology of pollen tubes of Chinese chestnut, with striations aligned parallel to the longitudinal axis, Proximal–distal regions (**e**), and undulating surface ornamentation in the apex (**f**)

longitudinal lines on their surface, but there were irregularly dispersed shallow foveolate depressions, and the abnormal pollen of Chinese chestnut lacked a germination groove (Fig. 1c and d). The pollen tube apex of Chinese chestnut exhibits undulating surface ornamentation, while the subapical and distal regions display striations aligned parallel to the longitudinal axis (Fig. 1e and f).

3.2 Preparation of Pollen Ultrathin Sections and Observation Under Transmission Electron Microscopy

1. Sample Fixation (Dual-Fixation Protocol): Primary fixation: Immerse anthers or pollen tubes in 2.5% glutaraldehyde (prepared in 0.1 M PBS, pH 7.2). Apply vacuum for 30 min to facilitate fixative penetration and ensure samples sink (avoid floating, which causes uneven fixation). Continue vacuum fixation for 1.5–2 h at room temperature, then transfer the sample-fixative mixture to a 4 °C refrigerator for overnight fixation (to maximize structural preservation).

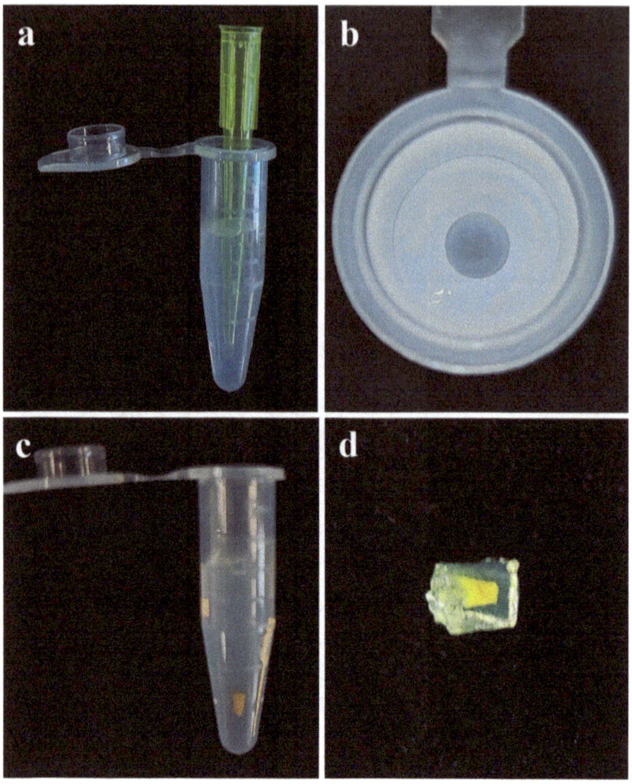

Fig. 2 The procedure for pollen encapsulation in agarose medium. (**a**) 1.5 mL centrifuge tube with gel, and a pipette tip inserted into the gel. (**b**) A tip-shaped hole in the gel for obtaining the pollen encapsulation mold. (**c**) Pollen or pollen tubes were injected into the tip-shaped hole in the gel. (**d**) Pollen or pollen tubes were wrapped in gel

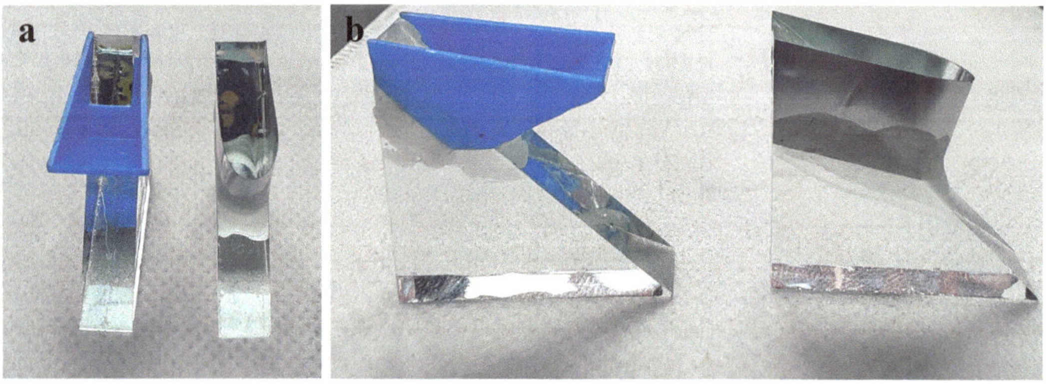

Fig. 3 Glass knife and liquid trough assembly for ultramicrotomy. (**a**) Two different sinks. (**b**) Seal the sink with paraffin wax to prevent leakage

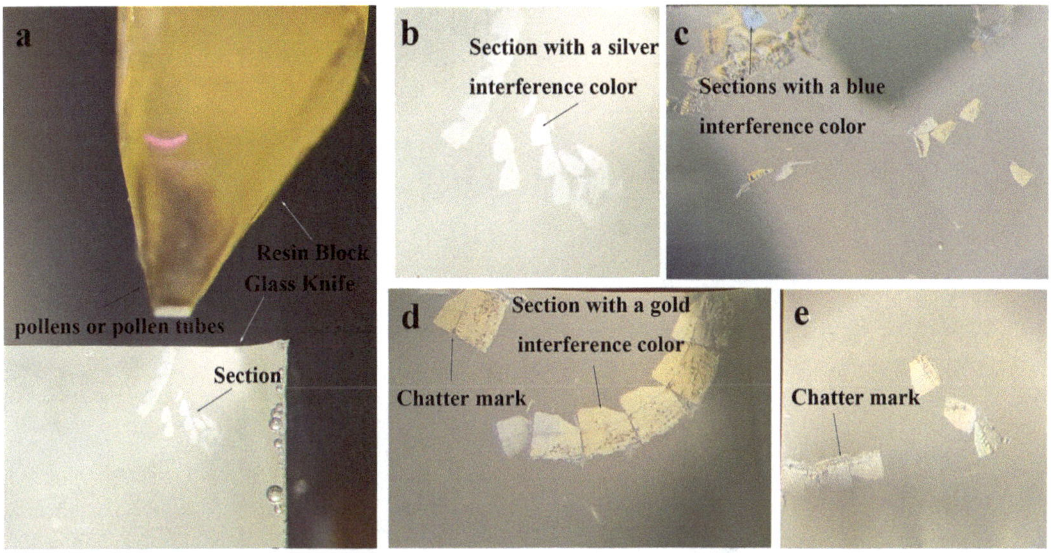

Fig. 4 Ultramicrotomy of pollen and pollen tubes at varying physiological states. (**a**) The specimen block, glass knives, and ultrathin sections were under the microscope field of view. (**b–e**) Sections with a silver interference color (**b**) and golden yellow color (**d**) were optimal, and the thickness was approximately 50–90 nm. But sections with a blue interference color must be discarded due to excessive thickness (**c**). The chatter marks in the sections in (**c**) to (**e**) cause localized obliteration of ultrastructural information in corresponding areas, thereby compromising morphological observation in these regions

2. Post-Fixation Rinsing: After primary fixation, remove the glutaraldehyde solution and rinse the samples with 0.1 M PBS (pH 7.2) 10 times, with each rinse lasting 10 min at room temperature. After each rinse, centrifuge the samples at low speed (e.g., 6000 rpm for 2 min) to collect pollen/pollen tubes, then discard the supernatant (PBS).

3. Pollen Encapsulation (for Small or Fragile Samples): Encapsulation in agarose gel prevents sample loss or damage during subsequent steps [18]. Mold preparation: prepare 1% agarose gel using deionized water and heat until fully dissolved. Pipette the molten gel into a 1.5 mL centrifuge tube. When the gel is nearly solidified (slightly warm to the touch), insert a sterile pipette tip (100–200 μL) vertically into the gel (Fig. 2a). Once the gel is completely solidified, gently remove the pipette tip to leave a tip-shaped cavity, which serves as the pollen encapsulation mold (Fig. 2b).

   Pollen tube encapsulation: resuspend the centrifuged pollen/pollen tubes (from **Step 2**) in a small volume of PBS and inject the suspension into the agarose mold cavity (Fig. 2c). Allow the samples to settle naturally, then aspirate the supernatant, leaving only the sedimented pollen/pollen tubes in the cavity. Trim excess agarose from the mold using a sterile blade,

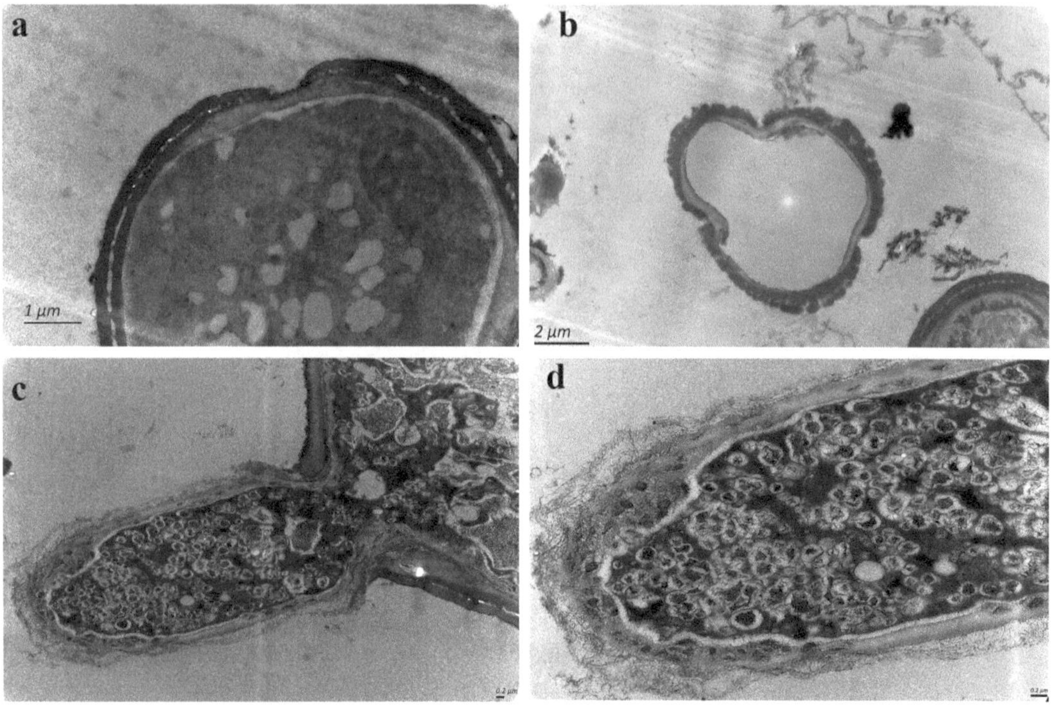

Fig. 5 Ultrastructural imaging of pollen grains and pollen tubes via TEM. (**a**) Normal pollen grain of Chinese chestnut. (**b**) Abnormal pollen grains of Chinese chestnut without cytoplasm. (**c**) Morphology of the pollen tube of the Chinese chestnut. (**d**) Morphology of the apex and proximal-distal region of the pollen tube in the Chinese chestnut

then seal the top of the cavity with a small amount of molten 1% agarose (to secure the sample) (Fig. 2d).

4. Secondary Fixation (Osmium Tetroxide Fixation): In a fume hood, prepare a 1% OsO_4 solution by mixing 2% OsO_4 stock solution with 0.1 M PBS (volume ratio 1:1) in a 1.5 mL centrifuge tube. Immerse the encapsulated samples (or unencapsulated pollen tubes, if robust) in the OsO_4 solution and cap the tube tightly immediately to prevent vapor leakage. Gently invert the tube 2–3 times to ensure uniform exposure to OsO_4, then incubate at room temperature for approximately 1.5 h (shake the tube once midway through incubation).
5. Post-Secondary Fixation Rinsing: Carefully discard the OsO_4 solution (follow hazardous waste disposal protocols) and rinse the samples 10 times with 0.1 M PBS (pH 7.2), 10 min per rinse. After the final rinse, discard the PBS and blot excess liquid from the sample surface using filter paper (avoid crushing the sample).
6. Gradient Dehydration: Use a sterile toothpick to transfer the sample to a new 2.0 mL centrifuge tube. Prepare a series of

ethanol solutions (10%, 30%, 50%, 70%, 80%, 90%, 95%, 100%, 100%) in 10 mL graduated centrifuge tubes using absolute ethanol and deionized water. Dehydrate the sample using a series of ethanol solutions (10%, 30%, 50%, 70%, 80%, 90%, 95%, 100%, 100%) at room temperature (20–25 °C), with each step lasting 30 min.

7. Transfer the sample using a sterile toothpick to avoid mechanical damage, and incubate on a gentle shaker (50–80 rpm) to ensure thorough dehydration. The final two 100% ethanol steps are critical to eliminate residual water, which is essential for subsequent resin infiltration.
8. Acetone Substitution: After dehydration, replace ethanol with acetone (to improve resin compatibility) by immersing the sample 3 times in fresh 100% acetone, 30 min per incubation at room temperature.
9. Resin Infiltration: To minimize structural shrinkage, infiltrate the sample with Spurr resin using a sequential acetone-resin gradient (prepare all solutions in 10 mL centrifuge tubes, sealed with parafilm to prevent solvent evaporation). Prepare three infiltration solutions, including acetone: resin = 3:1, 1:1, and 1:3. Immerse the sample in each solution sequentially, incubating for 8 h at room temperature per gradient. Finally, the sample was infiltrated with pure resin twice: first for 2 h, and then overnight.
10. Embedding: Pre-clean a silicone embedding mold (with 5–10 mm diameter wells) using acetone and dry it thoroughly. Inject fresh, degassed Spurr resin into each well until 2/3 full, carefully avoiding air bubbles (tap the mold gently on the bench to release trapped bubbles). Use a sterile toothpick to transfer the infiltrated sample into the resin-filled well, orienting it for optimal sectioning. Label each well: Write sample information (e.g., "Treatment 1 or T1") on a $2 \times 4\ mm^2$ piece of butter paper with a pencil (ink dissolves in resin), then place the paper adjacent to the sample in the well.
11. Resin Polymerization: Transfer the embedding plate to a 60 °C oven and incubate for 6 days or longer. Confirm complete polymerization using two criteria: the surface of the polymerized resin block with the embedded sample is hard and non-tacky to the touch; Scratching the surface with a fingernail produces no debris. After polymerization, cool the mold to room temperature (20–25 °C) for 1–2 h before demolding (gently peel the silicone mold to release the resin blocks).
12. Block Trimming: Mount the polymerized resin block (with embedded sample) on a block trimmer. Trim the block into a trapezoid shape with a trapezoidal cross-section using a sharp

double-edged razor blade, ensuring the sample is exposed at the pyramid's tip.

13. Glass Knife Preparation: Prepare a clean, sharp glass knife using a knife maker (follow manufacturer instructions) (Fig. 3a). Ensure the knife edge is free of nicks (checked under a stereomicroscope), as defects cause section tearing. The trough can be purchased as a finished product or can be formed by wrapping a rubber strip around a glass knife. Subsequently, use paraffin to adhere and seal the finished trough or rubber strip to the knife. Both types of troughs are shown in Fig. 3b. If the budget allows, a diamond knife is the best choice for cutting pollen, as it provides smoother sections and reduces chatter marks.
14. Water Trough Assembly: Assemble a water trough to float ultrathin sections (aids in flatting and collection). Seal the area around the glass knife edge with waterproof tape to form a shallow trough. Seal tape edges with molten paraffin wax to prevent water leakage (Fig. 3b) (*see* **Note 7**).
15. Ultrathin Sectioning (taking Leica UC6 ultramicrotome as an example). Block and Knife Installation: mount the trimmed resin block in the sample holder, fully exposing the sample. With the sample arm stationary, secure the holder to the sample arm's orientation head, ensuring the sample's cutting surface is parallel to the knife edge. Clamp the glass knife firmly in the knife holder and align the knife edge with the top of the sample's exposed surface.
16. Knife-Sample Alignment: Adjust the knife's rotation and translation knobs to align the knife edge with the sample's cutting surface, ensuring they are parallel.
17. Trough Preparation: Raise the sample arm to separate the sample from the knife. Use a 1 mL syringe to add deionized water to the trough until the liquid surface covers the knife edge and forms a white reflective area (indicates proper water level).
18. Automatic Sectioning: Press the "Run/Stop" button to initiate automatic cutting mode. Adjust the sectioning speed (typically 1–5 mm/s) and thickness (50 nm for ultrathin sections) as needed, based on section quality. Optimal sections appear silver-white to golden yellow when observed under reflected light (Fig. 4a, b, and d). Pollen grains exhibiting a scattered distribution pattern can be clearly observed in the sections (Fig. 4d). Sections with a blue interference color, however, must be discarded, as they are excessively thick and unsuitable for TEM imaging. Moreover, if knife marks are detected on the sections (Fig. 4c–e), the cutting position must be realigned

immediately to bypass notches on the knife edge—this mitigates the recurrence of such defects.

19. Section Collection: If sections have wrinkles, hold a piece of filter paper soaked in chloroform approximately 1 cm above the sections for 10 s to flatten the wrinkles. Use fine forceps to grip the edge of a formvar–/carbon-coated copper grid, ensuring the coated side faces downward. Align the grid horizontally with the floating section, then gently lower the grid onto the section. Lift the grid horizontally to pick up the section. Place the section-loaded grid on filter paper to dry completely before staining, as moisture can cause staining artifacts.

20. Section Staining (Double Staining for Contrast). Stain sections with uranyl acetate (for membrane contrast) and lead citrate (for cytoplasmic contrast), with strict minimization of CO_2 exposure—this prevents lead carbonate precipitation.

    (a) Uranyl Acetate Staining: Prepare 100 mL of boiled ultrapure water and cool it to room temperature (boiling removes CO_2 to avoid interference with staining). On a wax plate (a Petri dish lined with aluminum foil, with a glass slide covered in sealing film), place 8–10 μL drops of uranyl acetate solution (one drop per grid). Place the dried copper grids (section side down) onto the uranyl acetate drops, cover the wax plate with a lid, and stain for 30 min—protect the setup from light, as uranyl acetate is photosensitive. Use forceps to transfer the grids to a Petri dish lined with filter paper. Blot excess stain gently from the grids using filter paper, then rinse them 3–5 times with boiled ultrapure water (dispensed via a 2 mL syringe). Blot the grid surfaces with filter paper to remove residual water.

    (b) Lead Citrate Staining. On a new wax plate, place a circle of NaOH pellets around the glass slide (these absorb atmospheric CO_2 to maintain a CO_2-free environment). Add 8–10 μL drops of lead citrate solution to the slide (one drop per grid). Then immediately place the uranyl acetate-stained grids (section side down) onto the lead citrate drops, cover the plate tightly to minimize CO_2 infiltration, and stain for 30 min. Rinse the grids thoroughly with boiled ultrapure water 5–6 times to remove excess lead citrate, then blot them dry with filter paper.

    (c) Storage. Place the stained grids in a grid storage box (or gridded Petri dish) and store them in a desiccator to prevent moisture-induced damage until TEM observation.

21. TEM Observation: First, browse and examine the whole copper grid at low magnification (e.g., 1000–2000×) to locate

areas containing sections, and note positions of interest. Systematically progress from low to high magnification for observation: begin with the overall structure, then focus on specific regions (such as exine, intine, cytoplasm, nucleus, etc.). Before capturing high-magnification images (e.g., 40,000× or higher), perform precise focusing on adjacent areas at an even higher magnification, and then acquire images at the desired magnification.

22. When saving images, preserve the original format (e.g.,.dm4) and also convert and save a copy in TIFF format. Save images in real time to prevent data loss due to system freezes or power outages. Several ultrastructural images of Chinese chestnut pollen grains and pollen tubes are presented in Fig. 5. The intine, exine, plasma membrane, cytoplasm, and nucleus of a pollen grain are clearly visible in Fig. 5a. We can also observe an abnormal pollen grain with absent cytoplasm in Fig. 5b. In Fig. 5c and d, we can see the difference between the apex and proximal–distal region of the pollen tube. There is a single layer of fibrous structure at the tip and a callose layer at the distal end.

4 Notes

1. Osmium tetroxide (OsO_4) is highly toxic, volatile, and corrosive. All operations involving OsO_4—including preparation, sample immersion, and waste disposal—must be performed in a certified fume hood. Wear nitrile gloves, safety goggles, and a lab coat to avoid skin, mucous membrane, or ocular exposure.
2. Glutaraldehyde reacts with osmium tetroxide to form fine, electron-dense precipitates that can interfere with subsequent imaging. Thorough rinsing with PBS is essential to remove excess glutaraldehyde, ensuring the samples are compatible with secondary fixation.
3. Osmium tetroxide reacts with ethanol (used in dehydration) to form insoluble byproducts. Complete removal of residual OsO_4 is mandatory to avoid interference with the gradient dehydration step.
4. Dehydration must be thorough to eliminate all residual water from the samples. Incomplete dehydration can lead to ice crystal formation during freeze-drying, which damages ultrastructural features.
5. Ethanol volatilizes rapidly, which can cause sample collapse if allowed to dry naturally. Tert-Butanol, with a slower volatilization rate, acts as a cryoprotectant and facilitates uniform freeze-drying by preventing ice crystal formation.

6. Pollen and pollen tubes are non-conductive; the gold coating enhances electrical conductivity, preventing charge accumulation during SEM imaging and improving image resolution.
7. Ensure the trough's upper edge is parallel to the glass knife edge; misalignment will prevent continuous sectioning.

References

1. Ashfaq S, Zafar M, Ahmad M et al (2018) Microscopic investigations of palynological features of convolvulaceous species from arid zone of Pakistan. Microsc Res Tech 81(2):228–239
2. Lingzhu Y, Jian L (2019) The fundamental principles and applications of scanning electron microscopy. Exp Sci Technol 17(5):85–92
3. Devore ML, Zhou Z, Jansen RK et al (2007) Pollen morphology and ultrastructure of calyceraceae. Lundellia 10(1):32–48
4. Cai XY, Hou YQ, Wang H et al (2022) Pollen morphology of the genus *Gossypium* and its systematic implications. PREPRINT (version 1) available at Research Square. https://doi.org/10.21203/rs.3.rs-2174547/v1
5. Blackmore S, Wortley AH, Skvarla JJ et al (2007) Pollen wall development in flowering plants. New Phytol 174(3):483–498
6. Hesse M, Halbritter H, Weber M et al (2009) Pollen terminology: an illustrated handbook. Springer Vienna
7. Banks H, Klitgaard BB, Claxton F (2008) Pollen morphology of the family *Polygalaceae* (*Fabales*). Bot J Linn Soc 156(2):253–289
8. Wu JY, Tzeng HY, Yang TYA (2025) Taxonomic revision of *Polygalaceae* (*Fabales*) in Taiwan. PhytoKeys 262:73–103
9. Shi J, Cui M, Yang L et al (2015) Genetic and biochemical mechanisms of pollen wall development. Trends Plant Sci 20(11):741–753
10. Zhang F, Zhang YC, Zhang JP et al (2018) Rice UCL8, a plantacyanin gene targeted by miR408, regulates fertility by controlling pollen tube germination and growth. Rice (N Y) 11(1):60
11. Wang YB, Coomey J, Miller K et al (2022) Interactions between a mechanosensitive channel and cell wall integrity signaling influence pollen germination in *Arabidopsis thaliana*. J Exp Bot 73(5):1533–1545
12. Chebli Y, Kaneda M, Zerzour R et al (2012) The cell wall of the Arabidopsis pollen tube--spatial distribution, recycling, and network formation of polysaccharides. Plant Physiol 160(4):1940–1955
13. Zhang WW, Zhang Q, Xing Y et al (2022) Effect of boron toxicity on pollen tube cell wall architecture and the relationship of cell wall components of *Castanea mollissima* Blume. Front Plant Sci 13:946781
14. Wang J, Li D, Shang F et al (2017) High temperature-induced production of unreduced pollen and its cytological effects in *Populus*. Sci Rep 7(1):5281
15. Ma KB, Tang L, Cui JW (2017) Change of morphological structure during pollen development process in *Cunninghamia lanceolate*. Guihaia 37(10):1342–1347
16. Mondol PC, Xu D, Duan L et al (2020) Defective Pollen Wall 3 (DPW3), a novel alpha integrin-like protein, is required for pollen wall formation in rice. New Phytol 225(2):807–822
17. Nthakanio NP, Xue QZ (2011) Programmed cell death-like behavior in photoperiod-sensitive genic male sterile (PGMS) rice. Afr J Biotechnol 10(16):3027–3034
18. Honglan F (2004) Practical electron microscopy. Higher education press, Beijing

Chapter 5

Visualizing Pollen and Pollen Tube Development Using Fluorescence Microscopy

Ewa Szczuka and Agata Leszczuk

Abstract

There are two main methods of studying pollen grains and pollen tubes using fluorescence microscopy: immunocytochemical reaction with antibodies and techniques using fluorescent dyes. The protocols described cover issues related to the immunocytochemical labeling of structural components in pollen grains and pollen tubes, as well as the detection of individual components during in vivo observations of the processes of development and growth of pollen tubes. The first approach is based on sectioned and fixed material, while the other method involves in vivo observations of the pollen tube culture and germinating pollen grains. The protocol describes the step-by-step procedure of analyses, provides information on the required reagents and equipment, and summarizes basic troubleshooting during the performance of individual technical stages.

Key words Fluorescence, Imaging, Microscopy, Pollen, Pollen tube, Staining

1 Introduction

One of the characteristics of flowering plants, both gymnosperms and angiosperms, is the production of pollen grains. The most crucial of the many factors and processes ensuring the reproductive success of plants are the determinants of pollen grain quality. Hence, the successive cell divisions in the generative lineage (meiotic and mitotic divisions, or mitosis and meiosis), which take place in microsporangia and lead to the formation of microspores and the male gametophyte in plants, are frequently investigated by scientists [1–5].

The first observations of pollen grains were carried out by Grew in the late seventeenth century (1682), and these structures have continuously been a subject of scientific interest for almost three and a half centuries. This is associated with the extremely important role of pollen grains in the sexual reproduction of both gymnosperms and angiosperms. A groundbreaking finding that stimulated

Giampiero Cai et al. (eds.), *Pollen and Pollen Tube Analysis: Methods and Protocols*, Methods in Molecular Biology, vol. 3042, https://doi.org/10.1007/978-1-0716-5308-1_5,

research on pollen grains was the discovery of the double fertilization phenomenon in flowering plants in the early 1900s. Initially, researchers focused on factors determining pollen grain production, including physical (abiotic) factors, such as temperature and humidity, and biotic determinants, e.g., the condition of the maternal plant resulting from the availability of nutrients [6]. These studies provided crucial information on the pollen yield, which was the basis for further studies of these structures playing an essential role in the fertilization process. Particular attention was paid to issues related to haploid selection and pollen competition [6].

The fertilization process opens a long list of investigations into the importance of pollen grains and their germination into pollen tubes. This process is involved in the generation of offspring and species survival. Furthermore, a crucial factor is the ability to ensure genetic diversity in offspring plants as a basis for adaptation to changing environmental conditions. Hence, the quality of pollen grains, which is determined by such factors as the absence of genome damage, adequate moisture, presence of sufficient reserve substances, and a properly formed sporoderm protecting the interior of the male gametophyte, is essential [7].

Fluorescence microscopy studies of pollen grain formation focus primarily on the analysis of genetic material dividing during microsporogenesis and mitotic divisions leading to the formation of a two- or three-celled pollen grain, as well as the analysis of the sporoderm. In particular, analyses of sporoderm morphology provide valuable information about the pollen grain sculpture. Results of fluorescence microscopy studies facilitate rapid and precise determination of the pollen grain type and identification of the group of plants producing a specific type of pollen grain. This is associated with the sculpture of pollen grains, which is specific for the different taxonomic groups [8]. At the same time, the knowledge of pollen grains is the basis for experiments aimed at designing new varieties with desirable traits, e.g., resistance to diseases or such climatic factors as drought. Such experiments are particularly important for crop plants, as they offer the possibility to manipulate the size and quality of yields, as in the case of the spectacular outcomes in ornamental plants.

An important role in the taxonomy of flowering plants is also played by pollen morphology [9]. Both taxonomic and evolutionary investigations of these plants focus their attention on the remarkable diversity in pollen grain morphology [10, 11]. One of the numerous traits considered in both these disciplines is the characteristics of the sculpture of pollen grain surfaces. In addition to its sculpture traits in individual taxa, the exine, containing the stable biopolymer sporopollenin, exhibits extraordinary resistance to external factors [12]. To underscore the importance of research on pollen grains, the use of pollen grains in palynology should be mentioned. The sporoderm of pollen grains, in particular its sporopollenin component, is exceptionally resistant to environmental factors and substances contained in the outer layer of the Earth's crust. This property facilitates dating certain layers of the Earth's crust and analyzing the vegetation present in a specific area.

An important issue related to human health is the allergenicity of pollen grains. A growing number of people suffer from symptoms of hay fever, such as sneezing, itching, a runny nose, and watery eyes, which are caused by pollen allergens. Some studies have focused on Japanese cedar (*Cryptomeria japonica*) and Japanese cypress (*Chamaecyparis obtusa*) pollen suspended in the air every spring, which trigger allergies in humans [13]. Hence, pollen grain monitoring is commonly used in many countries to warn against the presence of plant pollen posing a particular risk to allergy sufferers. This allows appropriate measures to be taken to reduce allergy symptoms. One of the methods for classification of pollen species is the use of autofluorescence image analysis [13].

Since its accidental discovery on the *Portulaca oleracea* stigma by Giovanni Battista Amici in the nineteenth century, the pollen tube has been a frequent subject of research. The main reason for this great interest is its importance in the transfer of sperm cells (male units) produced in the pollen grain or the pollen tube to ovules located in the ovary. This function of the pollen tube has an impact on the proper fertilization process, leading to the formation of an embryo surrounded by reserve tissue and enclosed in the seed coat. In turn, proper seed formation contributes to, e.g., the generation of yields, which provide food for humans. Additionally, the pollen tube is an excellent model in studies of plant cell behavior, e.g., its extraordinary growth through elongation only at the apex, and in investigations of electrotropism [14].

Scientists widely use sensitive, specific, reliable, and extensive fluorescence methods in advanced studies of pollen grains and pollen tubes. The phenomenon in which organic and inorganic substances are capable of absorbing light and emitting photons at a longer wavelength was first named "fluorescence" by Stokes in 1852, laying the foundation for the development of fluorescence microscopy [15]. The turning points in the development of this technique included the first use of fluorophores in biological investigations, fluorescence labeling, and green fluorescent protein (GFP) cloning. These groundbreaking discoveries led to the advancement of the fluorescence microscopy technique, which is widely used for observation of the localization of molecules within cells and cells within tissues [16]. Currently, techniques based on new molecular probes are being developed for precise and specific labeling of individual components of the pollen tube wall [17–23]. These techniques are characterized by high sensitivity down to the level of single molecules [24] and provide valuable information for the design of optical probes [25] and labeling studies

[26, 27]. Fluorescence techniques are part of experimental methods that utilize the unique or specific properties of fluorescence, facilitating research and analysis of biological materials. These fluorescence properties include excitation wavelength, emission wavelength, intensity, lifetime, and anisotropy.

Considering the information presented above, two main methods for the analysis of pollen grains and pollen tubes will be described. One of them facilitates rapid assessment of the pollen grain type, whereas the other approach, consisting of the fixation of pollen grains and pollen tubes, is suitable for analyses at subsequent stages of development. The protocols described address issues related to the immunocytochemical labeling of structural components in pollen grains and pollen tubes (Fig. 1), as well as the rapid and easy staining of individual components during in vivo observations of pollen tube development and growth (Fig. 2).

The two techniques described involve numerous steps that contribute to a positive immunocytochemical reaction. Moreover, any deviation from the procedure may result in a completely different result. It should be emphasized that there is great species specificity, which influences the structure, germination time of pollen grains, and the structure of the pollen tube wall itself. An issue addressed in studies of pollen grains and germination into pollen tubes is the developmental stages of normal pollen grains. Appropriate identification of stages with demonstration of the complete sequence of events can pose problems whose solution may be time-consuming. A thorough analysis of consecutive stages is focused on the development of stamens and microsporangia with their tissues, which influences the archesporial cells and microsporocytes undergoing meiotic divisions. The divisions result in the formation of microspores, which undergo mitotic divisions leading to the generation of the male gametophyte. Disturbances may occur at each stage of development, leading to the formation of pollen grains and, ultimately, the participation of the gametes contained within the grains in the fertilization process. It has long been known that pollen grains and the process of their formation are highly sensitive to environmental factors, with such abiotic factors as temperature, humidity, and osmotic stress playing a key role. Undoubtedly, the temperature is a key determinant not only of the development, quantity, and quality of pollen grains but also of the achievement of reproductive success in plants, reflected by crop yields. The observation of the stages may also be challenging due to the duration of the individual phases of meiotic division and asymmetric mitotic divisions in the forming pollen grain. The division of the generative cell resulting in the emergence of two sperm cells may take place several hours after the cell formation.

Despite the difficulties indicated in this chapter, the structure of pollen grains and their germination process are an extremely rewarding object of research.

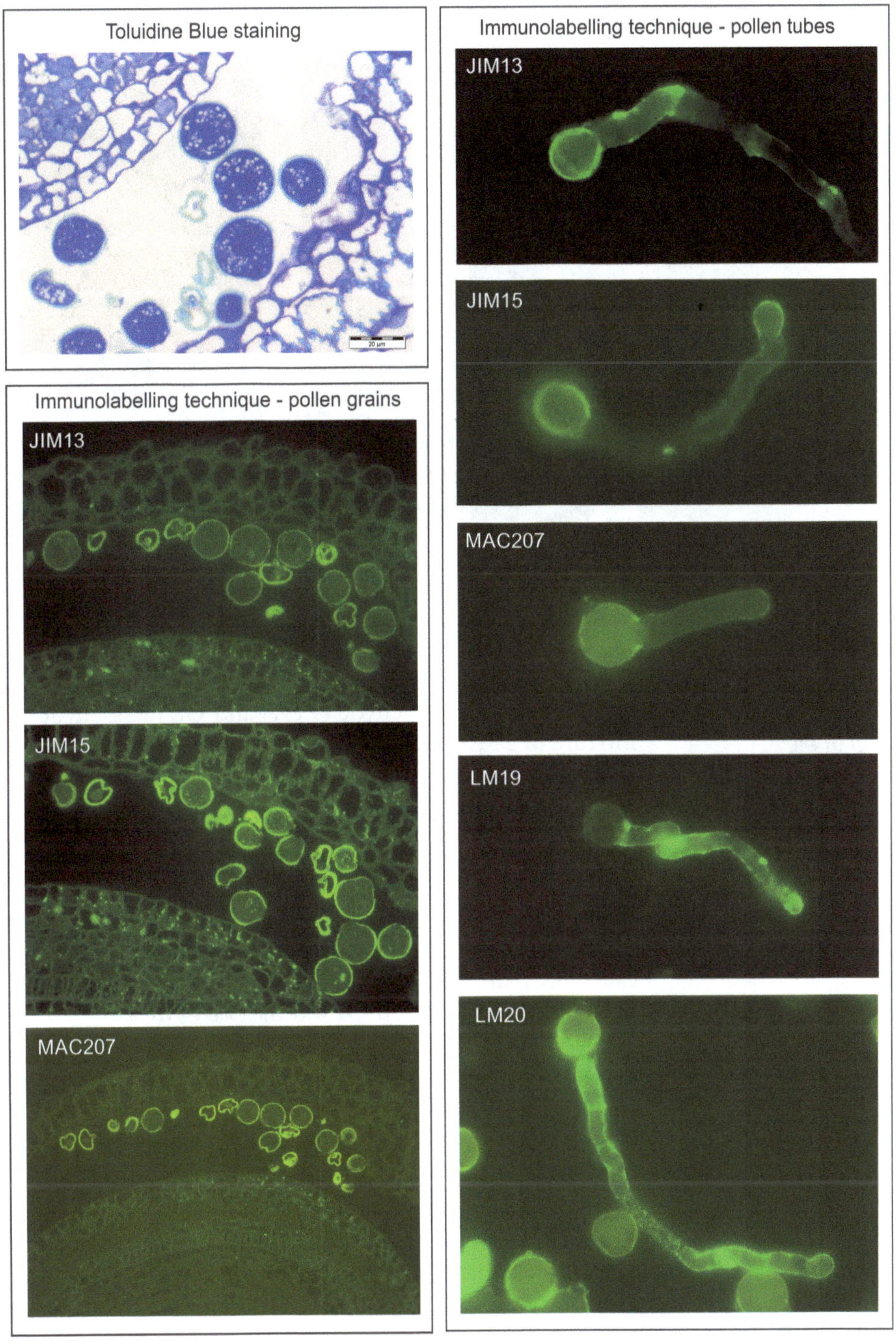

Fig. 1 Examples of the pollen grain and pollen tube immunostaining technique. Imaging of arabinogalactan proteins using antibodies recognizing different targets of carbohydrate chains and homogalacturonans—LM19, LM20. JIM13 (β-GlcA(1→3)-α-GalA(1→2)-α-Rha), JIM15 (GlcA-β(1-O-Me)), MAC207 ((β-GlcA (1→3)-α-GalA(1→2)-α-Rha), LM20 (α-MeGalA(1→4)-α-MeGalA(1→4)-α-MeGalA-(1→4)-α-MeGalA), LM19 (α-GalA(1→4)-α-GalA(1→4)-α-GalA(1→4)-α-GalA)

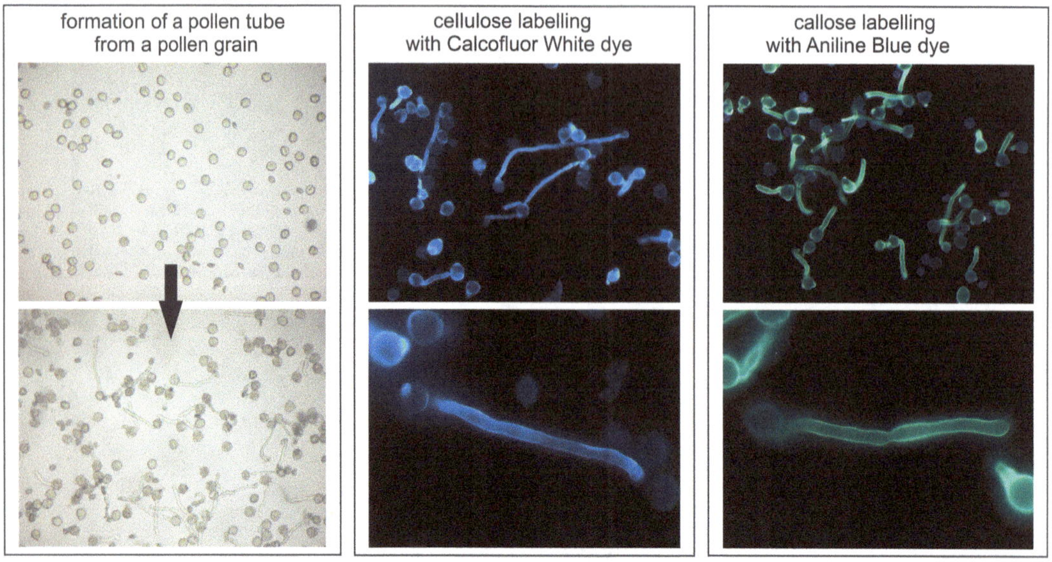

Fig. 2 Examples of pollen tube labeling with specific fluorescent dyes—Calcofluor White and Aniline Blue

2 Materials

2.1 Reagents for the First Step of the Preparative Part

1. Fixative: Paraformaldehyde (PFA) and Glutaraldehyde (GA) - mixture of 2–4% GA + 2% PFA in 1X phosphate-buffered saline (PBS) (*see* **Note 1**).
2. Resin: LR White. It comes in two parts: resin and hardener to initiate polymerization (dibutyl peroxydicarbonate (DBPC) or benzoin methyl). The standard resin-to-hardener ratio is about 10:1 by volume.
3. Blocking buffer: BSA stands for Bovine Serum Albumin. Typically, BSA in a buffer (commonly PBS or TBS) is used at concentrations between 1% and 5% (*see* **Note 2**).

2.2 Antibodies

1. Primary antibodies: the first step in the immunostaining process is to specifically bind and detect a particular target molecule. Table 1 lists the commonly used primary antibodies in pollen tube studies.
2. Secondary antibodies: antibodies that bind to the primary antibody in a secondary binding event and are conjugated to a detection molecule (such as fluorochromes (e.g., FITC, Alexa Fluor, Cy3, Cy5). Table 2 lists the commonly used secondary antibodies.

Table 1
Commonly used antibodies in pollen grain and pollen tube research

Name of antibody	Recognised epitope/specific molecules or structures within the pollen/tube	References
Antibodies against cytoskeletal components		
Anti-actin antibodies	Actin filaments	[16]
Anti-tubulin antibodies	Microtubules	[28]
Antibodies against pollen tube wall components		
Anti-Callose antibodies	Callose	[29]
Anti-cellulose antibodies	Cellulose	[30]
Anti-pectin antibodies	Pectins	[31]
Antibodies against membrane proteins and lipids		
Anti-protein disulfide isomerase (PDI) antibodies	Protein disulfide isomerase in the endoplasmic reticulum	[32]
ARF (ADP-ribosylation factor) and RAB	Vesicles containing cell wall materials	[33]
Anti-AGP antibodies	AGP as part of the extracellular continuum	[34]
Antibodies against pollen tube-specific proteins	Pollen tube-specific receptor kinases	[35]
Antibodies against signalling molecules		
Anti-calcium binding proteins	Calmodulin	[36]
Anti-Proline-rich proteins (PRPs)	Proline	[37]
Antibodies for investigating the role of hormones		
Anti-auxin binding proteins (ABPs)	Auxins	[38]

2.3 Fluorescent Dyes Used to Visualize Pollen Tube Growth in Intravital Observations (Table 3)

1. Calcofluor white: The typical working concentration of Calcofluor White is around 0.01% to 0.1% (w/v). To prepare 100 mL of a 0.1% solution, add 0.1 g of Calcofluor White to 100 mL of distilled water or PBS.
2. Aniline blue: The typical working concentration of Aniline blue is usually around 0.05% to 0.1% (w/v).
3. SYTO 9: The typical working concentration for SYTO 9 is around 1–5 μM.
4. DAPI: The typical working concentration of DAPI is around 0.1–1 μg/mL.
5. Fluorescein diacetate (FDA): The typical concentration for stock FDA is around 5 mg/mL or 10 mg/mL.
6. Propidium Iodide (PI): The typical final working concentration of PI is around 1–5 μg/mL.

Table 2
Common secondary antibodies conjugated with fluorochromes

Name of antibody	Properties	Advantages/disadvantages
FITC-conjugated antibodies	FITC fluoresces green when excited with blue light in the 488 nm range and emits fluorescence at 520 nm (green light).	1. Good for basic applications 2. Lower quantum yield 3. Produces a weaker fluorescence signal compared to Alexa Fluor 488 4. Sensitive to pH changes
Alexa Fluor 488-conjugated antibodies	Alexa Fluor 488 is excited with blue light (around 495 nm) and emits bright green fluorescence around 519 nm	Ideal for long-term imaging, multicolor experiments More photostable than FITC Low background fluorescence, which ensures that the signal-to-noise ratio is high and makes it easier to distinguish specific signals from nonspecific ones Higher fluorescence intensity More sensitive in detecting low-abundance targets Higher brightness and higher quantum yield than FITC Broader emission spectrum Less spectral overlap with other fluorophores, making it useful in multicolor experiments

7. Phalloidin coupled with FITC: The typical working concentration of Phalloidin-FITC for staining is 0.05–0.2 μg/mL.

2.4 Essential: Specific Equipment for the Immunolabeling Method

1. Vacuum chamber (*see* **Note 3**).
2. Poly-L-lysine-coated glass slides (*see* **Note 4**).
3. Gelatine capsules, pipettes, and incubation trays.
4. Glass knife maker.
5. Ultramicrotome: sections in the range of 30–100 nm to 1 μm.
6. Humidified chamber.
7. Fluorescence microscope with image analysis software (*see* **Note 5**).

3 Methods

3.1 Immuno-cytochemistry (ICC) with Fluorescence Detection

Fixation and Sectioning

1. Preparative step: Prepare the pollen grains and/or pollen tubes.
2. Fixation: Fix the samples in a glass falcon in a solution of 2.5% paraformaldehyde and 2.5% glutaraldehyde in 0.15 M PBS (*see* **Note 6**).

Table 3
Fluorescent dyes are used to visualize pollen tube growth in intravital observations

Name of dye	Labeled component	Application
Calcofluor white	Cellulose in the cell wall	Cell wall composition and tube elongation
SYTO 9	Nucleic acid stain	Cell viability in pollen tube development
DAPI	A-T-rich regions of double-stranded DNA	Nuclear behavior during pollen tube growth
Fluorescein diacetate (FDA)	Enters living cells and is hydrolyzed by cellular esterases to produce fluorescent fluorescein	Pollen viability and monitor growth patterns
Propidium iodide (PI)	Stains dead cells—Cells with damaged membranes	Discrimination between live and dead cells in pollen tubes
Phalloidin coupled with FITC	Actin filaments in the cytoskeleton	Actin network in growing pollen tubes in vesicle trafficking and polar growth
Aniline blue	Callose	Visualization of callose plugs that form in the tube during growth

3. De-airing: Put the falcons in the vacuum chamber under vacuum in the range of 0.5 to 0.7 atm for 3 h and keep overnight. Prolonged or excessive vacuum exposure can damage delicate samples.
4. Post-fixation: Rinse the samples three times with PBS to remove excess fixative solution.
5. Dehydration: Dehydrate the samples in a graded ethanol series: 30% ethanol at room temperature (RT) for 15 min.
6. Dehydration: Dehydrate the samples in a graded ethanol series: 50% ethanol at RT for 15 min.
7. Dehydration: Dehydrate the samples in a graded ethanol series: 70% ethanol at RT for 15 min.
8. Dehydration: Dehydrate the samples in a graded ethanol series: 90% ethanol at RT for 15 min.
9. Dehydration: Dehydrate the samples in a graded ethanol series: 96% ethanol at RT for 30 min.
10. Dehydration: Dehydrate the samples in a graded ethanol series: 99.8% ethanol at RT for 30 min.
11. Infiltration: Infiltrate with resin (e.g., Epon, LR White, Spurr's resin). Fix the samples in LR White according to the procedures described by Sigma Aldrich. Prepare an LR White working solution with benzoyl peroxide blended with dicyclohexyl phthalate, mix, and keep overnight.

12. Infiltration: Fix the samples in a solution of 99.8% ethanol: LR White (3:1) at RT for 3 h.
13. Infiltration: Fix the samples in a solution of 99.8% ethanol: LR White (1:1) at RT for 3 h.
14. Infiltration: Fix the samples in a solution of 99.8% ethanol: LR White (1:3) at RT for 3 h.
15. Infiltration: Fix the samples in LR White at 4 °C for 24 h.
16. Infiltration: Fix the samples in LR White at RT for 5 h.
17. Embedding: Put the samples directly into embedding moulds, i.e., gelatine capsules, and fill with resin. Close them!
18. Embedding: Keep the samples for polymerization at 55 °C for 48 h (*see* **Note 7**).
19. Sectioning: Tissue sectioning using an ultramicrotome with a glass knife.
20. Trimming the block: Trim the resin block down to a manageable size using a sharp scalpel or razor blade.
21. Mounting the block: Place the resin block into the ultramicrotome chuck or holder.
22. Sectioning; Cut semi-thin sections (1 μm) using an ultramicrotome with a glass knife.
23. Sectioning: Collect the sections onto poly-L-lysine coated slides using a fine paintbrush or forceps.
24. Sectioning: Dry the sections in a drop of water on a hot plate at 50 °C for 30 min.
25. Sectioning: Circle the localization of the sections on the slices with a liquid blocker Pen.

Immunocytochemical Reaction and Visualization of Fluorescence

1. Washing: Wash the sections in PBS 3 times at RT for 15 min (*see* **Note 8**).
2. Blocking: Pre-incubate the sections with 1% bovine serum albumin (BSA) in PBS for 30 min to avoid nonspecific binding of antibodies (*see* **Note 9**).
3. Primary Antibody Incubation: Incubate the sections with the primary antibody adjusted to a 1:50 dilution in 0.1% BSA at 4 ° C for 24 h (*see* **Note 10**).
4. Washing: Wash the sections in PBS 3 times at RT for 15 min.
5. Secondary Antibody Incubation: Incubate the sections with the secondary antibody diluted 1:50 in a blocking solution at 4 °C for 24 h in the dark!
6. Washing: Wash the sections in PBS 3 times at RT for 15 min in darkness!

7. Washing: Wash the sections in deionized water 3 times at RT for 15 min, in darkness.
8. Imaging: Imaging the fluorescence signal. All parameters (i.e., laser intensity, gain) are kept constant. The excitation wavelength for the FITC/Alexa Fluor488 is 475 nm, and the emission wavelength is 520 nm (*see* **Note 11**).

3.2 In Vivo Observations of the Pollen Tube and Biochemical Detection of Its Components

Pollen Tube Cultivation

In vivo observations are an experimental technique used to study the growth, development, and behavior of pollen tubes.

1. Preparation of culture medium: 1–10 mM $CaCl_2$, 1–10 mM KNO_3, 1–10 mM $MgSO_4$, 0.1–1 mM boric acid, 10–20% sucrose, pH of 6.0–7.0. For a solid medium, add agar (1–2%).
2. Research material: Collect pollen grains from mature anthers and add a small number of pollen grains to the prepared culture medium.
3. Incubation: Incubate the culture at the optimal temperature (20–25 °C).
4. Duration of growth: Depending on the species, pollen tube growth can take from 30 min to several hours.

Pollen Grain and Pollen Tube Labeling with Specific Fluorescent Dyes

The staining method for pollen tubes depends on the aspects of the pollen tube metabolic events, such as changes in the cytoskeletal structure, cell wall composition, viability, or gene expression. The commonly used dyes include fluorescent dyes (e.g., Calcofluor White, DAPI), vital dyes (e.g., FDA, PI), cytoskeletal markers (e.g., Phalloidin-FITC), and specific cell wall stains (e.g., Aniline Blue).

1. Calcofluor White: After preparing the Calcofluor White solution, apply it directly to the sample and let it incubate for 5–10 min. Then, rinse the sample with water to remove excess dye. The sample will now fluoresce under blue light.
2. Aniline blue: Apply the working solution to the section, incubate for the required time (usually 5–10 min), rinse with water, and proceed with further callose visualization.
3. SYTO 9: Add the SYTO 9 working solution directly to the cells or tissue sections. SYTO 9 binds to nucleic acids and emits blue light (excitation around 480 nm, emission around 500–530 nm).
4. DAPI: Add the DAPI working solution to tissue samples. Incubate for 5–10 min at RT. DAPI binds to A-T-rich regions of DNA and fluoresces when exposed to UV light (excitation around 358 nm and emission around 461 nm).

5. Fluorescein diacetate (FDA): Incubate the sample with the FDA solution for 15–30 min at RT. FDA penetrates live cells and produces fluorescein, which emits green fluorescence upon exposure to light (excitation at ~490 nm and emission at ~515 nm).
6. PI: Incubate the working solution of PI with the sections for 10–15 min at RT. PI will only enter cells with compromised membranes and intercalate into the DNA, emitting red fluorescence upon excitation (excitation at 535 nm and emission at 617 nm).
7. Phalloidin coupled with FITC: Incubate the sections with the Phalloidin-FITC solution for about 20–30 min at RT in a dark environment. Phalloidin-FITC will bind to F-actin filaments, and the fluorescence emission will be in the green range (excitation around 495 nm, emission around 520 nm).
8. Auramine O: Incubate the sections with a 0.01–0.05% Auramine O solution in distilled water or PBS. After incubation for 5–15 min at RT, rinse them gently with PBS or distilled water. Observe under a fluorescence microscope using blue excitation light (around 450–490 nm). Auramine O binds to β-1,3-glucans, i.e., callose plugs or cell walls.
9. Eosine: Incubate the germinated pollen tubes in 0.1% to 1% Eosin in distilled water or PBS for 1–5 min at RT. Observe under a fluorescence microscope (excitation ~525–540 nm). Eosin staining allows the assessment of cell viability.

4 Notes

1. PFA is a commonly used fixative that crosslinks proteins by forming methylene bridges. In turn, GA crosslinks proteins more strongly and is used for preserving ultrastructural details.
2. LR white is used to block nonspecific binding sites on slides to prevent background interference during antibody binding and to stabilize antibodies in experiments by preventing denaturation, aggregation, or degradation.
3. The vacuum chamber helps the fixative solution penetrate the tissue by removing any air from the cells and allowing the fixative to fill the spaces, ensuring thorough infiltration.
4. The coating provides a positively charged surface, which helps sections/pollen tubes adhere better to the material. A necessary solution during a large number of steps involving solution exchanges and a high possibility of sample detachment.

5. Images captured with fluorescence or confocal microscopes are analyzed with software like ImageJ (https://imagej.net/ij/) or software conjugated with a microscope. The tools allow quantification of fluorescence intensity, spatial protein localization, and other relevant data.
6. Fixative concentration and tissue fixation time: Aldehyde-based fixatives, such as formaldehyde, formalin, and glutaraldehyde, are most commonly used. Aldehydes react with and cross-link cellular proteins by crossing the plasma membrane; however, some epitopes can be masked by cross-linking reactions, thereby reducing target antigenicity. These effects can be minimized by shortening the incubation times.
7. The resin block must be hard enough for cutting and free of air bubbles or other imperfections. For large samples, accelerated polymerization can be done at 40–60 °C.
8. In all steps, put the slices in the humidified chamber to prevent their evaporation, ensuring that the sample remains consistently exposed to the reagents for accurate results. Without a humidified chamber, the samples can dry out, which leads to uneven staining.
9. Preincubation with BSA is a crucial step that enables the specific reaction to occur. A blocking step is required to limit fluorescence, known as false positives. Shortening or omitting the step or using too low a concentration enhances the fluorescence signal. Furthermore, the omission of this step also causes a high background reaction. The "gold standard" of blocking solutions should contain a 5–10% solution of serum.
10. Optimization of antibody incubation time: Our proposed reactions with antibodies have been developed and validated for optimal results when incubated at 4 °C overnight. Reactions at elevated temperatures, e.g., at room temperature, accelerate the immunocytochemical reaction, and a proportional shortening of the incubation time is required.
11. Imaging and fluorescence signal observation parameters: Autofluorescence samples should be eliminated by using unstained samples as controls. Imaging should be performed appropriately soon after the reaction is completed. This is also related to photobleaching, which is a phenomenon that diminishes fluorophore fluorescence following prolonged exposure to excitation radiation. To avoid errors during fluorescence imaging, mounting media with anti-fade reagents can be used, which minimizes photobleaching effects in the samples. In the event of a very weak fluorescence signal or very specific epitope localization, counterstaining is often used to define individual cells and cell structures. The most common are DAPI, phalloidin, and Calcofluor White.

References

1. Cheung AY, Duan QH, Costa SS, De Graaf BH, Di Stilio VS, Feijo J, Wu HM (2008) The dynamic pollen tube cytoskeleton: live cell studies using actin-binding and microtubule-binding reporter proteins. Mol Plant 1(4):686–702
2. Liao F, Wang L, Yang LB, Peng X, Sun M (2010) NtGNL1 plays an essential role in pollen tube tip growth and orientation likely via regulation of post-Golgi trafficking. PLoS One 5(10):e13401
3. Hao H, Chen T, Fan L, Li R, Wang X (2013) 2, 6-dichlorobenzonitrile causes multiple effects on pollen tube growth beyond altering cellulose synthesis in *Pinus bungeana* Zucc. PLoS One 8(10):e76660
4. Daghma DES, Hensel G, Rutten T, Melzer M, Kumlehn J (2014) Cellular dynamics during early barley pollen embryogenesis revealed by time-lapse imaging. Front Plant Sci 5:675
5. Weng X, Shen Y, Jiang L, Zhao L, Wang H (2023) Spatiotemporal organization and correlation of tip-focused exocytosis and endocytosis in regulating pollen tube tip growth. Plant Sci 330:111633
6. Williams JH, Mazer SJ (2016) Pollen—Tiny and ephemeral but not forgotten: new ideas on their ecology and evolution. Am J Bot 103(3):365–374
7. Adhikari PB, Liu X, Kasahara RD (2020) Mechanics of pollen tube elongation: a perspective. Front Plant Sci 11:589712
8. Urbanczyk J, Casado MAF, Díaz TE, Heras P, Infante M, Borrego AG (2014) Spectral fluorescence variation of pollen and spores from recent peat-forming plants. Int J Coal Geol 131:263–273
9. Taia WK (2022) Pollen grain diversity and application in taxonomy and evolution. Taeckholmia 42(1):27–42
10. Pacini E, Franchi GG (2020) Pollen biodiversity–why are pollen grains different despite having the same function? A review. Bot J Linn Soc 193(2):141–164
11. Almosawi MBH (2025) A review in pollen grains for plants. Biol Pharm Sci 30(1):21–30
12. Piffanelli P, Ross JH, Murphy DJ (1998) Biogenesis and function of the lipidic structures of pollen grains. Sex Plant Reprod 11:65–80
13. Mitsumoto K, Yabusaki K, Aoyagi H (2009) Classification of pollen species using autofluorescence image analysis. J Biosci Bioeng 107(1): 90–94
14. Aloisi I, Cai G, Faleri C, Navazio L, Serafini-Fracassini D, Del Duca S (2017) Spermine regulates pollen tube growth by modulating Ca^{2+}-dependent actin organization and cell wall structure. Front Plant Sci 8:1701
15. Dunst S, Tomancak P (2019) Imaging flies by fluorescence microscopy: principles, technologies, and applications. Genetics 211(1):15–34
16. Bove J, Vaillancourt B, Kroeger J, Hepler PK, Wiseman PW, Geitmann A (2008) Magnitude and direction of vesicle dynamics in growing pollen tubes using spatiotemporal image correlation spectroscopy and fluorescence recovery after photobleaching. Plant Physiol 147(4): 1646–1658
17. Wang H, Jiang L (2011) Transient expression and analysis of fluorescent reporter proteins in plant pollen tubes. Nat Protoc 6(4):419–426
18. Hartman E, Levy C, Kern DM, Johnson MA, Basu A (2014) A rapid, inexpensive, and semi-quantitative method for determining pollen tube extension using fluorescence. Plant Methods 10:1–6
19. Potocký M, Pleskot R, Pejchar P, Vitale N, Kost B, Žárský V (2014) Live-cell imaging of phosphatidic acid dynamics in pollen tubes visualized by S po20p-derived biosensor. New Phytol 203(2):483–494
20. Vogler F, Konrad SS, Sprunck S (2015) Knock-in'on pollen's door: live cell imaging of early polarization events in germinating Arabidopsis pollen. Front Plant Sci 6:246
21. Qu H, Xing W, Wu F, Wang Y (2016) Rapid and inexpensive method of loading fluorescent dye into pollen tubes and root hairs. PLoS One 11(4):e0152320
22. Dunker S, Motivans E, Rakosy D, Boho D, Mäder P, Hornick T, Knight TM (2021) Pollen analysis using multispectral imaging flow cytometry and deep learning. New Phytol 229(1): 593–606
23. van den Wildenberg SM, Prevo B, Peterman EJ (2024) A brief introduction to single-molecule fluorescence methods. In: Single molecule analysis: methods and protocols. Humana Press, Totowa, pp 81–99
24. Tholen M, Yim JJ, Groborz K, Yoo E, Martin BA, van den Berg NS et al (2020) Design of Optical-Imaging Probes by screening of diverse substrate libraries directly in disease-tissue extracts. Angew Chem Int Ed 59(43): 19143–19152
25. Wilson SM, Bacic A (2012) Preparation of plant cells for transmission electron microscopy to optimize immunogold labeling of carbohydrate and protein epitopes. Nat Protoc 7(9): 1716–1727

26. Dean KM, Palmer AE (2014) Advances in fluorescence labeling strategies for dynamic cellular imaging. Nat Chem Biol 10(7):512–523
27. Wang X, Li T, Xu J, Zhang F, Liu L, Wang T et al (2024) Distinct functions of microtubules and actin filaments in the transportation of the male germ unit in pollen. Nat Commun 15(1): 5448
28. Wilsen KL, Lovy-Wheeler A, Voigt B, Menzel D, Kunkel JG, Hepler PK (2006) Imaging the actin cytoskeleton in growing pollen tubes. Sex Plant Reprod 19:51–62
29. Kapoor K, Geitmann A (2023) Pollen tube invasive growth is promoted by callose. Plant Reprod 36(2):157–171
30. Zhang M, Zhang R, Qu X, Huang S (2016) Arabidopsis FIM5 decorates apical actin filaments and regulates their organization in the pollen tube. J Exp Bot 67(11):3407–3417
31. Luo N, Yan A, Liu G, Guo J, Rong D, Kanaoka MM et al (2017) Exocytosis-coordinated mechanisms for tip growth underlie pollen tube growth guidance. Nat Commun 8(1): 1687
32. Feldeverd E, Porter BW, Yuen CY, Iwai K, Carrillo R, Smith T et al (2020) The Arabidopsis protein disulfide isomerase subfamily M isoform, PDI9, localizes to the endoplasmic reticulum and influences pollen viability and proper formation of the pollen exine during heat stress. Front Plant Sci 11:610052
33. Yoo CM, Wen J, Motes CM, Sparks JA, Blancaflor EB (2008) A class I ADP-ribosylation factor GTPase-activating protein is critical for maintaining directional root hair growth in Arabidopsis. Plant Physiol 147(4):1659–1674
34. Leszczuk A, Kozioł A, Szczuka E, Zdunek A (2019) Analysis of AGP contribution to the dynamic assembly and mechanical properties of cell wall during pollen tube growth. Plant Sci 281:9–18
35. Schneidereit A, Scholz-Starke J, Sauer N, Büttner M (2005) AtSTP11, a pollen tube-specific monosaccharide transporter in Arabidopsis. Planta 221(1):48–55
36. Cárdenas L, Lovy-Wheeler A, Kunkel JG, Hepler PK (2008) Pollen tube growth oscillations and intracellular calcium levels are reversibly modulated by actin polymerization. Plant Physiol 146(4):1611–1621
37. Sede AR, Wengier DL, Borassi C, Ricardi M, Somoza SC, Aguiló R et al (2024) Arabidopsis pollen prolyl-hydroxylases P4H4/6 are relevant for correct hydroxylation and secretion of LRX11 in pollen tubes. J Exp Bot 75(14): 4415–4427
38. Zhu L, Zhang Y, Kang E, Xu Q, Wang M, Rui Y et al (2013) MAP 18 regulates the direction of pollen tube growth in Arabidopsis by modulating F-actin organization. Plant Cell 25(3): 851–867

Chapter 6

Automated Measurements of Fluorescence in Spatially Defined Regions of Pollen Tubes

Jeremy Adler and Ingela Parmryd

Abstract

Pollen tubes are a widely used model for the extension and development of polarized cells. Their relatively standard shape simplifies automated identification and image analysis of subregions. The key is reliable identification of the cell plasma membrane and the pollen tube tip. This chapter describes how regions of interest can be created and used to assess distributions in tubular cells using two ImageJ macros. The first macro enables efficient and precise identification of the pollen tube plasma membrane and tip, based on a few imprecisely marked points. Subsequent steps are automated, with users prompted to confirm their accuracy, a balance between automation and supervision. The plasma membrane and the tip are the bases of five distance maps: three starting from the tip, one from the plasma membrane, and one from the pollen tube midline. Measurements of fluorescence intensity and generalized polarization gradients are then made from the tip along the plasma membrane. The second macro uses the distance maps to measure fluorescence intensities and generalized polarization from user-specified regions derived from a single or a combination of the five distance maps. Highly reproducible measurements are achieved by automation with minimal user interaction, while every analysis step is recorded for complete traceability.

Key words Distance map, Fluorescence gradient, Generalized polarization, ImageJ, Macro, Plasma membrane identification, Pollen tubes, Region of interest

1 Introduction

A pollen tube extends from a pollen grain to deliver sperm cells to fertilize egg cells in the pistil [1]. Pollen tubes rapidly elongate and expand unidirectionally. To enable extension, cell wall material is delivered to the pollen tube tip in vesicles that undergo exocytosis, as are membrane components. However, the need for cell wall material exceeds the need for membrane expansion, resulting in extensive membrane recycling by endocytosis [2, 3]. This makes pollen tubes an excellent model system for studies of directional cell growth and membrane dynamics. Given the fast pollen tube growth and membrane remodeling, gradients of membrane lipids and proteins appear both along the plasma membrane and intracellularly

Giampiero Cai et al. (eds.), *Pollen and Pollen Tube Analysis: Methods and Protocols*, Methods in Molecular Biology, vol. 3042,
https://doi.org/10.1007/978-1-0716-5308-1_6,

[4]. To increase our understanding of the biological mechanisms underlying growth and cell development, it is important to map the distribution of these components, an area where we have recently contributed by developing and using the software described in detail in this chapter [5].

Accurately demarcating a region of interest (ROI) at the plasma membrane by hand is extremely difficult, any inaccuracy degrades subsequent measurements. Precise freehand lines following thin structures like the plasma membrane are almost impossible to draw, and even marking and joining sequential points is, while better, not very reproducible and also time-consuming. When cells are stained with a membrane marker, the pixels best representing the plasma membrane are those with the highest local intensity at the cell border. Based on this notion, we recently developed an efficient method to identify the plasma membrane of spherical cells like immune cells, where the user only needs to select a few points in the vicinity of the plasma membrane [6]. The position of each point is optimized using a local search for the most intense pixel. Then, additional points are interpolated and similarly repositioned. This approach has now been applied to tubular cells like pollen tubes in the ImageJ macro PollenTube_DisMaps.ijm. This macro also automatically identifies the position of the pollen tube tip and has the capacity to measure gradients from the pollen tube tip along the plasma membrane. The regional specificity is derived from a distance map with its origin at the pollen tube tip that extends exclusively along the plasma membrane. Several additional distance maps are created in the intracellular direction, both from the plasma membrane and the tip. In addition, the pollen tube midline is identified and used as the starting point for a distance map. All images, ROIs, and distance maps are saved and available for repeated analysis by a second macro performing data analysis.

To reproducibly draw intracellular ROIs by hand is even more challenging than identifying the plasma membrane because there are no readily identifiable reference points. To this end, distance maps are useful [7]. Another problem is that the position of organelles and other intracellular structures may vary between imaging planes. However, in pollen tubes, organelles and other cellular components are symmetrically distributed around the midline, making pollen tubes an ideal system for studies of intracellular distributions [4]. The distance maps generated in PollenTube_DisMaps.ijm are used to define and analyze ROIs at any user-specified position(s) within the pollen tubes in the ImageJ macro Pollen_Tube_Distributions.ijm, exemplified in our recent study [5].

The biophysical properties of the plasma membrane of eukaryotic cells are not homogeneous but vary with the relative proportions of two co-existing liquid phases, the liquid ordered and liquid disordered phase, that differ in their lipid packing [8, 9]. The more ordered domains are also known as lipid rafts and have roles in multiple cellular processes like adhesion, signal transduction, secretion, and pathogen entry [10, 11]. By collecting the photons emitted by solvatochromic probes at two wavelength ranges, membrane order can be quantified using generalized polarization (GP) calculations, where higher GP-values report a higher proportion of ordered membrane domains [12].

Our two macros reproducibly report the spatial fluorescence in user-defined ROIs. Using two fluorescent channels or more, ratiometric measurements like GP-values can be obtained. This makes the macros versatile tools for quantification of the spatial fluorescence intensity and membrane order differences along the plasma membrane and intracellularly relative to the tip, inwards from the plasma membrane, and outwards from the pollen tube midline.

2 Materials

The harvesting of anthers and the media optimized for pollen grain germination have been described by Fritz & Kost [13].

2.1 Slides

1. A uniformly fluorescent slide, e.g., 474,029–9080 from Zeiss (Jena, Germany) (*see* **Note 1**).
2. Specimen with no added fluorophore.
3. Specimen with only a single fluorophore, one for each fluorophore used (*see* **Note 2**).
4. Specimen with all fluorophores.

2.2 Imaging Systems

If more than one channel is used, simultaneous acquisition is recommended to enable image alignment despite growth-related movement. The following microscopes meet the requirements:

1. Wide-field.
2. Confocal.
3. Multiphoton.
4. Stimulated emission depletion.
5. Structured illumination.
6. Lattice sheet.

2.3 Software

1. Optional: deconvolution, e.g., Huygens from SVI (Hilversum, The Netherlands) or AutoDeblur from Media Cybernetics (Maryland, US) (*see* **Note 3**).
2. ImageJ/Fiji (https://imagej.net/ij/download.html) [14].
3. ImageJ Macros PollenTube_DisMaps.ijm and Pollen_Tube_Distributions.ijm and the accompanying file Default_ROI_Settings.tif (https://github.com/Parmryd/Pollen_tubes).

4. ImageJ Plugins StackReg, Translation alignment [15], and MorphoLibJ [16]. Please cite these references when the macros described in this chapter are used.

3 Methods

3.1 Selection of Fluorophores and Filters

1. If more than one fluorophore is required, examine their excitation and emission spectra together with excitation sources and filter setups to ensure that each channel collects photons originating from only one of the fluorophores. This can be done with online spectral viewers (https://www.bdbiosciences.com/en-us/resources/bd-spectrum-viewer, https://www.biolegend.com/en-gb/spectra-analyzer, https://www.thermofisher.com/order/fluorescence-spectraviewer/#!/ and others) (*see* **Note 4**).
2. Ensure that the optical filters selected are appropriate for the combination of fluorophores and illumination source(s) used. Filter specifications should be obtainable from the filter and/or microscope manufacturer.

3.2 Microscope Setup

1. Check for even illumination using a uniformly stained test slide and examine the images using a false color look-up table. Adjust the illumination of the light source and use a shading correction if required.
2. Optimize the image acquisition and excitation light intensity settings for each fluorophore used so that the signal remains within the response range of the system. Failing to optimize the settings will degrade any quantitative image analysis (*see* **Note 5**).
3. Check the background in an unlabeled specimen for autofluorescence using the settings established in **step 2**. When imaging an area of the specimen, the signal should not exceed that from areas where no specimen is present. If a signal within the specimen is observed, either the sample preparation protocol, the fluorophore(s) used, the emission filters, or the excitation range should be changed (*see* **Note 6**).

3.3 Image Acquisition

1. Establish inclusion criteria. For tobacco pollen tubes, a growth rate of 3.0 μm/min is a recognized threshold for healthy cells [17]. A method to determine the growth rate can be found in Fritz & Kost [18].
2. If possible, blind the operator from the experimental identity of the specimen(s) to reduce operator bias.

3. Make an experimental plan, deciding on how many images/image sets from how many different specimens per group are required.
4. Check for consistent staining patterns at random positions on the slide to ascertain that the images acquired are representative of the staining/treatment group.
5. Acquire images using the settings established in Subheading 3.2, **step 2** to ensure that the fluorescence signal is within the dynamic range of the microscope.
6. Minimize exposure times and use illumination at low intensity to reduce photobleaching, photodamage, and blurring caused by movement (*see* **Note 7**).
7. Acquire the required number of images/image sets from different areas of the coverslip by randomly moving the stage and avoiding areas where neighboring pollen tubes are close (*see* **Note 8**).

3.4 Image Deconvolution (Optional)

For widefield images, deconvolution is an essential processing step prior to image analysis. If working with images from other imaging systems, deconvolution is advantageous but less crucial (*see* **Note 3**).

1. Perform image deconvolution to improve the image clarity and resolution. An illustration of the power of deconvolution can be seen in [19].
2. Name the images by adding identifiers for the deconvolved images from the shorter and longer wavelength channels, respectively, if GP-measurements are required. For instance, use shorter/longer or emission wavelengths (*see* **Note 9**).

3.5 Identifying the Plasma Membrane and the Pollen Tube Tip, Creating Distance Maps, and Making Measurements Along the Plasma Membrane Using PollenTube_DisMaps.ijm

1. By trial and error, optimize the variables after the line "EDITABLE SETTINGS" for your specimens; explanations are given in Fig. 1. All the settings appear in the Results table, along with the date, name(s), and location of the input file(s) (*see* **Note 10**).
2. Click "Run." This closes any open images, clears the ROI manager, and starts the macro.
3. Open the image/z-stack from the first channel. Open the first image, named in the first position in the settings in **step 1**. If a z-stack, the center slice is opened by default, but this can be changed in the settings in **step 1**.
4. Open any images/z-stacks from additional channels. The user is prompted to open any subsequent images, named in the settings in **step 1**.

```
// EDITABLE SETTINGS -------------------------------------------------------------
// Default values for variables -  adjust to suit images.
// Then save the macro (with a new name) to reuse the new settings.
// For traceability the settings are shown in the Results Table.
AnalysisBy="Jeremy";  // name of analyst
macroVersion="PollenTube_DisMaps01.ijm";// CHANGE IF MACRO SAVED WITH NEW NAME
//      INTERACTION - images shown to the user as the macro runs
shw=1; // show overlays during analysis  0-off    1-show
shwStep=1;  // periodically stop macro to show user current state    1-on,0-off
//       IMAGES
nChan=2;// number of channels used
defaultSlc=0; // the default slice - if z-series,  0-selects the middle slice
//       CHANNEL NAMES, in sequence when loaded separately or sequence in a stack
ChannelNamesArray=newArray("DiI_shorter","DiI_longer","Transmission","Alexa568");
pxlSze=0.090;// pixel size in ums
//       ALIGNMENT
AlignDo=1; // 1 align images   0 do not align, not relevant for a single channel
rngAlgn=28;// max range in pixels to check images for misalignment
//       MEMBRANE ROI  Initial(marked) points to identify PM & optimise
ChanForPMArray=newArray(1,1,0,0);// channels to use/combine for PM, 1-use
sig=0.90; // sigma for Gaussian when smoothing the image for optimising points
checkInit=1; // search for max intensity in:  0-circular area   or 1-perpendicular
rng=17; // search distance from initial (those marked) points in pixels
//       INTERPOLATION & Search around initial points
spacing=4; // interpolation spacing (in pixels)
checkInterp=1;// search for max intensity in: 0-circular area  or  1-perpendicular
InterplRng=8; // search distance around interpolated points
//        FIND TIP
FndTipMethod=1;  //  1-from centre, 2-from ends
//        BACKGROUND ROI,expanded outwards from PM then extended by the band
BKoffset=30; // offset from between the plasma membrane ROI (pixels)
BKband=15;  // width of the background ROI, extending from inner edge (pixels)
//        MEASUREMENTS
PM_bin=1; // size of increments for measurements along PM (ums)
GPmeasure=1; // option to measure GP, 1-yes, 0-no
```

Fig. 1 User-defined settings in PollenTube_DisMaps.ijm. Explanatory text is shown in green

5. Identify the area of interest. Move and resize the default rectangular selection tool to enclose the area of interest around a single pollen tube with sufficient space outside the pollen tube for the software to define a background ROI. If no image cropping is not required, simply click "OK" (*see* **Note 11**).
6. Image alignment. Alignment of the cropped images from **step 5** is, if selected, automatically performed using the ImageJ plugin *Translation alignment*, which is a modified version of the ImageJ plugin *StackReg* [15]. Examples of images before and after successful and failed alignment are shown in Fig. 2 (*see* **Note 12**). The alignment option can be turned off in the settings in **step 1**. If only one channel is used, the alignment is automatically turned off.
7. Check the alignment of the images by flipping the slider between the different channels in a stack consisting of the single, cropped input images and the intensity equalized image combination of the two. Click "OK" if the images are aligned.

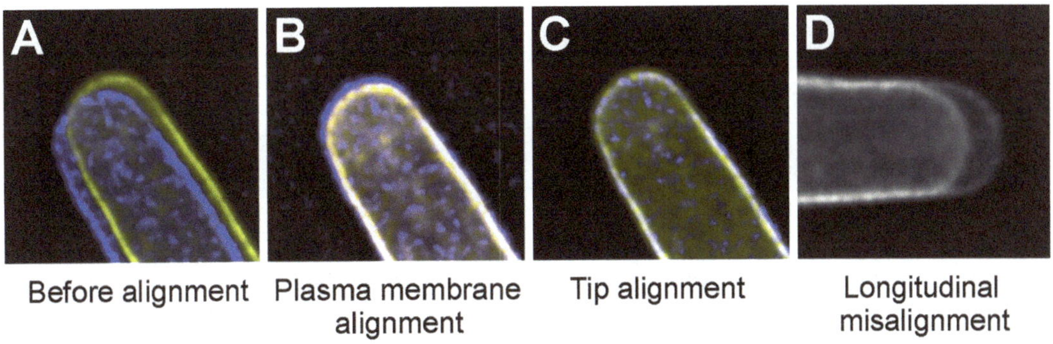

Fig. 2 Image alignment. (**A**) Original unaligned images. (**B**) Automatic alignment based on the images after cropping. (**C**) Backup alignment using only the tip region of the pollen tube. (**D**) Failed image alignment along the longitudinal axis of the pollen tube. Scale bar = 8 μm. The scale bar works as a reference also for Figs. 3, 4, and 5 since the same pollen tube is used in subsequent illustrations. From the macro PollenTube_DisMaps.ijm

8. Use the backup alignment tool if the initial alignment fails. The tool prompts the user to mark only the curved tip of the pollen tube, which is used for the subsequent alignment (*see* **Note 13**).
9. Mark the plasma membrane. Select whether to use one of the single images or the merged image by flipping the slider. Crudely mark a few sequential points near the plasma membrane (Fig. 3A). The start and the end points should be roughly equidistant from the tip. A start point, a few points around the tip, and an end point are usually sufficient (*see* **Note 14**).
10. Identification of the plasma membrane. The accuracy of the marked points is improved using an automated local search for the brightest pixel in the vicinity of the marked point. The search is performed either along a line or within a circle over a range editable under the settings in **step 1** and the positions of the points are adjusted (Fig. 3A). A spline is then fitted to the optimized points (Fig. 3B). Next, new points are automatically interpolated along the plasma membrane from the start point at intervals set by the user in **step 1** (Fig. 3C). Finally, the positions of the interpolated points are also optimized by a local search, to create the plasma membrane ROI (Fig. 3D) (*see* **Note 15**).
11. Correct any mislocated points. The user is shown and given the opportunity to manually relocate or remove any aberrant points. Individual points can be either moved by right-clicking on a point and holding the button down or deleted by right-clicking on a point while pressing the Alt-key if they are

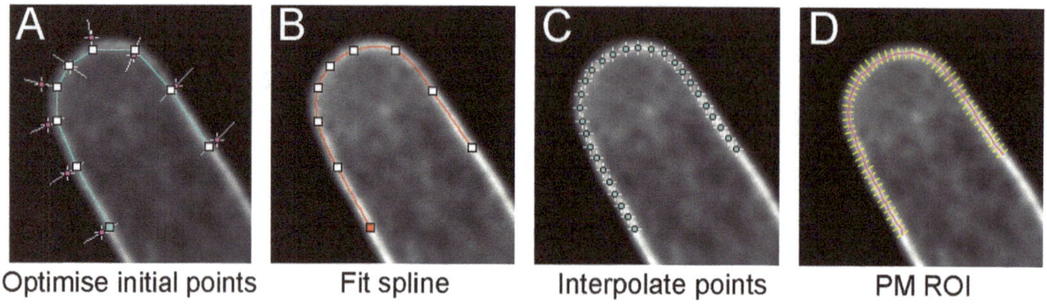

Fig. 3 Plasma membrane identification. (**A**) A few sequential points (purple rings) are imprecisely marked along the plasma membrane, the majority at the pollen tube tip. An automatic search for the brightest pixel is performed along a line extending, here set at 13 pixels, in both directions at an angle perpendicular to a line between the two adjacent points, and each point is repositioned to the most intense pixel along the search line (white squares). (**B**) A spline is fitted between the optimized points. (**C**) New points are interpolated (blue circles) from the start point along the spline at fixed intervals, here five pixels. (**D**) The position of each interpolated point is optimized like the original points, but at a shorter search range, here five pixels. From the macro PollenTube_DisMaps.ijm

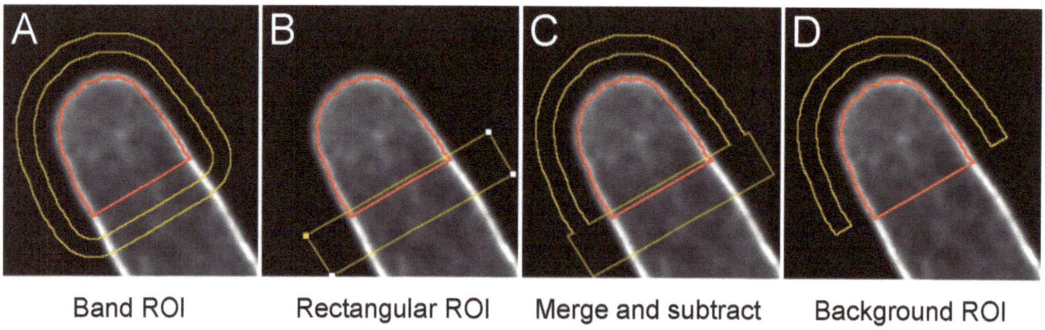

Fig. 4 Automatic creation of a background ROI. (**A**) The endpoints of the plasma membrane ROI are joined to form an area. This area is enlarged, becoming the inner border of a band. (**B**) A rectangular ROI just overlapping the area at the intracellular end of the band is created. (**C**) The band area and the rectangle are merged, and the parts of the band area that coincide with the rectangle are removed. (**D**) The background ROI is formed. From the macro PollenTube_DisMaps.ijm

assessed to be in inappropriate positions. For this step, it can be helpful to zoom in.

12. Generation of an ROI for measuring the background intensities. The background ROI is created automatically. First, the two endpoints are joined to form an area. The area is enlarged to exclude the immediate vicinity of the pollen tube and a band with width specified in **step 1** is created (Fig. 4A). To remove the part of the band that includes the pollen tube, a rectangular area is created (Fig. 4B) and the two areas are merged

(Fig. 4C). The band overlap with the rectangle is removed leaving a horseshoe-shaped area (Fig. 4D). For each fluorophore, the mode of the background ROIs is then subtracted from the image(s).

13. Calculation of GP-values. If desired, GP-values are calculated for every pixel in the background-subtracted images, and a GP-image is automatically created. The formula used for the calculation is:

$$GP = \frac{I(\text{shorter wavelength range}) - I(\text{longer wavelength range})}{I(\text{shorter wavelength range}) + I(\text{longer wavelength range})}$$

For the GP-values to be calculated according to the formula above, the image from the shorter wavelength range needs to be opened in **step 3**, and the image from the longer wavelength range needs to be opened in **step 5** (*see* **Note 16**).

14. Identification of the tip position. To enable tip identification, the images must include the cell midplane. The tip position can be automatically derived by two methods and selected in the settings from **step 1**. The first method identifies the point of the plasma membrane that is furthest away from the midpoint of the two end points. The second method identifies the point in the plasma membrane ROI with the greatest combined distance from the two ROI end points (Fig. 5). The software displays the nominal tip, which can be overridden by the user (*see* **Note 17**).

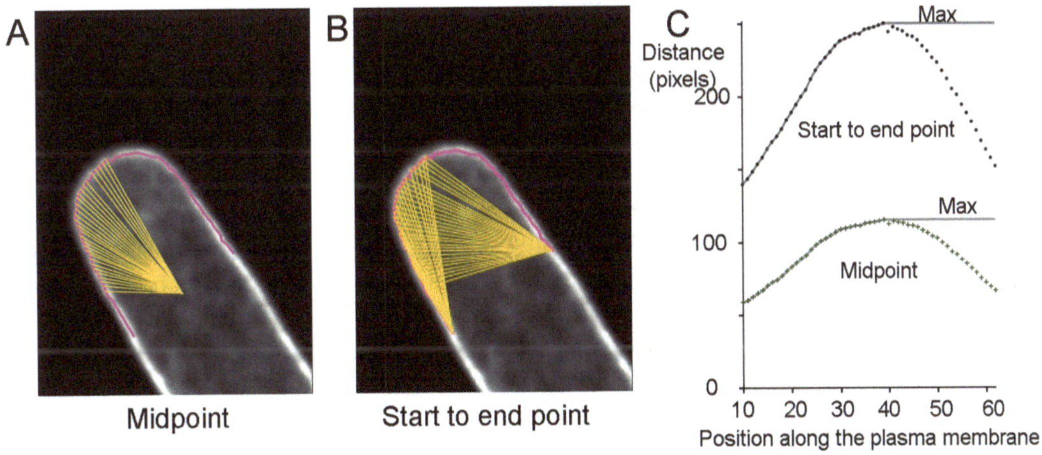

Fig. 5 Identification of the position of the pollen tube tip. Two automatic search methods are offered: (**A**) the point in the plasma membrane ROI with the longest distance from the midpoint between the two endpoints, or (**B**) the point on the plasma membrane ROI furthest from the two endpoints using the sum of their combined distance. (**C**) Graphical illustration of the distances obtained by the two search methods. From the macro PollenTube_DisMaps.ijm

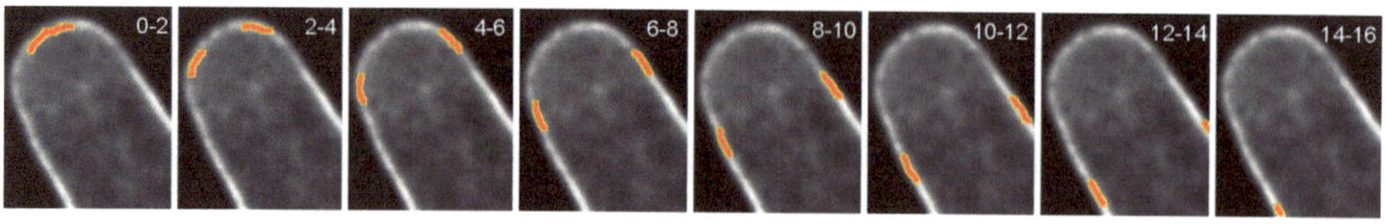

Fig. 6 Illustration of ROIs along the plasma membrane. Sequential 2 μm ROIs along the plasma membrane from the pollen tube tip. The outline of each ROI has been enlarged from the single-pixel line for better visibility, and the interval settings end at 1.99, 3.99 μm, etc.. From the macro PollenTube_DisMaps.ijm

15. Generation of a distance map from the tip along the plasma membrane. A distance map with the geodesic distance from the tip along the plasma membrane is automatically created (from Tip along PM).
16. Generation of intracellular distance maps. Intracellular distance maps are automatically created from (i) the entire plasma membrane ROI (from PM), (ii) the plasma membrane ROI excluding the tip region (used to define the midline), and (iii) Euclidean distance from the tip (from Tip Euclidean).
17. Generation of a distance map from the pollen tube midline. First, the pollen tube midline is automatically identified from the intensity ridge in a distance map from the plasma membrane. To allow the ridge to reach the tip, the area around the tip is removed from the plasma membrane distance map. Then, a distance map from the midline to the plasma membrane is created (from Midline). An additional intracellular distance map perpendicular to the midline starting from the tip (from Tip transverse).
18. Intensity values (I) and GP-values (if required) are measured from sequential ROIs along the plasma membrane, starting at the tip at increments set in **step 1**. An illustration of sequential ROIs in the plasma membrane is shown in Fig. 6, where each ROI covers a distance range from the tip.
19. A Results file, all the ROIs, and a folder containing the images and distance maps are automatically generated and, by default, saved in the folder from which the input images originated. The Results file includes all the adjustable settings, the date, the original name and the source of the images, the mean background for the channel(s), the slice selected from an input image z-stack, and the measurements of fluorescence intensity and GP-values (Fig. 7). The number of pixels in each ROI is also reported (*see* **Note 18**).

Category	Value	fromTip	nPixels	GPmean	GPstd	iShorter	Sstd	iLonger	Lstd
Image	8.5_decon06_ch01	0	19	-0.704	0.036	2499	340	14444	1290
Folder	C:PollenTubes21	1	19	-0.745	0.049	2488	595	16973	1738
Pixel size ums	9.000E-2	2	20	-0.695	0.043	3042	328	17048	1994
Analysis by	Jeremy	3	19	-0.732	0.028	2596	385	16751	1668
MacroVersion	PollenTube_DisMaps_V01	4	20	-0.732	0.053	2492	740	15758	1358
Year month day	2026_05_10	5	18	-0.771	0.042	2096	356	16276	1345
Channels	2	6	17	-0.762	0.035	2081	305	15467	1386
Slice in stack	7	7	18	-0.778	0.015	2212	316	17617	1558
Chn1	Dil_shorter	8	19	-0.779	0.037	2358	581	18764	1470
Chn2	Dil_longer	9	17	-0.777	0.029	2588	548	20451	1555
Initial search	perpendicular	10	18	-0.755	0.028	3165	558	22549	1775
Initial search pxls	17	11	17	-0.727	0.037	3355	578	21215	2027
Interpolated search	perpendicular	12	18	-0.777	0.041	2806	485	22438	1318
Interpolation interval pxls	4	13	17	-0.776	0.030	2965	409	23476	936
Interpolation search pxls	8	14	18	-0.778	0.015	2855	271	22832	1050
Blur sigma pxls	0.9	15	18	-0.736	0.044	3439	560	22611	951
BKgrnd_PMexpand_pxls	30	16	18	-0.735	0.039	3337	687	21608	1845
BKgrnd_Band_width_pxls	15	17	18	-0.762	0.043	3184	931	23167	3056
Channels use for PM	1 2	18	17	-0.751	0.070	3494	627	25372	4365
Background shorter	3463	19	18	-0.740	0.039	3786	544	25436	1914
Background longer	3657	20	18	-0.766	0.021	3438	387	25886	748
PM_incr um	1.0	21	18	-0.765	0.058	3127	586	23943	3229
GP	1	22	18	-0.746	0.045	3744	653	25768	1677
GPchan	chan 1 2	23	17	-0.773	0.039	3284	717	25495	2717

Fig. 7 Layout of the results table from PollenTube_DisMaps.ijm. Screenshot of the results table

3.6 Making Measurement from Intracellular ROIs Using Pollen_Tube_Distributions.ijm

1. Open images. The user is prompted to open the stack of 11 images created by PollenTube_DisMaps.ijm in Subheading 3.5, **step 19**.
2. Open the Dialog with the settings. The user is prompted to open the file Default_ROI_Settings.tif or a similar file with alternative settings (*see* **Note 19**).
3. Define a ROI for measurements. The ROI is specified by the user from a subset of the five distance maps in a Dialog (Figs. 8 and 9). The distance maps Tip Transverse and Tip Euclidean are similar and cannot be usefully combined to define an ROI. For each distance map, a range from the smallest distance (min) to the largest distance (max) can be specified. When more than one distance map is used, the final measurement area is the overlap (Boolean AND), and within this measurement ROI, subregions at incremental spatial steps can be specified to measure gradients (Fig. 9E). The ranges can be specified in pixels or μm (*see* **Note 18**).

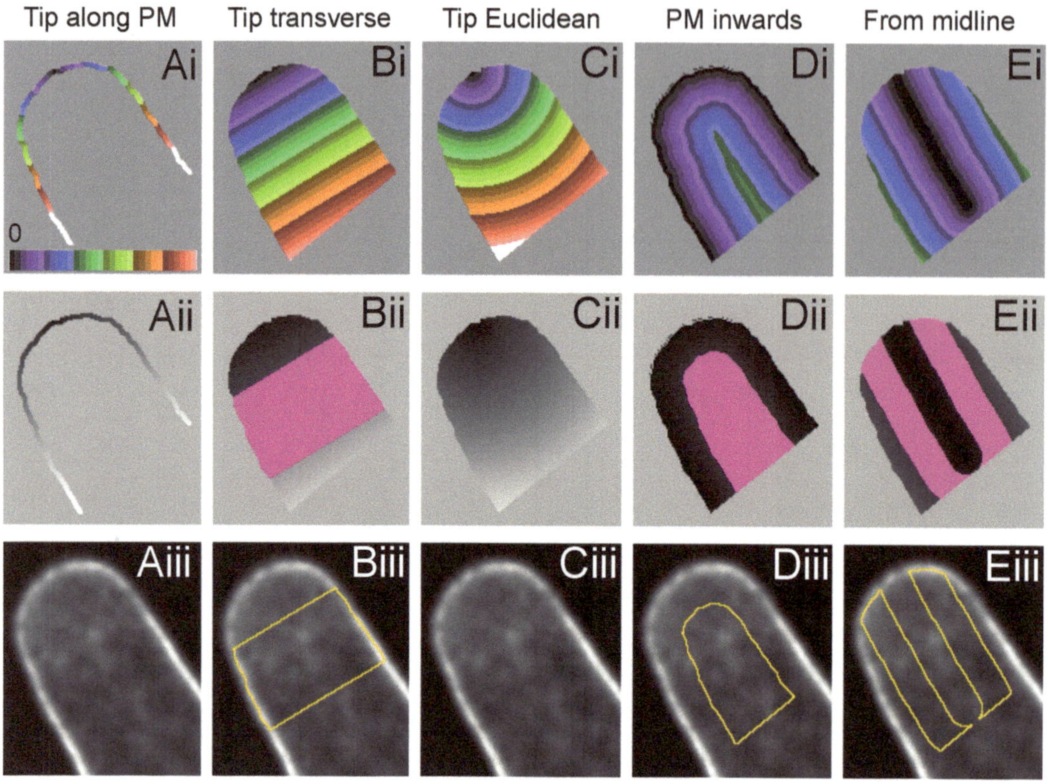

Fig. 8 Using distance maps to specify areas within the pollen tube. First row (**i**) distance maps, (**A**) (**Ai**) From the tip along the plasma membrane. (**B**) (**Bi**) Transverse from the tip. (C) Euclidean from the tip. (D) From the plasma membrane inwards. (E) From the midline towards the plasma membrane. Second row (**ii**), the areas specified by the Dialog settings in Fig. 9. Third row (**iii**), the areas superimposed on the pollen tube image. The lookup table displays distances from the respective starting points. From the macro Pollen_Tube_Distributions.ijm

4. A Results file containing the fluorescence intensity and GP-measurements in the selected ROI is automatically generated and saved in the folder from which the input images were opened.

4 Notes

1. Uniformly fluorescent slides can be prepared by sealing a layer of a high concentration of a fluorophore between a slide and a coverslip. The fluorophore(s) should be selected based on the excitation wavelengths and filters used.
2. When using one fluorophore only, this control is not necessary since there can be no crosstalk apart from autofluorescence.
3. In image deconvolution, out-of-focus light is reassigned to a position closer to its origin, which improves the image

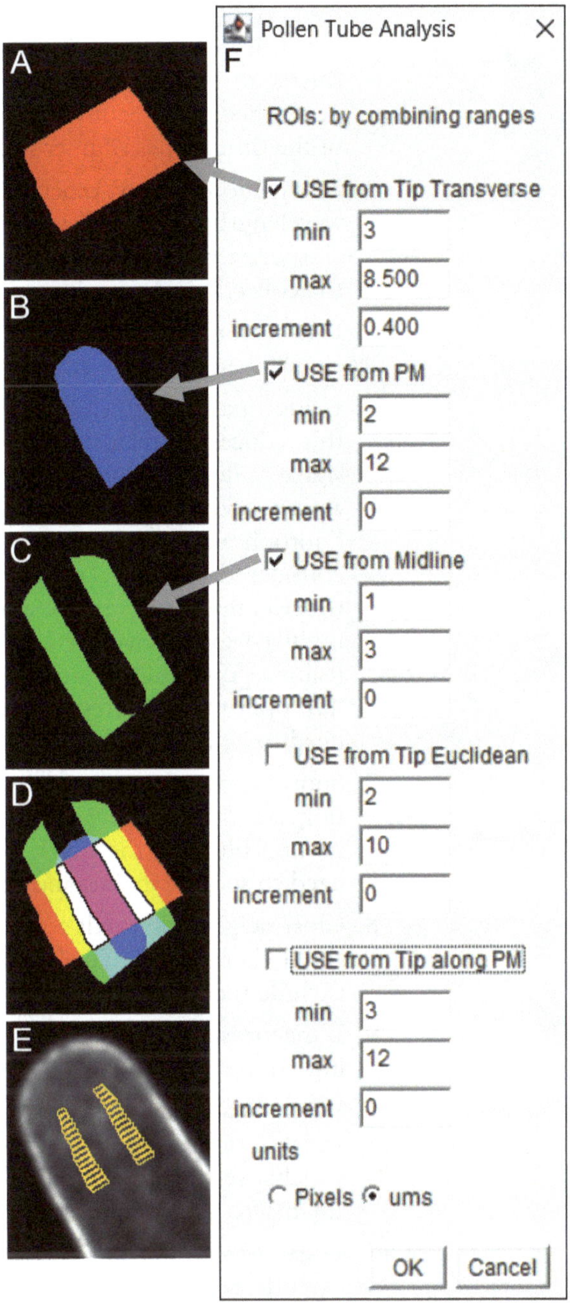

Fig. 9 Creating an intracellular measurement ROI. (**A–C**) Areas are specified in three distance maps. (**D**) Overlay of the three areas, where white indicates the area common to all three, the desired ROI. (**E**) Subdivision of the measurements in increments set along the Euclidean distance from the tip allows gradient measurements. (**F**) Screenshot of the user Dialog where usage, range, and increment are selected for each distance map. Arrows link the areas with their respective distance map specifications. From the macro Pollen_Tube_Distributions.ijm

resolution, partially reversing the convolution (distortion) inherent in image acquisition [20]. For wide-field microscopy, this is an essential step before quantitative image analysis is performed, but deconvolution will also improve the resolution of the other types of microscopes listed in Subheading 2.2.

4. Many fluorophores, especially those with emission in the red wavelength region, have multiple excitation peaks. Ensure that all spectra cover all wavelengths used for excitation as well as the collection of emission.
5. False color look up tables display intensity differences more clearly than greyscale and monochromatic look up tables.
6. Using quantum-efficient fluorophores is recommended since this reduces the relative contribution of autofluorescence to the signal. Since autofluorescence, e.g., originating from aromatic amino acids, generally appears at short wavelengths, using fluorophores with longer emission wavelengths should also be considered. A spectral scan of any autofluorescence can be used to select fluorophores that emit over a different range. If studying fixed specimens, the autofluorescence can be bleached out using a powerful excitation source before the addition of the fluorophores of interest. In addition, the fixation protocol may need to be adjusted since it could cause autofluorescence.
7. Short exposure times are particularly important for deconvolution, which is improved when using a z-stack, but also for not motion blurring the borders of membrane components localized to specific membrane zones.
8. Close neighboring cells could compromise the measurement of the background signal since the macro makes no provision to exclude their fluorescence from the background ROI.
9. If the measurement(s) involve only a single channel, the renaming can be ignored.
10. When the macro is saved, this includes the current (edited) variable settings. Saving the macro thus allows the adjusted variable settings to be reused. After editing the settings, save the macro with a new name to avoid overwriting the original.
11. Local deconvolution artifacts that can appear where a cell extends beyond the image border should be excluded from the cropped image, both to facilitate the alignment and not to corrupt any measurements.
12. For most imaging systems, the microscope should produce aligned images if recorded by the same detector. However, small adjustments may still be required. In addition, chromatic aberration resulting from large differences in the emission

wavelengths of the selected fluorophores may cause misalignment. Aim to keep this difference as small as possible without causing crosstalk between channels. For images generated by dual or quad viewers that record two or four areas with different emission filters, respectively, image alignment is usually required. The same applies when multiple detectors are used.

13. The long and almost parallel sides of the pollen tube can, when labeled with a membrane probe, have very similar intensities throughout their lengths, which may interfere with the alignment (Fig. 2d). Since a shift along the axis of the pollen will still produce a strong cross-correlation, the offset images have substantial alignment. After marking the tip, the macro aligns the images based only on this part of the image. Alignment may also be improved by cropping the image(s) differently.
14. If the pollen tube is not straight, more points along its sides may be required.
15. Bright intracellular structures close to the plasma membrane can disrupt the automatic repositioning of points to the plasma membrane. If these structures are present, try decreasing the search range and marking the initial points just outside the cells before moving or deleting points. For optimization of the interpolated points, a smaller search range is recommended than for the initial points. If the distance between the interpolated points is narrow, a line search is preferable so that search areas for adjacent points do not overlap.
16. Depending on what solvatochromic fluorophore and imaging setups are used, the actual wavelength ranges collected for the respective shorter and longer channels will differ.
17. Failure to precisely locate the tip position could be caused by the two marked parts of the pollen tube sides having appreciably different lengths. The failure can also be caused by the pollen tube being curved or the tip not being completely round. The user can then override the automatic tip identification to adjust the location of the tip.
18. If the consecutive pixels are orthogonal, the distance between them is the same as the xy-pixel size, but if the consecutive pixels are diagonal, the distance will be 1.41 times the xy-pixel size. To ensure every ROI includes pixels, the increment in distance should be a multiple of the pixel size, and it is recommended not to have a distance smaller than twice the pixel size. Note, acquiring images with pixels smaller than Nyquist sampling does not increase the resolution.
19. The last settings used in the Dialog can be saved and reused for efficient analysis of ROIs from multiple pollen tubes.

Acknowledgments

This work was supported by funding to IP from the Swedish Research Council (2015-04764) and Magnus Bergvall's Foundation.

References

1. Bedinger P (1992) The remarkable biology of pollen. Plant Cell 4(8):879–887
2. Ketelaar T, Galway ME, Mulder BM, Emons AM (2008) Rates of exocytosis and endocytosis in Arabidopsis root hairs and pollen tubes. J Microsc 231(2):265–273
3. Picton JM, Steer MW (1983) Membrane recycling and the control of secretory activity in pollen tubes. J Cell Sci 63:303–310
4. Grebnev G, Cvitkovic M, Fritz C, Cai G, Smith AS, Kost B (2020) Quantitative structural organization of bulk apical membrane traffic in pollen tubes. Plant Physiol 183(4): 1559–1585
5. Fritz C, Reimann TM, Adler J, Knab J, Schulmeister S, Kriechbaum C, Muller S, Parmryd I, Kost B (2024) Plasma membrane and cytoplasmic compartmentalization: a dynamic structural framework required for pollen tube tip growth. Plant Physiol 197(1)
6. Adler J, Huang A, Parmryd I (2023) Find_plasma_membrane and measure_plasma_membrane: ImageJ macros for efficient identification of and measurements at and around the plasma membrane. SoftwareX 24: 101570
7. Adler J, Melia CD (1994) Using a distance transform function in image processing. Microsc Anal 41:31–33
8. Owen DM, Williamson DJ, Magenau A, Gaus K (2012) Sub-resolution lipid domains exist in the plasma membrane and regulate protein diffusion and distribution. Nat Commun 3:1256
9. Dinic J, Ashrafzadeh P, Parmryd I (2013) Actin filaments attachment at the plasma membrane in live cells cause the formation of ordered lipid domains. Biochim Biophys Acta 1828(3): 1102–1111
10. Levental I, Levental KR, Heberle FA (2020) Lipid rafts: controversies resolved, mysteries remain. Trends Cell Biol 30(5):341–353
11. Kulkarni R, Wiemer EAC, Chang W (2021) Role of lipid rafts in pathogen-host interaction – a mini review. Front Immunol 12:815020
12. Parasassi T, De Stasio G, d'Ubaldo A, Gratton E (1990) Phase fluctuation in phospholipid membranes revealed by Laurdan fluorescence. Biophys J 57(6):1179–1186
13. Fritz C, Kost B (2020) Testing pollen tube proteins for in vivo binding to phosphatidic acid by n-butanol treatment and confocal microscopy. Methods Mol Biol 2160:307–325
14. Schindelin J, Arganda-Carreras I, Frise E, Kaynig V, Longair M, Pietzsch T, Preibisch S, Rueden C, Saalfeld S, Schmid B, Tinevez JY, White DJ, Hartenstein V, Eliceiri K, Tomancak P, Cardona A (2012) Fiji: an open-source platform for biological-image analysis. Nat Methods 9(7):676–682
15. Thevenaz P, Ruttimann UE, Unser M (1998) A pyramid approach to subpixel registration based on intensity. IEEE Trans Image Process 7(1):27–41
16. Legland D, Arganda-Carreras I, Andrey P (2016) MorphoLibJ: integrated library and plugins for mathematical morphology with ImageJ. Bioinformatics 32(22):3532–3534
17. Sun J, Eklund DM, Montes-Rodriguez A, Kost B (2015) In vivo Rac/Rop localization as well as interaction with RhoGAP and RhoGDI in tobacco pollen tubes: analysis by low-level expression of fluorescent fusion proteins and bimolecular fluorescence complementation. Plant J 84(1):83–98
18. Fritz C, Kost B (2026) Quantitatively analyzing the structural organization at the tip of growing tobacco pollen tubes. In: Cai G, Piccini C, Conti V, Parri S (eds) Pollen and pollen tube analysis: methods and protocols (pp 175–198). Methods in molecular biology. Springer, New York
19. Huang A, Adler J, Parmryd I (2023) Optimised generalized polarisation analysis of C-laurdan reveals clear order differences between T cell membrane compartments. Biochim Biophys Acta Biomembr 1865(2): 184094
20. van der Voort H, Strasters K (1995) Restoration of confocal images for quantitative image analysis. J Microsc 178:165–181

Chapter 7

Quantitative Profiling of Metabolites, Sugars, and Lipids in Pollen Tubes by Mass Spectrometry

Maneesh Lingwan, Neeta Lohani, and Erin Cullen

Abstract

Pollen tube germination and elongation are energy-demanding, metabolically active processes essential for plant reproduction. Recent research combining metabolomics, imaging, and molecular genetics has highlighted the crucial roles of sucrose synthase activity, starch breakdown, and lipid signaling in maintaining pollen tube health and promoting precise growth, particularly under stress conditions. This chapter presents a mass spectrometry-based workflow for high-resolution profiling of metabolites, sugars, and lipids in pollen and pollen tubes. The workflow includes metabolite extraction, chemical derivatization, and chromatographic separation steps compatible with both gas chromatography–mass spectrometry (GC–MS) and liquid chromatography–mass spectrometry (LC–MS). These techniques enable simultaneous quantification of polar metabolites, lipids, and fatty acids, capturing the metabolic complexity that supports pollen tube germination, elongation, and fertilization. It also covers the application of internal standards and isotopic normalization to ensure precise quantification and reproducibility across experiments. These analytical approaches collectively provide a foundation for comparative metabolite profiling in reproductive biology and can be adapted to various plant species, lipid signaling, and membrane biogenesis, thereby offering a mechanistic understanding of how pollen tubes sustain polarized growth under both developmental and stress conditions.

Key words Pollen tube, Metabolites, Lipid, Sugar, Chromatography, Mass spectrometry

1 Introduction

Pollen tube growth represents one of the most remarkable cellular expansion processes in nature. These highly polarized structures serve as the essential conduit through which sperm cells navigate from the stigma to the ovule during sexual reproduction in flowering plants. The metabolic processes supporting this rapid growth require precise coordination of lipid biosynthesis for membrane expansion, sugar metabolism for energy production, and specialized metabolite accumulation for stress tolerance and cellular signaling [1]. Elongation rates can reach extraordinary speeds of up to 1 cm per hour in maize (*Zea mays*), with tubes extending over

Giampiero Cai et al. (eds.), *Pollen and Pollen Tube Analysis: Methods and Protocols*, Methods in Molecular Biology, vol. 3042, https://doi.org/10.1007/978-1-0716-5308-1_7,

30 cm through silk tissue to achieve fertilization [2]. Since pollen viability directly impacts crop yields and reproductive tissues are particularly vulnerable to environmental stresses [3], understanding pollen tube metabolism has become increasingly critical for ensuring agricultural productivity under changing climatic conditions.

The transition from dormant pollen grain to actively growing pollen tube involves profound metabolic reprogramming, with over 2000 distinct metabolomic features altering dynamically during hydration, germination, and elongation [4]. Pollen grains possess unique lipid-rich structures, both extracellular and intracellular, that are essential for reproduction and environmental interactions. The extracellular pollenkitt, a sticky matrix composed primarily of neutral lipids, fatty acids, carotenoids, and proteins, aids in dispersal, adhesion, pollinator attraction, and pollen-stigma interactions [5]. Within the pollen grain, lipids are stored as triacylglycerols in lipid droplets and maintained as membrane glycerolipids, sphingolipids, and sterols to support rapid tube growth and membrane expansion [5]. Further, the pollen wall derives its chemical durability from a sporopollenin scaffold formed from lipid and phenolic precursors synthesized in the tapetum [6].

Notably, pollen lipid profiles differ substantially from those of vegetative tissues [7]. For example, extraplastidial galactolipids are synthesized in growing tubes, and sphingolipids and sterols organize into unique membrane microdomains. Lipids in pollen grain's function both as structural components of the expanding membrane and as signaling molecules. Phosphatidic acid and phosphoinositides localize at the apex to regulate tip polarity, while sterol-rich membrane microdomains contribute to exocytosis and vesicle targeting [8]. This complex lipid environment, spanning the pollenkitt coat to intracellular signaling networks, necessitates careful methodological approaches to accurately profile the pollen lipidome without introducing artifacts or contamination.

Sugar metabolism provides energy and biosynthetic precursors essential for pollen tube growth while exhibiting unexpected regulatory complexity. In many crop species, starch accumulated during pollen maturation is the primary storage carbohydrate, and mutations impairing starch biosynthesis cause male sterility due to insufficient energy and carbon for germination and tube elongation [9]. Imported sucrose traverses the apoplast of the anther and is converted to starch in developing grains via coordinated activity of sucrose transporters, proton ATPases, and starch biosynthetic enzymes. Upon germination, this starch is rapidly hydrolyzed by amylases to release glucose and maltose that fuel glycolysis and mitochondrial respiration [10, 11]. Interestingly, pollen tubes also resynthesize sucrose via sucrose-phosphate synthase, generating UDP-glucose for callose plug formation that compartmentalizes the cytoplasm during extension

[11, 12]. Hexokinase-mediated phosphorylation of glucose integrates sugar sensing with metabolic flux, and glucose uptake through specific transporters can paradoxically inhibit tube growth via HEXOKINASE1-dependent signaling pathways, indicating that sugar sensors modulate growth independently of substrate supply [13]. Beyond sugars, pools of organic acids, amino acids, and nucleotides reflect mitochondrial activity, biosynthetic demands, and stress-response readiness during tube emergence [4, 14]. Characterizing these metabolite classes requires complementary extraction protocols and chromatographic-ionization platforms to address each compound class's unique physicochemical and ionization properties and achieve comprehensive metabolome coverage.

This chapter details methods for quantifying and localizing these metabolites, combining traditional biochemical assays with advanced metabolomics and lipid analysis. Gas chromatography–mass spectrometry (GC–MS) provides robust quantification of metabolites following chemical derivatization. In contrast, liquid chromatography tandem mass spectrometry (LC–MS/MS) enables structural characterization of complex lipids through high-resolution accurate mass measurement and diagnostic fragmentation. The chapter outlines data acquisition and processing methods, including spectral deconvolution, retention index alignment, and metabolite annotation using curated spectral libraries.

2 Materials

2.1 Reagents

1. 1.5 mL, 2.0 mL, and 5 mL microcentrifuge tubes.
2. Sterile metal spatulas.
3. Liquid nitrogen.
4. Mortar and pestle.
5. 8 mL glass tubes (8 mL).
6. 2 mL analytical glass vials with inserts.
7. Cryogenic gloves.
8. Safety goggles.
9. Micropipettes and sterile tips.
10. 0.8 μm Polyethersulfone (PES) clarifying filters.

2.2 Equipment

1. Vortex.
2. Centrifuge.
3. Lyophilizer (freeze dryer).
4. −80 °C freezer.
5. Ice bucket.

6. Rotary mixer or orbital shaker (4 °C compatible).
7. Speed vacuum concentrator (SpeedVac).
8. Liquid chromatography–mass spectrometry (LC–MS).
9. LC–MS columns (C18, C8).
10. Vacuum filtration unit.
11. Sterile filters.
12. Gas chromatography–mass spectrometry (GC–MS).
13. Gas chromatography-flame ionization detection (GC–FID).
14. GC capillary columns (RTX-5, DB-5MS, DB23).
15. Fume hood for reagent preparation.

2.3 Software and Spectral Libraries

1. Agilent MassHunter, ChemStation, or Xcalibur for data analysis.
2. NIST Library (National Institute of Standards and Technology).
3. Fiehn Metabolomics Library.
4. Freestyle (Thermo Fisher Scientific).
5. Skyline (University of Washington).
6. LipidSearch 4.0 (Thermo Fisher Scientific).
7. MS-DIAL (RIKEN, Japan).
8. LipidXplorer (Max Planck Institute of Biochemistry).

2.4 Solvents and Metabolite Standard for Mass Spectrometry

1. All solvents used for extraction and chromatography must be of UHPLC grade to prevent contamination and improve ionization efficiency.
2. Commonly used solvents include chloroform, methanol, isopropanol (IPA), acetonitrile (ACN), ethanol, and formic acid.
3. For GC–MS analysis, methoxylamine hydrochloride, anhydrous pyridine, and N-methyl-N-(trimethylsilyl)-trifluoroacetamide (MSTFA) are required for derivatization steps, which enhance the volatility and thermal stability of a wide range of compounds, including sugars, amino acids, and organic acids (*see* **Note 1**).
4. To achieve accurate quantification and normalization across samples, internal standards such as ribitol (for polar metabolites), piperazine-N, N′-bis 2-ethanesulfonic acid (PIPES) for organic acids, and norvaline (for amino acid and nitrogen metabolism) should be added before extraction or derivatization.
5. Internal standards serve as recovery controls, helping to correct for extraction efficiency, injection variability, and derivatization yield during GC–MS and LC–MS analyses.

6. For absolute quantification, use metabolite standards such as sugars (glucose, sucrose, fructose), organic acids (malate, citrate, fumarate), and amino acids (glutamate, alanine, serine) for calibration and validation of compounds.
7. Lipid standards representing each major lipid class or similar compounds should be included in all analyses to monitor extraction efficiency and allow for accurate quantitative normalization. Common standards include phosphatidylcholine (PC), phosphatidylethanolamine (PE), phosphatidylglycerol (PG), phosphatidylserine (PS), lysophosphatidylcholine (LPC), ceramide (Cer), diacylglycerol (DG), triacylglycerol (TG), and free fatty acids (FA).
8. Lipid standards are available in Avanti Polar Lipids or in Sigma-Aldrich and are crucial for evaluating extraction recovery, matrix effects, and reproducibility of analytical results.
9. For quantification of total lipid content in pollen tube using fatty acid methyl esters (FAMEs), sulfuric acid (H_2SO_4), butylated hydroxytoluene (BHT), TAG internal standards, C15: 0 (tripentadecanoin), C17:0 (triheptadecanoin), Sodium chloride (NaCl), and Hexane (HPLC grade) are required.

3 Method

3.1 Pollen Tube Collection and Metabolite Quenching

1. Collect pollen tube samples in a clean, dry environment to prevent contamination from dust or environmental lipids [15].
2. Transfer pollen tube samples carefully into pre-chilled 5 mL Eppendorf tubes using sterile, cold tools like metal spatulas or fine brushes.
3. Minimize processing time anticipated because enzymatic hydrolysis and oxidative changes can happen within minutes after collection.
4. Immediately after collection, immerse the collection tubes in liquid nitrogen to quench metabolism and preserve volatile or labile metabolites [15, 16].
5. While frozen, grind the samples into a fine powder using a pre-cooled mortar and pestle under liquid nitrogen.
6. Transfer the homogenized powder into cryovials using chilled tools.
7. Lyophilize samples overnight for long-term storage.
8. Store lyophilized material at −80 °C until metabolite or lipid extraction.

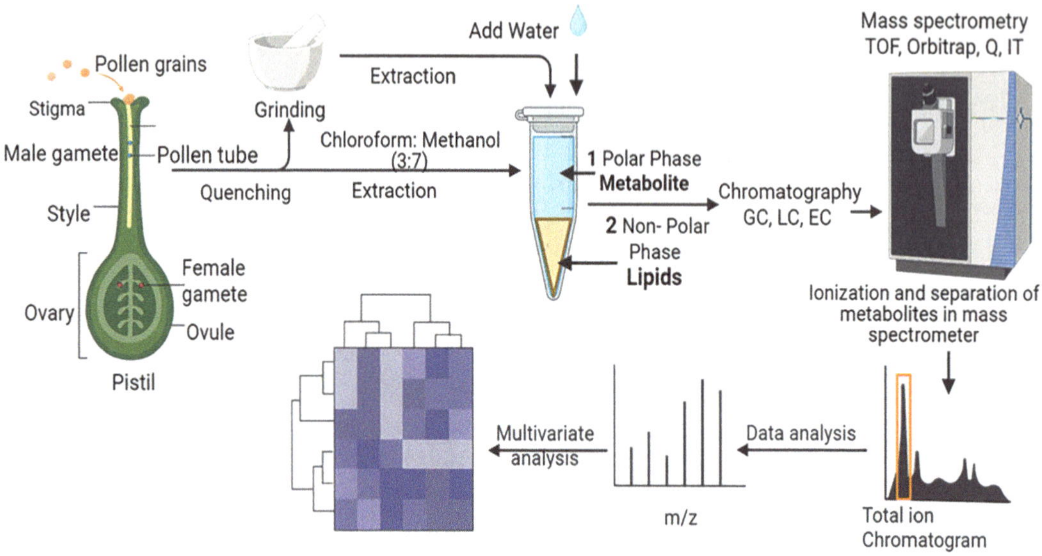

Fig. 1 Typical workflow for metabolite profiling in pollen biology. The steps include collecting tissues, quenching, extracting polar and nonpolar metabolites with appropriate solvents, profiling using analytical platforms, and performing statistical analysis for biological interpretation, All the illustration was created using biorender

3.2 Metabolite Extraction from Pollen Tubes

1. Weigh approximately 30 mg of lyophilized pollen powder or 100 mg of freshly ground pollen tube tissue into a pre-chilled 2 mL microcentrifuge tube (*see* **Note 2**).
2. Samples were maintained on ice throughout the extraction process unless specified (*see* **Note 3**).
3. Add 1.2 mL of pre-chilled methanol: chloroform (7:3, v/v) to each sample tube [4, 17, 18] (Fig. 1).
4. Add 15 μL of 100 μg/mL PIPES-ribitol-norvaline internal standard to each sample.
5. Set the samples on a rotary mixer at 4 °C for 2 h to facilitate efficient extraction of both polar and semi-polar metabolites while minimizing enzymatic degradation.
6. This extraction method allows the simultaneous recovery of small-molecule metabolites, sugars, and lipid intermediates from small amounts of pollen material.
7. After 2 h, add 480 μL chilled double-distilled water (ddH_2O) to induce phase separation between the aqueous (polar) and organic (lipid) fractions.
8. After adding water, vortex the samples vigorously and then centrifuge at 14,000 rpm at 4 °C for 10 min, allowing the samples to phase separate.

9. Carefully collect the upper aqueous phase containing sugars, amino acids, and central metabolites into a new 1.5 mL microcentrifuge tube.
10. For lipid profiling, carefully transfer the lower organic phase into a separate 1.5 mL centrifuge tube.
11. Dry both fractions overnight in a speed vacuum concentrator to evaporate residual solvent. And sample extracts can be stored at −80 °C for analysis on different analytical instruments (e.g., GC–MS or LC–MS).

3.3 GC–MS-Based Metabolites Analysis and Data Acquisition

For GC–MS analysis, take ~35–50 μL of the metabolite extract from the polar supernatant obtained after centrifugation.

1. Dry the aliquot completely in a speed vacuum concentrator to remove all residual solvent and ensure accurate derivatization (*see* **Note 4**).
2. Ensure that the residue is fully dried; incomplete drying can interfere with silylation efficiency.
3. Prepare a 20 mg/mL methoxylamine hydrochloride solution in anhydrous pyridine freshly inside a fume hood to prevent moisture absorption.
4. Add 35 μL of the methoxylamine solution to each dried sample tube.
5. Incubate the mixture for 2 h at 37 °C and 900 rpm in a thermomixer. This ximation step transforms carbonyl groups into stable methoxime derivatives, reducing sugar degradation and enhancing chromatographic consistency.
6. After oximation, add 49 μL of N-methyl-N-(trimethylsilyl)-trifluoroacetamide (MSTFA) to each sample. Methoxamine-trimethylsilylation (MeOx-TMS) derivatization enables detection of a broad range of pollen metabolites, including sugars, amino acids, and organic acids [19–21] (*see* **Note 5**).
7. Incubate again at 37 °C and 900 rpm for 30 min to achieve complete silylation of hydroxyl, carboxyl, and amine functional groups, enhancing volatility and thermal stability of the metabolites.
8. Centrifuge the derivatized samples at 13,000× *g* for 10 min to pellet any particulates or insoluble material.
9. Carefully transfer the clear supernatant into GC–MS autosampler vials or glass inserts, ensuring no debris is carried over.
10. Seal the vials immediately to prevent evaporation and deterioration of volatile derivatives.
11. For data acquisition, ensure the GC–MS system is calibrated correctly and tuned using standard tuning mixtures to verify mass accuracy and detector sensitivity before sample analysis.

12. Install a suitable capillary column based on the metabolite profile of interest. RTX-5 or DB-5MS columns are ideal for nonpolar or derivatized metabolites (sugars, amino acids, organic acids) (*see* **Note 6**).
13. Load derivatized samples into the automatic liquid sampler (ALS) to ensure consistent and contamination-free injections.
14. Inject 1–2 μL of each sample into the front inlet of the GC system in split or splitless mode, depending on sample concentration and instrument sensitivity.
15. Maintain the injector temperature at 250 °C to ensure complete volatilization of derivatized metabolites.
16. Separate metabolites using the following oven temperature program [20, 21]. An initial hold at 50 °C for 1 min, followed by a 10 °C min^{-1} ramp to 200 °C, where the temperature is held for 4 min. The temperature is then further increased at the same rate (10 °C min^{-1}) to a final temperature of 300 °C, maintained for 10 min to ensure elution of high-molecular-weight and late-eluting metabolites (*see* **Note 7**).
17. Use helium as the carrier gas at a constant flow rate of 0.6–1.0 mL/min, adjusted according to column dimensions and sample complexity.
18. Recheck the resulting total ion chromatograms (TICs) for well-resolved peaks, each corresponding to an individual metabolite or co-eluting compound.

3.4 LC–MS Setup for Polar Metabolites, Phosphorylated Metabolites, and Lipids in Pollen Tubes

1. Reconstitute dried polar extracts in 50 μL of 50% methanol containing 0.2% formic acid before LC–MS analysis (*see* **Note 8**).
2. Reconstitute the organic dried phase in chloroform: methanol (2:1, v/v) for lipids in LC–MS analysis.
3. Filter the reconstituted extracts through 0.8 μm PES clarifying filters (Sartorius) prior to LC–MS injection to remove particulates.
4. Transfer clean extract to a glass vial for injection into an LC–MS (*see* **Note 9**).
5. Prepare all LC solvents freshly prior to analysis and ensure they are filtered through a 0.22 μm membrane filter, degassed, and stored in amber or inert-gas-flushed bottles to avoid oxidation or contamination (*see* **Note 10**).
6. Equilibrate the liquid chromatography system with the initial mobile phase conditions for at least 10 min before sample injection to stabilize the baseline (*see* **Notes 11** and **12**).

7. Always maintain the required column temperature at 25–30 °C throughout the analysis to ensure consistent chromatographic performance.
8. For the central metabolites and sugars, analyze the extracted aqueous fraction using a reverse-phase C18 (100 × 0.3 × 5 μm) column [4].
9. Inject 3–5 μL of the aqueous metabolite extract into the LC–MS system (*see* **Note 13**).
10. For reverse-phase chromatography, use mobile phases consisting of Solvent A, 0.1% formic acid in ddH_2O, and 0.1% formic acid in Acetonitrile (Solvent B).
11. Set the flow rate to 15 μL/min.
12. Run the following gradient conditions for the C18 method: 0–3 min at 2% B, 3–13 min linear ramp to 100% B, 13–16 min hold at 100% B, 16–19 min ramp back to 2% B, and 19–30 min equilibration at 2% B.
13. For high-resolution separation of phosphorylated sugar, sugar, and organic acid, use the Imtakt Intrada Organic Acid column (150 × 2 mm, 3 μm; Kyoto, Japan) [17, 18].
14. For the Imtakt Intrada Organic Acid column, prepare mobile phase A, which consists of 100 mM ammonium formate in 10% acetonitrile (ACN), and mobile phase B, made up of 75% ACN and 1% formic acid.
15. Program the chromatographic step with an initial concentration of 0–1 min at 100% B, 1–5 min linear decrease to 88% B, 5–7 min hold at 88% B, 7–8 min ramp to 84% B, 8–10 min hold at 84% B, 10–13 min ramp to 75% B, 13–15 min ramp to 0% B, 15–19.5 min hold at 0% B for column cleaning, 19.5–20 min ramp back to 100% B, and 20–25 min hold at 100% B for column equilibration. The total run time was 25 min, with a flow rate of 0.225 mL/min.
16. Perform lipid separation by taking a nonpolar fraction [see extraction section] and set up a C8 column (100 × 0.5 × 1.7 μm) from Higgins Analytical Inc. (Mountain View, CA, United States). The C8 column provides high resolution and reproducibility across diverse lipid classes [4, 22].
17. Prepare solvent A mobile phases as 1% 1 M ammonium acetate, 0.1% acetic acid in ddH_2O, and solvent B prepared by 1% 1 M ammonium acetate, 0.1% acetic acid in 7:3 (v/v) acetonitrile: isopropanol. Set the flow rate to 40 μL/min (*see* **Notes 14** and **15**).
18. Set the gradient as 0–1 min at 55% B, 1–3 min ramp to 75% B, 3–8 min ramp to 89% B, 8–10 min ramp to 99% B, 10–11 min hold at 99% B, 11–12 min return to 55% B, and 12–18 min equilibrate at 55% B [4, 22].

3.5 Mass Spectrometry Parameters

The following methods were optimized in Thermo Scientific quadrupole–Orbitrap mass spectrometers.

1. Connect the UPLC system to the Q Exactive Orbitrap mass spectrometer through a heated electrospray ionization (HESI) source [4, 23].
2. Acquire data in both positive and negative ionization modes using a data-dependent MS/MS acquisition method to capture broad metabolite coverage.
3. Set the mass scan range to 70–1000 m/z for untargeted metabolomics across all chromatographic methods.
4. Acquire full MS scans at a resolution of 70,000 (at m/z 200) in both positive and negative ion modes to ensure accurate mass measurement.
5. Optimize source parameters such as spray voltage, sheath gas, auxiliary gas, and capillary temperature through pilot runs to achieve stable ionization and maximal signal intensity.
6. Adjust sample injection volumes and dilutions to prevent ion suppression and minimize matrix effects during data acquisition.
7. Optimize the heated electrospray ionization (HESI) source for sheath gas, auxiliary gas, capillary temperature, auxiliary gas heater temperature, and S-lens RF level.
8. Set the spray voltage to 4.2 kV and 3.9 kV in positive and negative modes, respectively.
9. Monitor the total ion chromatograms (TICs) after each injection to confirm stable retention times, reproducible peak shapes, and consistent ion intensities.

3.6 Quantification of Total Lipid Content in Pollen Tube Using FAMES

Total lipid content is generally analyzed by converting total lipids into fatty acid methyl esters (FAMEs) [19, 24, 25] (*see* **Note 16**).

1. Weigh approximately 5–10 mg lyophilized biomass in an 8 mL glass tube.
2. Add 2 mL freshly prepared 5% (v/v) sulfuric acid: methanol. For example, to prepare 20 mL total, add 1 mL H_2SO_4 to 19 mL MeOH.
3. Add 25 μL of 0.2% (w/v) butylated hydroxytoluene (BHT) in methanol to prevent oxidation.
4. Add triheptadecanoin (0.1 mg) and tripentadecanoin (0.025 mg) as internal standards for quantification of fatty acid (*see* **Notes 12** and **15**).
5. Cap tubes securely and incubate at 110 °C for 3 h, with intermittent vortexing every 30–45 min to promote complete lipid transesterification.

6. After incubation, allow samples to cool to room temperature (~5–10 min).
7. Add 3 mL of 0.9% NaCl (w/v) to each tube to quench the reaction and facilitate phase separation.
8. Add 2 mL hexanes. Vortex thoroughly and spin down briefly to get good separation of phases.
9. Briefly centrifuge the tubes (2 min at low speed) to separate the organic (hexane) and aqueous layers.
10. Carefully collect 1.5 mL of the upper hexane layer containing FAMEs and transfer to a clean 2 mL GC vial.
11. Prepare a hexane blank using the same solvent but without a sample to monitor background noise and contamination.
12. The FAMEs can be quantified by gas chromatography–flame ionization detection (GC–FID) using a DB23 column (30 m, 0.25-mm i.d., 0.25-μm film; J&W Scientific).
13. Load the vials onto the GC-FID autosampler and analyze using a DB-23 column (30 m × 0.25 mm × 0.25 μm).
14. Operate the GC system in split mode (30:1). Set the flame ionization detector temperature to 250 °C.
15. Set the oven temperature ramp profile from 180 to 260 °C at a rate of 20 °C min^{-1} followed by a hold time of 7 min.
16. Maintain helium or nitrogen as the carrier gas at the manufacturer-recommended flow rate (typically ~1.0 mL min^{-1}).
17. Identify peaks corresponding to known FAMEs based on retention times of standard compounds.
18. Quantify total fatty acids by comparing integrated peak areas of sample-derived FAMEs against those of the C15:0 and C17:0 internal standards.

3.7 Metabolite Identification and Data Processing

1. Import raw chromatographic data (total ion chromatograms; TICs) from the GC–MS system into a suitable preprocessing software.
2. Use MetAlign [26] for automated baseline correction, noise filtering, and mass spectral alignment, which are particularly suited for large-scale plant metabolomics datasets.
3. Following preprocessing, transfer the corrected chromatograms to Instrument-specific data analysis software such as MassHunter, ChemStation, or equivalent GC–MS processing suites for metabolite identification and profiling.
4. Match individual mass-to-charge (m/z) fragments against reference spectra available in commercial libraries such as the NIST (National Institute of Standards and Technology) and Fiehn Metabolomics Library [27].

5. Assign metabolite identities based on the combined evaluation of retention time, mass spectral similarity index, and probability of match obtained from spectral libraries.
6. Confirm critical metabolite annotations using authentic chemical standards analyzed under identical chromatographic and ionization conditions to enhance identification confidence.
7. Integrate peak areas of the identified metabolites to derive quantitative information, ensuring all integrations are performed under consistent threshold settings.
8. Compile the resulting metabolite abundance table, including compound names, retention times, m/z fragments, peak areas, and spectral match scores.
9. Export the finalized dataset in.csv or.xlsx format for downstream statistical and multivariate analyses such as normalization, principal component analysis (PCA), and clustering [20, 28].
10. Similarly, for LC–MS data, import raw LC–MS data files into the designated data processing software (e.g., Freestyle, Skyline, LipidSearch, MS-DIAL, or LipidXplorer) [4, 20, 29, 30].
11. Define the appropriate ionization mode (positive or negative) and set parameters for peak detection (minimum intensity threshold, retention time range, and noise level) to ensure accurate identification of chromatographic features.
12. Perform automated retention-time alignment across all chromatograms to correct for minor shifts between runs.
13. Execute automated peak detection to identify chromatographic features corresponding to individual lipid species.
14. Match detected peaks based on accurate mass (m/z values), isotopic distribution patterns, and, where available, MS/MS fragment ions for structural confirmation.
15. Generate a comprehensive list of candidate lipid features, including molecular formula, retention time, and intensity values.
16. Quantify each identified lipid species relative to its corresponding internal standard, preferably a compound belonging to the same lipid class (e.g., PC, TAG, DAG, PE) or a structurally similar surrogate.
17. Always retain both raw and processed files for data traceability and validation in future analyses. Import raw chromatographic data (total ion chromatograms; TICs) from the GC–MS system into a suitable preprocessing software.
18. Use MetAlign [26] for automated baseline correction, noise filtering, and mass spectral alignment, which are particularly suited for large-scale plant metabolomics datasets.

19. Following preprocessing, transfer the corrected chromatograms to Instrument-specific data analysis software such as MassHunter, ChemStation, or equivalent GC–MS processing suites for metabolite identification and profiling.
20. Match individual mass-to-charge (m/z) fragments against reference spectra available in commercial libraries such as the NIST (National Institute of Standards and Technology) and Fiehn Metabolomics Library [27].
21. Assign metabolite identities based on the combined evaluation of retention time, mass spectral similarity index, and probability of match obtained from spectral libraries.
22. Confirm critical metabolite annotations using authentic chemical standards analyzed under identical chromatographic and ionization conditions to enhance identification confidence.
23. Integrate peak areas of the identified metabolites to derive quantitative information, ensuring all integrations are performed under consistent threshold settings.
24. Compile the resulting metabolite abundance table, including compound names, retention times, m/z fragments, peak areas, and spectral match scores.
25. Export the finalized dataset in.csv or.xlsx format for downstream statistical and multivariate analyses such as normalization, principal component analysis (PCA), and clustering [20, 28].
26. Similarly, for LC–MS data, import raw LC–MS data files into the designated data processing software (e.g., Freestyle, Skyline, LipidSearch, MS-DIAL, or LipidXplorer) [4, 20, 29, 30].
27. Define the appropriate ionization mode (positive or negative) and set parameters for peak detection (minimum intensity threshold, retention time range, and noise level) to ensure accurate identification of chromatographic features.
28. Perform automated retention-time alignment across all chromatograms to correct for minor shifts between runs.
29. Execute automated peak detection to identify chromatographic features corresponding to individual lipid species.
30. Match detected peaks based on accurate mass (m/z values), isotopic distribution patterns, and, where available, MS/MS fragment ions for structural confirmation.
31. Generate a comprehensive list of candidate lipid features, including molecular formula, retention time, and intensity values.
32. Quantify each identified lipid species relative to its corresponding internal standard, preferably a compound

belonging to the same lipid class (e.g., PC, TAG, DAG, PE) or a structurally similar surrogate.

33. Always retain both raw and processed files for data traceability and validation in future analyses.

3.8 Visualization of Callose Plugs (β-1,3-Glucan) by Aniline Blue Staining

To evaluate total carbohydrate stores, the phenol-sulfuric acid assay measures neutral sugars colorimetrically, providing a high-throughput estimate of overall carbohydrate pools. In addition to soluble sugars, callose deposition can be examined with aniline blue staining, which visualizes β-1,3-glucan plugs that form behind the growing tip [31, 32]. During pollen tube growth through the pistil tissues, callose (β-1,3-glucan) is periodically deposited to form callose plugs, which maintain cytoplasmic pressure and compartmentalize the growing tube (*see* **Note 16**). Visualization of callose provides valuable insights into pollen tube growth dynamics and guidance. The fluorescent dye decolorized aniline blue specifically binds to β-1,3-glucan polymers, allowing detection under UV excitation [31, 32].

1. All steps should be performed using glassware such as screw-cap bottles to prevent dye adsorption and contamination.
2. Pollinated pistils are collected and immediately fixed in acetic alcohol (1:3, v/v) and incubated overnight at room temperature.
3. Following fixation, pistils are washed three times with purified double-distilled water to remove residual fixative.
4. To soften tissues and facilitate dye penetration, transfer pistils to 1 N NaOH or 1 N KOH solution and incubate at 50–60 °C for 1–3 h. The softening process should be checked periodically to prevent over-digestion or tissue disintegration.
5. After incubation, wash samples twice with double-distilled water to remove residual alkali.
6. For staining, immerse the softened pistils in decolorized aniline blue solution (0.005% in 0.05 M Na_2HPO_4, pH 11). The dye solution should be freshly filtered to remove precipitates and stored in a dark refrigerator to maintain stability.
7. Incubate the samples in the staining solution at 4 °C for 24 h, and protect them from light by wrapping the containers in aluminum foil.
8. After staining, mount pistils on microscope slides using the same aniline blue staining solution as the mounting medium.
9. Perform visualization under a fluorescence microscope equipped with a UV excitation filter (450–460 nm, blue range). Callose deposits and plugs will appear as bright blue or whitish fluorescence along the pollen tube, allowing for the assessment of pollen tube growth and integrity within pistil tissues.

4 Notes

1. Always work inside a chemical fume hood when using volatile solvents like chloroform, methanol, hexane, and derivatization chemicals.
2. Metabolites can be extracted from lyophilized pollen powder or the fresh weight of finely powdered pollen tubes.
3. Keep all steps-grinding, extraction, and concentration-on ice or at 4 °C. This helps prevent enzymatic degradation and lipid oxidation.
4. Always use solvent-rinsed borosilicate glassware for all extractions and derivatizations. Plastic materials can easily leach fatty acids or adsorb analytes, which can contaminate samples and distort signals.
5. This two-step MeOx-TMS derivatization method provides consistent and comprehensive coverage of central metabolites and sugars from pollen tissues. The workflow described here has been successfully adapted for multiple systems and validated in recent studies [19–21].
6. DB-2MS or DB-Wax columns may be used for polar compounds and volatile organics.
7. GC-MS acquisition protocol enables robust quantitative profiling of sugars, amino acids, organic acids, and small lipid derivatives from pollen and pollen tube samples, supporting high-throughput metabolic analyses across developmental and environmental contexts.
8. To enhance ionization efficiency and metabolite recovery, 0.2% formic acid can be added to the reconstitution solvent, particularly when analyzing acidic metabolites or phosphorylated intermediates.
9. Various chromatographic methods can be adopted with different LC–MS instruments, mainly depending on which instruments (TOF, Orbitrap, Q, IT) are available (Fig. 1).
10. Prepare all aqueous reagents and mobile phases using ultrapure water. This can minimize background ions and improve MS signal quality.
11. Include procedural blanks and solvent-only blanks in every batch. These can help identify background contamination from the lab environment, solvents, or glassware.
12. Inject blank samples between runs to confirm that no carryover of analytes remains in the injector or column.
13. Store all lipid and metabolite extracts in amber glass vials under nitrogen or argon at −20 °C or below. This can preserve sample integrity and prevent oxidation.

14. Always include at least four biological replicates for each experimental condition. This guarantees statistical confidence and reproducibility. Replicate injections can be performed as needed to evaluate reproducibility, using internal standard responses to monitor analytical accuracy.
15. Spike known internal standards before and after extraction. This can be used to monitor extraction efficiency, ion suppression, and recovery rates.
16. Consider including time-course experiments. Pollen lipid and metabolite profiles can change rapidly during hydration, germination, and tube elongation, so sampling at multiple time points can reveal dynamic metabolic trends.

Acknowledgments

ML expresses gratitude to the Doug Allen Laboratory for their support and to the Donald Danforth Plant Science Center for providing mass spectrometry facilities.

References

1. McCormick S (2004) Control of male gametophyte development. Plant Cell 16(suppl_1): S142–S153
2. Bedinger P (1992) The remarkable biology of pollen. Plant Cell 4(8):879
3. De Storme N, Geelen D (2014) The impact of environmental stress on male reproductive development in plants: biological processes and molecular mechanisms. Plant Cell Environ 37(1):1–18
4. Wang J, Kambhampati S, Allen DK, Chen L-Q (2022) Comparative metabolic analysis reveals a metabolic switch in mature, hydrated, and germinated pollen in Arabidopsis thaliana. Front Plant Sci 13:836665
5. Pacini E, Hesse M (2005) Pollenkitt-its composition, forms and functions. Flora-Morphol Distrib Funct Ecol Plants 200(5):399–415
6. Matveyeva N, Polevova S, Smirnova A, Yermakov I (2012) Sporopollenin accumulation in Nicotiana tabacum L. microspore wall during its development. Cell Tissue Biol 6(3): 293–301
7. Ischebeck T (2016) Lipids in pollen-they are different. Biochim Biophys Acta 1861(9): 1315–1328
8. Monteiro D et al (2005) Phosphoinositides and phosphatidic acid regulate pollen tube growth and reorientation through modulation of [Ca2+] c and membrane secretion. J Exp Bot 56(416):1665–1674
9. Liu S, Li Z, Wu S, Wan X (2021) The essential roles of sugar metabolism for pollen development and male fertility in plants. Crop J 9(6): 1223–1236
10. Lee S-K, Lee J, Jo M, Jeon J-S (2022) Exploration of sugar and starch metabolic pathway crucial for pollen fertility in rice. Int J Mol Sci 23(22):14091
11. Seitz J et al (2023) How pollen tubes fight for food: the impact of sucrose carriers and invertases of Arabidopsis thaliana on pollen development and pollen tube growth. Front Plant Sci 14:1063765
12. Hong Z, Zhang Z, Olson JM, Verma DPS (2001) A novel UDP-glucose transferase is part of the callose synthase complex and interacts with phragmoplastin at the forming cell plate. Plant Cell 13(4):769–779
13. Harrington GN, Bush DR (2003) The bifunctional role of hexokinase in metabolism and glucose signaling. Plant Cell 15(11): 2493–2496
14. Lingwan M (2024) Fats influencing flowering: pistil-derived lipids affect pollen tube growth and fertility in Arabidopsis thaliana. Plant Physiol 196:670–671
15. Feijó JA, Sainhas J, Hackett GR, Kunkel JG, Hepler P (1999) Growing pollen tubes possess

a constitutive alkaline band in the clear zone and a growth-dependent acidic tip. J Cell Biol 144(3):483–496
16. Connor KF, Towill LE (1993) Pollen-handling protocol and hydration/dehydration characteristics of pollen for application to long-term storage. Euphytica 68(1):77–84
17. Ma F, Jazmin LJ, Young JD, Allen DK (2017) Isotopically nonstationary metabolic flux analysis (INST-MFA) of photosynthesis and photorespiration in plants. In: Photorespiration: methods and protocols. Springer New York, New York, pp 167–194
18. Koley S, Chu KL, Gill SS, Allen DK (2022) An efficient LC-MS method for isomer separation and detection of sugars, phosphorylated sugars, and organic acids. J Exp Bot 73(9): 2938–2952
19. Obermeyer G, Fragner L, Lang V, Weckwerth W (2013) Dynamic adaption of metabolic pathways during germination and growth of lily pollen tubes after inhibition of the electron transport chain. Plant Physiol 162(4): 1822–1833
20. Shree M, Lingwan M, Masakapalli SK (2019) Metabolite profiling and metabolomics of plant systems using 1H NMR and GC-MS. In: OMICS-based approaches in plant biotechnology, pp 129–144
21. Lingwan M, Masakapalli SK (2022) A robust method of extraction and GC-MS analysis of monophenols exhibited UV-B mediated accumulation in Arabidopsis. Physiol Mol Biol Plants 28(2):533–543
22. Hummel J, Segu S, Li Y, Irgang S, Jueppner J, Giavalisco P (2011) Ultra performance liquid chromatography and high resolution mass spectrometry for the analysis of plant lipids. Front Plant Sci 2:54
23. Gathungu RM, Larrea P, Sniatynski MJ, Marur VR, Bowden JA, Koelmel JP, Starke-Reed P, Hubbard VS, Kristal BS (2018) Optimization of electrospray ionization source parameters for lipidomics to reduce misannotation of in-source fragments as precursor ions. Anal Chem 90(22):13523–13532
24. Allen DK, Young JD (2013) Carbon and nitrogen provisions alter the metabolic flux in developing soybean embryos. Plant Physiol 161(3): 1458–1475
25. Koley S, Jyoti P, Lingwan M, Wei M, Xu C, Chu KL, Williams RB, Koo AJ, Thelen JJ, Xu D, Allen DK (2025) Persistent fatty acid catabolism during plant oil synthesis. Cell Rep 44(4):115492
26. Lommen A, Kools HJ (2012) MetAlign 3.0: performance enhancement by efficient use of advances in computer hardware. Metabolomics 8(4):719–726
27. Fiehn O, Kind T, Barupal DK (2011) Data processing, metabolomic databases and pathway analysis. In: Annual plant reviews volume 43: biology of plant metabolomics, vol 43, pp 367–406
28. Pang Z, Zhou G, Ewald J, Chang L, Hacariz O, Basu N, Xia J (2022) Using MetaboAnalyst 5.0 for LC-HRMS spectra processing, multi-omics integration and covariate adjustment of global metabolomics data. Nat Protoc 17(8):1735–1761
29. Cullen E, Lingwan M (2025) Go with the flux: modeling accurately predicts phenotypes of Arabidopsis lipid mutants. Plant Physiol 197
30. Hoffmann N, Mayer G, Has C, Kopczynski D, Al Machot F, Schwudke D, Ahrends R, Marcus K, Eisenacher M, Turewicz M (2022) A current encyclopedia of bioinformatics tools, data formats and resources for mass spectrometry lipidomics. Metabolites 12(7):584
31. Mangla Y, Khanduri P, Gupta CK (2023) Reproductive biology of angiosperms: concepts and laboratory methods. Cambridge University Press, Cambridge
32. Qin P, Ting D, Shieh A, McCormick S (2012) Callose plug deposition patterns vary in pollen tubes of Arabidopsis thaliana ecotypes and tomato species. BMC Plant Biol 12(1):178

Chapter 8

Analyzing Pollen by Infrared and Raman Spectroscopy

Boris Zimmermann and Achim Kohler

Abstract

Mid-infrared (mid-IR) and Raman vibrational spectroscopy have been widely adopted in pollen research due to their unique capability for simultaneous taxonomic identification and comprehensive chemical profiling in a single measurement. The integration of both bulk (macro-sampling) and microscopy-based spectroscopic techniques provides complementary insights, where bulk measurements enable efficient, high-throughput screening of pollen samples, whereas microspectroscopies enable detailed analysis of single-grain heterogeneity and substructural chemical composition. Here, we present one Raman and three mid-IR methods for bulk pollen measurements, as well as one Raman and one mid-IR microspectroscopy method. Additionally, we provide guidelines for the preprocessing of spectral data, including advanced spectral preprocessing approaches related to Mie-type scattering interferences in mid-IR microspectroscopy.

Key words Mid-infrared spectroscopy, Raman spectroscopy, Fourier-transform infrared (FTIR), Fourier-transform Raman, Microspectroscopy, Chemical profiling, Taxonomic identification, Spectral preprocessing

1 Introduction

In the past 20 years, mid-infrared and Raman vibrational spectroscopy has progressed rapidly and been widely adopted in pollen research because it enables simultaneous taxonomic identification and comprehensive chemical profiling in a single measurement [1–4]. Together, bulk (macro-sampling) and microscopy-based techniques provide complementary insights, where bulk measurements enable efficient screening studies, while microspectroscopy provides information on single-grain heterogeneity and substructural chemistry. Mid-infrared spectroscopy of pollen is mostly performed using Fourier-transform infrared (FTIR) spectrometers [1, 2, 5–7], while Raman spectroscopy is performed using either FT or dispersive Raman spectrometers [1, 5, 8]. Bulk vibrational spectroscopy measurements (of approximately 1 mg of material) are relatively fast and economical, supporting high-throughput screening of large

Giampiero Cai et al. (eds.), *Pollen and Pollen Tube Analysis: Methods and Protocols*, Methods in Molecular Biology, vol. 3042, https://doi.org/10.1007/978-1-0716-5308-1_8,

sample sets when spatially resolved information is not required [6, 9]. Vibrational microspectroscopy enables analysis across spatial scales, from whole individual grains to discrete substructures such as exine regions, apertures, and internal contents [3–5, 10–13], and it is essential for analyzing complex environmental samples from sediments and air samplers that contain mixtures of multiple pollen taxa along with other biological and inorganic particles. In general, vibrational spectroscopy provides a precise and comprehensive assessment of pollen's biochemical composition, with distinct signals from lipids, proteins, carbohydrates, and other constituents, such as pigments [2, 5, 14]. Since Raman and mid-IR provide chemically complementary information, employing both techniques on the same samples yields exceptionally detailed biochemical characterization [1, 2, 5].

In bulk FTIR spectroscopy, spectra are typically acquired in reflectance mode using an attenuated total reflectance (ATR) accessory via an evanescent wave, and this technique is convenient for quick analysis with minimal sample preparation [1, 2, 5, 7, 14, 15]. Pollen FTIR spectra can also be acquired in transmission mode, either from pellets formed by embedding the pollen in an alkali halide matrix [1, 5] or from thin films (formed from pollen suspensions) deposited on silicon plates for high-throughput screening (FTIR–HTS) [6, 9]. Bulk Raman spectra are commonly acquired with FT–Raman spectrometers by loading the sample into aluminum cups or glass vials [1, 2, 8], which also enables high-throughput analysis (FT–Raman–HTS). In vibrational microspectroscopy, an infrared or Raman spectrometer is coupled to a microscope, allowing spectra to be acquired with high spatial resolution (typically lateral resolution of approx. 2–10 μm in FTIR and 1–2 μm in Raman microspectroscopy) [3, 5]. FTIR microscopy spectra of pollen are usually acquired in transmission mode [3–5, 11], although reflectance-based approaches such as ATR–FTIR microspectroscopy are also feasible. Generally, microspectroscopy techniques enable measurement of individual pollen grains and even their substructures [3, 5, 12, 16].

The choice of technique and protocol depends first on the amount of sample available. For example, whether there is enough for bulk measurements by rapid screening techniques or microspectroscopy is required [5]. It also depends on the type of information sought. If the focus is on specific chemical attributes (e.g., lipid-to-protein ratios, pigment signatures, etc.), bulk measurements can provide rapid biochemical assessments, whereas identification and classification of taxa, assessment of chemical heterogeneity within populations, or analysis of substructures are better served by single-grain microspectroscopy. In practice, FTIR and Raman can be applied at both scales, and combining bulk and microscopic measurements often yields the most robust insight by integrating comprehensive chemical characterization with precise taxonomic resolution [3, 5, 16].

Spectral preprocessing in FTIR and Raman of pollen is as crucial as spectral acquisition itself. Spectral preprocessing is the set of data processing steps applied to raw spectra to remove instrumental and sample-related artifacts and interferences, correct and scale data, and enhance chemically meaningful features for analysis [17, 18]. Common operations include baseline correction, spectral smoothing and scatter corrections, normalization or scaling, and transformation into derivative forms [17, 19]. In vibrational spectroscopy of pollen, spectral preprocessing depends not only on the technique and sampling mode (for example, infrared vs Raman, bulk vs single-grain) but also on the study objective. It is sometimes advisable to preprocess the same spectral dataset using several different approaches to highlight distinct spectral features for subsequent classification or chemical characterization [2, 4, 11, 12]. For identification/classification studies, workflows often favor steps that emphasize subtle spectral differences and improve class separability, whereas for chemical characterization/quantitation, preprocessing is typically focused on preserving absolute or relative band intensities (for example, physically motivated spectral corrections) so that biochemical features remain interpretable [4, 11]. Advanced spectral preprocessing is especially valuable in FTIR microspectroscopy of pollen because infrared data from microparticles are frequently corrupted by interferent signals arising from anomalous dispersion and strong IR-light scattering when particle sizes are comparable to mid-IR wavelengths (2–20 μm). Such distorted spectral features are often observed in IR spectra of smaller pollen grains with diameters of approx. 5–30 μm [20, 21]. Mie-type scattering is the dominant contributor, producing characteristic wiggles and ripples in spectra [20, 21]. A range of experimental and analytical strategies has been developed to mitigate these effects [4, 22, 23]. The specifics of these approaches are beyond the scope of this protocol; readers are referred to the following studies for detailed guidance [4, 12, 21, 24–26]. Another approach is spectral augmentation, which has recently gained popularity as an alternative to spectral pre-processing. Originating from image analysis (where techniques like rotation and background changes improve model robustness), augmentation in spectral analysis involves introducing instrumental and sample-related artifacts to spectra [27]. This increases training data and improves the robustness of calibration and classification models, with studies showing compatibility with standard chemometric and machine learning methods.

The use of different techniques can result in varying levels of information, linking pollen chemistry to both biotic parameters (such as taxonomic identity, population or genotype, and pollination mode) and abiotic parameters (such as temperature, nutrient levels, solar radiation levels, pollution, and annual variations). Bulk infrared and Raman spectroscopic measurements provide comprehensive information on pollen chemistry through fast, cheap, and simple methods. These techniques enable the detection of phylogenetic variations, including the separation of congeneric species [2, 8, 14, 15] and even population-level differences within the same species [6], as well as chemical differences related to pollination strategy (wind-pollinated vs. insect-pollinated species) [14]. Furthermore, the methodology allows for the assessment of phenotypic plasticity in response to different environmental conditions, such as temperature and nutrient levels [6], including inter-annual variations within populations [9]. Hundreds of samples were measured in these referenced studies, demonstrating the high-throughput capability of bulk measurements. The spectral differences related to environmental factors and taxonomy are interpreted biochemically, specifically as variations in lipids, proteins, carbohydrates, cell wall chemistry, and pigments. Similar to bulk measurements, microspectroscopy enables the assessment of biochemical differences between pollen samples related to phylogenetic variations [3, 5, 12], including for congeneric species [10, 11], as well as environmental effects on phenotypes [16]. This demonstrates that microspectroscopy can be used for the identification of single pollen grains to an extremely high taxonomic level, which is useful for applications such as aeroallergenic monitoring [11, 12] or palaeoecological studies [4]. Microspectroscopy can also provide chemical imaging of pollen, unveiling the spatial distribution and concentration of proteins, carbohydrates, sporopollenins, and lipids within pollen substructures [3].

2 Materials

As mentioned earlier, there are several different techniques available for IR and Raman spectroscopy of pollen, ranging from bulk analysis methods to advanced microscopy-based approaches. Each technique offers distinct advantages depending on the required analytical detail. Below, these techniques will be listed separately for clarity.

2.1 Bulk IR Measurements

1. FTIR spectrometer for routine laboratory applications (*see* **Note 1**). Some IR spectrometers require liquid nitrogen for cooling their detectors (*see* **Note 2**).
2. For ATR measurements: an ATR accessory (*see* **Note 3**) and a pestle and mortar (optional, *see* **Note 4**).
3. For transmission pellet measurements: an alkali halide powder as a transparent matrix, a pestle and mortar for grinding and homogenizing the pollen sample and alkali halide powder, and

a press die and pellet press for forming pellets (*see* **Note 5**). Optionally, a vacuum pump for evacuable pellet dies (*see* **Note 5**).

4. For transmission thin film measurements: a tissue homogenizer (ultrasonic or bead beating), 1.5–2.0 mL tubes for homogenizer (empty for ultrasonic homogenizer or pre-filled with glass (or ceramics) beads for bead beating homogenizer), deionized water, a high-speed benchtop centrifuge, and an IR-transparent microtiter plate.

2.2 Bulk Raman Measurements

1. Raman spectrometer for routine laboratory applications (*see* **Note 6**). Some Raman spectrometers require liquid nitrogen for cooling their detectors (*see* **Note 7**).
2. Aluminum cups or glass vials for samples (*see* **Note 8**).

2.3 Microscopy IR Measurements

1. IR microspectrometer for routine laboratory applications (*see* **Note 9**). Some IR spectrometers require liquid nitrogen for cooling their detectors (*see* **Note 2**).
2. For measurements on microscope slides: IR-transparent microscope slides (*see* **Note 10**).
3. For measurements in paraffin: soft paraffin (Vaseline-type petroleum jelly), polyethylene foils (approx. 10 μm thickness), X-ray fluorescence sample cups (*see* **Note 11**).

2.4 Microscopy Raman Measurements

1. Raman microspectrometer for routine laboratory applications (*see* **Note 12**). Some Raman spectrometers require liquid nitrogen for cooling their detectors (*see* **Note 7**).
2. Raman-compatible microscope slides (*see* **Note 13**).

3 Methods

3.1 Bulk IR Measurements with ATR Accessory

1. Ensure the ATR crystal is clean and free of any contaminants. Clean the crystal surface before and between measurements using cotton or paper and an appropriate solvent, such as water, isopropanol, or acetone.
2. Some IR spectrometers require liquid nitrogen for cooling their detectors (*see* **Note 2**). If this is the case, cool the detector to the required operating temperature by carefully following the protocols and safety procedures provided by the instrument manufacturer. Warning: Due to the extremely low temperature of liquid nitrogen (−196 °C), any skin contact can cause severe frostbite. Therefore, wear protective clothes, shoes, gloves, and glasses while handling liquid nitrogen.
3. Select spectral acquisition parameters: spectral region (usually within spectral range 4000–600 cm^{-1}), spectral resolution

(usually between 8 and 4 cm^{-1}), number of scans (usually between 16 and 64), and apodization function (*see* **Note 14**).

4. Record the single-channel spectrum of the clean sample-free ATR crystal (so-called background measurement) (*see* **Notes 15** and **16**).
5. Place the sample on the ATR crystal, ensuring that the entire crystal area is covered (*see* **Note 17**). Close the clamp and apply pressure according to the protocols provided by the ATR accessory manufacturer. Make sure there is good contact between the sample and the ATR crystal by using the clamp and an appropriate compression head anvil.
6. Record the single-channel spectrum of the sample. The final spectrum is obtained as the ratio of the single-channel sample spectrum to the empty ATR plate single-channel spectrum (*see* **Note 18**).
7. After the measurement, first remove the pressure from the sample, and afterward unlock the clamp. Clean the crystal surface. Usually, each sample is measured in triplicate (i.e., three technical replicates).
8. After spectral acquisition, perform spectral preprocessing before proceeding to data analysis (*see* **Notes 19–21**).

3.2 Bulk IR Measurements with Alkali Halide Pellets

1. Some IR spectrometers require liquid nitrogen for cooling their detectors (*see* **Note 2**). If this is the case, cool the detector to the required operating temperature by carefully following the protocols and safety procedures provided by the instrument manufacturer. Warning: Due to the extremely low temperature of liquid nitrogen (−196 °C), any skin contact can cause severe frostbite. Therefore, wear protective clothes, shoes, gloves, and glasses while handling liquid nitrogen.
2. Weigh 1 part pollen sample with 100 parts of dry alkali halide (*see* **Note 22**), for example, by weighing 1 mg of the sample with 100 mg of KBr powder (*see* **Note 23**).
3. Thoroughly mix the sample and alkali halide powder using a pestle and mortar to achieve a homogeneous mixture.
4. Transfer the mixture into a pellet die.
5. Cold-press the die using a laboratory press (typically 5–10 tons), following the protocols and safety guidelines provided by the press and die manufacturers. Optionally, a vacuum pump can be used during pressing if the pellet die is evacuable, to help remove air and moisture and improve pellet quality (*see* **Note 5**).
6. Select spectral acquisition parameters: spectral region (usually within spectral range 4000–600 cm^{-1}), spectral resolution

(usually between 8 and 4 cm^{-1}), number of scans (usually between 16 and 64), and apodization function (*see* **Note 14**).

7. Record the single-channel spectrum of the sample-free setup (so-called background measurement) (*see* **Notes 15** and **24**).
8. Place the prepared pellet into the pellet holder.
9. Record the single-channel spectrum of the sample. The final spectrum is obtained as the ratio of the single-channel pellet spectrum to the sample-free single-channel spectrum (*see* **Note 18**). Usually, each sample is measured in triplicate (i.e., three technical replicates).
10. After spectral acquisition, perform spectral preprocessing before proceeding to data analysis (*see* **Notes 19**, **21**, and **25**).

3.3 Bulk IR Measurements with Microplate-Readers (High-Throughput Screening)

1. Weigh 1 mg of pollen sample, and transfer the sample into a 1.5–2.0 mL microcentrifuge tube.
2. Add 500 μL of deionized water to the tube.
3. Sonicate the pollen suspension using a probe-based ultrasonic homogenizer (*see* **Note 26**), following the protocols and safety guidelines provided by the homogenizer manufacturer (*see* **Note 27**).
4. Centrifuge the suspension at 13,000 rpm for 10 min.
5. Carefully remove 400 μL of supernatant to concentrate the suspension (*see* **Note 28**).
6. Pipette the concentrated suspension into a clean IR-transparent silicon microtiter plate (*see* **Note 29**). Usually, three aliquots are transferred to three individual wells for triplicate measurements (i.e., three technical replicates).
7. Allow the microtiter plate to dry at room temperature for 1 h, forming adequate thin films for FTIR measurement (*see* **Note 30**).
8. Select spectral acquisition parameters: spectral region (usually within spectral range 4000–600 cm^{-1}), spectral resolution (usually between 8 and 4 cm^{-1}), number of scans (usually between 16 and 64), and apodization function (*see* **Note 14**).
9. Place the prepared microplate into the pellet holder and start the measurement (*see* **Notes 15** and **31**).
10. After spectral acquisition, perform spectral preprocessing before proceeding to data analysis (*see* **Notes 19**, **21**, and **32**).

3.4 Bulk Raman Measurements

1. Some Raman spectrometers require liquid nitrogen for cooling their detectors (*see* **Note 7**). If this is the case, cool the detector to the required operating temperature by carefully following the protocols and safety procedures provided by the instrument manufacturer. Warning: Due to the extremely low temperature

of liquid nitrogen (−196 °C), any skin contact can cause severe frostbite. Therefore, wear protective clothes, shoes, gloves, and glasses while handling liquid nitrogen.

2. Put pollen sample (approx. 1 mg) into a small aluminum cup, and compress the sample gently with a pestle to ensure uniform packing. Alternatively, the pollen sample can be placed in a glass vial; in this case, a larger amount of sample is typically required (5–20 mg), depending on the size and geometry of the vial (*see* **Note 8**).
3. Place the sample in the sample stand, and adjust laser focus if possible (*see* **Note 33**).
4. Select spectral acquisition parameters: collection geometry, spectral region (usually within spectral range 3500–400 cm^{-1}), spectral resolution (usually between 8 and 2 cm^{-1}), and laser power (*see* **Note 34**). For FT-Raman spectrometers, select a number of scans (usually between 64 and 256) and apodization function (*see* **Note 35**). For dispersive Raman instruments, select exposure time (*see* **Note 36**).
5. Record the Raman spectrum (*see* **Note 37**). Usually, each sample is measured in triplicate (i.e., three technical replicates).
6. After spectral acquisition, perform spectral preprocessing before proceeding to data analysis (*see* **Notes 19**, **21**, and **38**).

3.5 Infrared Microspectroscopy Measurements

1. FTIR microspectrometers require liquid nitrogen for cooling their detectors (*see* **Note 2**). Cool the detector to the required operating temperature by carefully following the protocols and safety procedures provided by the instrument manufacturer. Warning: Due to the extremely low temperature of liquid nitrogen (−196 °C), any skin contact can cause severe frostbite. Therefore, wear protective clothes, shoes, gloves, and glasses while handling liquid nitrogen. In addition to FTIR microspectroscopy, pollen measurements can also be acquired using mid-infrared microspectroscopic imaging with quantum cascade lasers (*see* **Note 39**).
2. Place pollen sample onto an IR-transparent microscope slide (*see* **Note 10**). Alternatively, pollen grains can be embedded in paraffin (*see* **Note 11**). In this case, place one sheet of polyethylene foil flat on a clean surface and evenly spread approximately 1 mg of soft paraffin per 25 cm^2 area onto the foil. Distribute the pollen samples onto the paraffin layer without any pretreatment, then cover the pollen and paraffin with a second sheet of polyethylene foil. Insert the assembled foil sandwich into an X-ray fluorescence sample cup designed to stretch the polyethylene foils, and carefully stretch the foils to minimize the formation of air bubbles within the sample matrix. Ensure that the pollen is well embedded and evenly

distributed within the paraffin matrix, and examine the embedded sample visually to confirm uniform paraffin distribution and the absence of air bubbles.

3. Select spectral acquisition parameters: spectral region (usually within spectral range 4000–600 cm^{-1}), spectral resolution (usually between 8 and 4 cm^{-1}), number of scans (usually between 64 and 256), and apodization function (*see* **Note 14**).
4. Choose an appropriate objective (usually 15×) and focus the microscope on the pollen grain (i.e., measurement area) of interest. For point measurements with a single-element detector, adjust the aperture size according to the pollen grain size (typically 10–100 μm in diameter), using either rectangular or circular apertures. This adjustment is not required when using a focal plane array detector for spatially resolved chemical imaging.
5. Move focus to an empty microscope slide area and record a single-channel spectrum (spectra in case of imaging) of the sample-free setup (so-called background measurement) (*see* **Note 15**). In case of paraffin-embedded samples, conduct background measurement by measuring the sample-free setting before the sample cup is placed under the microscope (*see* **Note 15**).
6. Focus the microscope back on the measurement area of interest, and record the single-channel spectrum of the sample. The final spectrum is obtained as the ratio of the single-channel pellet spectrum to the sample-free single-channel spectrum (*see* **Note 18**).
7. After spectral acquisition, perform spectral preprocessing before proceeding to data analysis (*see* Notes **19**, **21**, **40**, and **41**).

3.6 Raman Microspectroscopy Measurements

1. Place pollen sample onto a microscope slide suitable for Raman microscopy (*see* **Note 13**).
2. Place the microscope slide inside the closed measurement chamber to minimize environmental interference (including ambient light interference).
3. Choose an appropriate objective (usually 50× or 100×) and focus the microscope on the pollen grain (i.e., measurement area) of interest (*see* **Note 42**).
4. Select spectral acquisition parameters: spectral region (usually within spectral range 3500–400 cm^{-1}), spectral resolution (usually between 4 and 2 cm^{-1}), and laser power (*see* **Note 34**). For FT–Raman spectrometers, select the number of scans and apodization function (*see* **Note 35**). For dispersive Raman instruments, select exposure time (*see* **Note 36**).

5. In case of point measurement, record the Raman spectrum. In case of spectral mapping, select measurement area and define step size in both x and y directions (*see* **Note 43**).
6. After spectral acquisition, perform spectral preprocessing before proceeding to data analysis (*see* **Notes 19**, **21**, and **38**).

4 Notes

1. Entry-level FTIR spectrometers allow for transmission measurements of pellets and reflection measurements using ATR accessories. In addition, advanced-level FTIR spectrometers often include automated plate-readers, enabling high-throughput analysis in both reflection and transmission modes.
2. Some infrared spectrometers require liquid nitrogen for cooling their detectors, which enhances sensitivity and reduces thermal noise during measurements. This is particularly common for systems equipped with highly sensitive detectors such as mercury cadmium telluride (MCT).
3. The acquired spectra will vary depending on the type of ATR accessory used. An ATR accessory uses a crystal, commonly made from diamond, zinc selenide (ZnSe), or germanium Ge), through which the infrared beam is internally reflected in contact with the sample. The number of reflections within the crystal (single or multiple) and the crystal type both influence the sensitivity and depth of penetration for different sample types and analysis needs.
4. Pollen samples can be measured without any sample preprocessing. Optionally, the samples can be grind to a fine homogeneous powder by using pestle and mortar. There is a considerable difference between the ATR spectra of ground and intact pollen, particularly for larger pollen grains (diameter above 50 μm). The infrared light measured in reflection measurements penetrates the pollen grains only up to 0.5–5 μm in depth, depending on the wavelength. As a result, the spectrum of intact pollen is dominated by signals from sporopollenins in the outer wall, while the spectrum of ground grains shows predominant signals from internal components such as lipids and proteins. The readers are referred to the following study for detailed guidance [5].
5. A pressing die and a pellet press is used to shape the mixture of pollen sample and alkali halide powder into a uniform, thin pellet for FTIR transmission measurements. Typical alkali halide powders used for this purpose include potassium bromide (KBr), sodium chloride (NaCl), potassium chloride (KCl), and cesium iodide (CsI). These alkali halide powders

have slightly different optical properties, such as their refractive index and transmission range for IR radiation. Each provides a distinct accessible spectral range, with NaCl transmitting down to about 650 cm^{-1}, KBr to around 400 cm^{-1}, and CsI extending further to approximately 200 cm^{-1}. Keep in mind that the acquired spectral range will depend not only on the type of alkali halide powder used as the matrix but also on the optical elements of the spectrometer itself. The pellet press applies high pressure to the die to remove air and moisture. Optionally, the pellet press system can be coupled to a vacuum pumping system to ensure optimal removal of air and moisture, thus providing optimal pellet transparency.

6. Raman spectrometers use a focused excitation laser (commonly in the visible or near-infrared range), and the excitation laser frequency significantly affects Raman measurements of biological materials like pollen. Shorter wavelength lasers (such as ultraviolet and green lasers) can increase Raman signal intensity but often enhance fluorescence background and are more strongly absorbed, leading to potential sample heating or damage. In contrast, longer wavelength lasers (such as near-infrared lasers) typically reduce both fluorescence interference and the risk of heating, though often requiring longer spectral acquisition times. Raman systems for bulk measurements (macro-sampling) employ either a dispersive or an FT spectrometer, and often include automated plate readers, enabling high-throughput analysis of well plates and vials.

7. Some Raman spectrometers require liquid nitrogen for cooling their detectors, which enhances sensitivity and reduces thermal noise during measurements. This is particularly common for systems equipped with highly sensitive detectors such as germanium (Ge).

8. Some glass vials may fluoresce, causing interference with the Raman peaks. When using a new type of glass vial, it is recommended to check an empty vial for potential fluorescence or background interference before proceeding with sample analysis.

9. IR microscopes are usually coupled to FTIR spectrometers with a thermal infrared source, such as globar, in combination with either a single-element detector for point measurements or a focal plane array detector for spatially-resolved chemical imaging. In addition, IR microscopes based on quantum cascade laser (QCL) technology offer rapid imaging and enhanced spectral brightness, making them particularly suitable for high-throughput measurements. The specifics of QCL-based micro-spectroscopy approach is beyond the scope of this protocol,

and readers are referred to the following study for detailed guidance [28].

10. Plane parallel IR-transparent microscope slides are available in a variety of materials, such as zinc selenide (ZnSe), zinc sulfide (ZnS), calcium fluoride (CaF_2), barium fluoride (BaF_2), potassium bromide (KBr), and sodium chloride (NaCl), and come in different sizes, typically with diameters of 10–25 mm and thicknesses of 1–3 mm. These window materials have slightly different optical properties, including refractive index and infrared transmission range. Each provides a distinct accessible spectral range, with CaF_2 transmitting down to about 1100 cm^{-1}, ZnS to about 900 cm^{-1}, and ZnSe to about 650 cm^{-1}.

11. Pollen grains strongly scatter mid-infrared light in IR microspectroscopy of individual pollen grains. The prominent Mie-type scattering artifacts occur due to pollen size and morphology, and these spectral interferences complicate biochemical interpretation of pollen spectra. Embedding pollen in a medium with a similar refractive index, such as paraffin, can significantly reduce these scattering effects by minimizing refractive index mismatches at the particle surface.

12. Raman microspectrometers use a focused excitation laser (commonly in the visible or near-infrared range) and high-quality microscope objectives to achieve spatially resolved chemical analysis. The excitation laser frequency significantly affects Raman measurements of biological materials like pollen, as shorter wavelength lasers (such as ultraviolet and green lasers) can increase Raman signal intensity but often enhance fluorescence background and are more strongly absorbed, leading to potential sample heating or damage. In contrast, longer wavelength lasers (such as near-infrared lasers) typically reduce both fluorescence interference and the risk of heating, though often requiring longer spectral acquisition times. Most Raman microscopy systems employ a dispersive spectrometer, though FT configurations also exist, and can be configured for confocal measurements with either upright or inverted microscope geometries. The choice of microscope objectives and spectrometer grating in Raman microspectroscopy is closely matched to the excitation laser wavelength to optimize spatial resolution, signal collection, and spectral dispersion. Proper alignment ensures efficient focusing of the laser on the sample and accurate detection of the resulting Raman scattered light.

13. Microscope slides for Raman spectroscopy are typically made of glass, unlike the specialized IR-transparent materials used for IR microscopy. In addition to glass, IR-transparent microscope slides can also be used in Raman spectroscopy if the user wishes

to analyze the same samples with both IR and Raman microspectrometers. The characteristic Raman peaks of microscope slides can serve as internal standards for calibrating spectral frequency and ensuring measurement accuracy. However, not all glass and IR-transparent microscope slides are suitable for Raman spectroscopy, as impurities in some optical materials can produce strong interference signals in the Raman spectra. Therefore, it is recommended to check a sample-free microscope slide for background interference before proceeding with sample analysis.

14. The spectral acquisition parameters in FTIR spectroscopy depend on the type of spectrometer available (entry-level or advanced-level FTIR spectrometers) as well as the specific requirements of the study, so the parameters provided in this protocol should be considered as general guidelines. A spectral range of 4000–600 cm^{-1} is typically sufficient for pollen measurements, as it covers the major vibrational modes relevant to the chemical characterization of biological samples. The number of scans refers to how many times the spectrometer collects data and averages the results, with higher numbers of scans improving the signal-to-noise ratio but increasing the total measurement time. Spectral resolution refers to the instrument's ability to distinguish between closely spaced absorption bands. Higher resolution (lower wavenumber values, such as 4 cm^{-1}) allows for finer detail but increases measurement time and data size, while lower resolution (higher values, such as 8 cm^{-1}) provides faster, less detailed spectra. For biological samples such as pollen, using a spectral resolution finer than 4 cm^{-1} is generally unnecessary, as it does not significantly improve the quality or interpretability of the spectra for most applications. Systems equipped with highly sensitive detectors, such as mercury cadmium telluride (MCT), require fewer scans to obtain high-quality spectra and can achieve higher spectral resolution within the same acquisition time compared to systems with less sensitive detectors, such as deuterated triglycine sulfate (DTGS) detectors. The apodization function in FTIR spectroscopy is used to modify the interferogram to optimize spectral resolution and minimize artifacts. Since the optimal choice of apodization depends on various experimental parameters, it is advisable to follow the recommendations provided by the instrument manufacturers.

15. The background spectrum contains information on IR-absorbing gases present in the optical path, such as water vapor and carbon dioxide. Unless the optical path is fully controlled (for example, under vacuum, nitrogen, or dry air atmosphere), it is advisable to perform background measurements frequently, preferably before each sample measurement,

as changes in atmospheric composition can introduce unwanted signals into the sample spectra. This is why it is recommended to perform background measurements immediately before, or as shortly as possible prior to, sample measurements to minimize the impact of atmospheric fluctuations. The atmospheric interference can be reduced by applying spectral pre-processing using an atmosphere correction algorithm, which is often part of spectral acquisition software. This pre-processing enables the suppression of water vapor and carbon dioxide bands.

16. ATR accessory has a pressure arm or clamp to ensure good contact between the sample and the crystal. All background measurements should be performed when the crystal surface is clean and with the open clamp.
17. The amount of pollen sample required to obtain high-quality IR spectra depends on the specifics of the ATR accessory being used. For single-reflection ATR accessories with a diamond crystal, as little as 0.5–1.0 mg of sample is often sufficient.
18. In FTIR spectroscopy, spectral acquisition involves first recording a single-channel spectrum of the sample-free setup (background spectrum) and then collecting a single-channel spectrum of the sample. These are subsequently ratioed to produce the final absorbance or transmittance spectrum.
19. Spectral preprocessing can often be performed using the spectral acquisition software provided by the instrument manufacturer. Alternatively, dedicated spectral preprocessing software, such as Orange (University of Ljubljana, Slovenia), Origin (OriginLab Corporation, USA), or Aspen Unscrambler (Aspen Technology, Inc., USA), as well as software packages for R (R Foundation for Statistical Computing, Austria), software libraries for Python (Python Software Foundation, USA) or toolboxes for Matlab (The Mathworks, Inc., USA), can be used for more advanced or customized data processing.
20. Spectral preprocessing of FTIR ATR spectra typically involves basic baseline correction and, if needed, conversion to derivative form. Since scattering issues are minimal, attention is often given to removing spectral regions with low signal-to-noise, such as between 2300 and 1900 cm^{-1} when using diamond ATR crystals. A typical preprocessing procedure involves smoothing (or smoothing and transforming to the second derivative form) using the Savitzky–Golay algorithm, truncating the data to the 1900–700 cm^{-1} spectral region, and applying multiplicative scatter correction (MSC). The readers are referred to the following article for detailed guidance [17].
21. The window size in the Savitzky–Golay algorithm is a critical parameter, and its optimal value depends mainly on the

number of recorded data points, which can be influenced by spectral resolution and post-acquisition processing such as zero-filling in Fourier transform spectrometers. As a result, even spectra acquired with the same resolution may require different window sizes. Along with window size, the polynomial degree of the fitting curve (i.e., polynomial order) should also be considered. Typically, lower-degree polynomials are favored as they help reduce noise and minimize the risk of overfitting. The readers are referred to the following article for detailed guidance [17].

22. A 1:100 (w/w) ratio typically yields optimal absorbance values (0.5–1.5) for most samples. However, if the sample exhibits very strong absorption, a lower sample-to-matrix ratio (down to 0.5% w/w) should be used, whereas for weakly absorbing samples, the ratio can be increased slightly (up to 2% w/w). A higher sample-to-matrix ratio can result in poor physical properties of the pellets, such as crumbling or reduced transparency, which may negatively affect the quality of the FTIR measurements.
23. This example is based on a pellet die with a 10 mm diameter. For dies of other sizes, the quantities of pollen sample and alkali halide should be adjusted proportionally to the volume and area of the die to ensure the pellet has appropriate thickness and transparency for FTIR analysis.
24. In principle, a sample-free pellet composed solely of pure alkali halide may serve as a background measurement. However, due to inherent variability in the optical properties and thickness of individual pellets, this practice is not recommended. The use of pure alkali halide pellets for background measurements can degrade spectral data by affecting the spectrum baseline, causing it to shift from simple features such as constant offset and slope to more complex baseline distortions, including various curvatures, which complicate spectral preprocessing, interpretation, and analysis. Instead, a pellet-free configuration (i.e., aconfiguration where no pellet is present in the beam path) should be employed for background acquisition to ensure optimal reproducibility and accuracy.
25. Spectral preprocessing of FTIR pellet spectra typically involves baseline correction and, if needed, conversion to derivative form. Baseline issues in FTIR spectra of pellets arise due to several factors, including variations in pellet thickness, inhomogeneous mixing of the sample and matrix, differences in refractive indices between the sample and alkali halide, light scattering within the pellet, and instrumental artifacts, all of which can distort the baseline and complicate spectral preprocessing. A typical preprocessing procedure involves smoothing

(or smoothing and transforming to the second derivative form) using the Savitzky–Golay algorithm, truncating the data to the 1900–700 cm^{-1} spectral region, and applying extended multiplicative scatter correction (EMSC) with linear and quadratic components. The readers are referred to the following article for detailed guidance [17].

26. Instead of ultrasonic homogenization, bead beating homogenization can be performed by using a bead beating tissue homogenizer. Readers are referred to the following articles for general procedures on bead beating homogenization of biological samples [29, 30].
27. Ultrasonication can lead to a significant increase in temperature. Therefore, it is recommended to pause sonication periodically to minimize temperature rise, and to conduct the process in an ice bath for additional temperature control. Incomplete fragmentation of pollen occurs with short sonication times, resulting in low-quality spectra, whereas proper sonication improves sample homogeneity and spectral quality. The exact homogenization parameters will depend on both the sample and the homogenizer used. Readers are referred to the following article for guidance on assessing sample homogeneity and spectral quality.
28. The optimal concentration of the suspension depends on the sample; the described procedure typically yields optimal absorbance values (0.5–1.5) for most samples. However, if the sample exhibits very strong absorption, a lower sample-to-water ratio (or less removal of supernatant) should be used, whereas for weakly absorbing samples, the ratio can be increased slightly.
29. Microplates are available in different configurations, commonly with 96, 384, or 1536 wells. The aliquot volume depends on the well size, and for example, 8–12 μL is typically pipetted into each well of a 384 well plates.
30. Drying can be accelerated by directing an air flow over the microtiter plate, such as placing the plate in a chemical fume hood, which significantly shortens the drying time.
31. One well on the plate is usually left empty (sample-free) to serve as a background measurement for the analysis. The plate measurement process is automated, and typically, the acquisition software allows recording the single-channel spectrum of the sample-free setup before each sample-well measurement.
32. Spectral preprocessing of FTIR–HTS spectra typically involves baseline correction and, if needed, conversion to derivative form. Baseline issues in FTIR spectra of pellets arise due to several factors, including interference fringes (caused by thin film effects), variations in film thickness and uniformity,

scattering from surface imperfections, and reflections at the air–film and film–silicon interfaces, and instrumental artifacts, all of which can distort the baseline and complicate spectral preprocessing. A typical preprocessing procedure involves smoothing (or smoothing and transforming to the second derivative form) using the Savitzky–Golay algorithm, truncating the data to the 1900–700 cm^{-1} spectral region, and applying extended multiplicative scatter correction (EMSC) with linear and quadratic components. The readers are referred to the following article for detailed guidance [17].

33. Some Raman instruments allow optimization of the sample position relative to the laser focus, enabling improved signal quality.

34. The spectral acquisition parameters in Raman spectroscopy depend on the type of spectrometer available (entry-level or advanced-level spectrometers) as well as the specific requirements of the study, so the parameters provided in this protocol should be considered as general guidelines. The two most common collection geometries in Raman spectroscopy are 180° geometry (backscattering), where the Raman signal is collected from the same side as the excitation laser, and 90° geometry, where the signal is collected at a right angle to the excitation laser. Since the optimal choice of collection geometries depends on various experimental parameters, it is advisable to follow the recommendations provided by the instrument manufacturers. A spectral range of 3500–400 cm^{-1} is typically sufficient for pollen measurements, as it covers the major vibrational modes relevant to the chemical characterization of biological samples. Spectral resolution refers to the instrument's ability to distinguish between closely spaced absorption bands. Higher resolution (lower wavenumber values, such as 2 cm^{-1}) allows for finer detail but increases measurement time and data size, while lower resolution (higher values, such as 8 cm^{-1}) provides faster, less detailed spectra. For biological samples such as pollen, using a spectral resolution finer than 2 cm^{-1} is generally unnecessary, as it does not significantly improve the quality or interpretability of the spectra for most applications. Higher laser power generally improves the signal-to-noise ratio of the spectra. However, excessive power can cause thermal degradation to sensitive pollen samples. Therefore, it is recommended to balance laser power to achieve optimal spectral quality without compromising sample chemistry. For example, overheating during Raman measurement of pollen samples containing carotenoid pigments can cause thermal degradation and photobleaching, leading to altered spectral features.

35. The number of scans refers to how many times the spectrometer collects data and averages the results, with higher numbers of scans improving the signal-to-noise ratio but increasing the total measurement time. Systems equipped with highly sensitive detectors, such as germanium (Ge), require fewer scans to obtain high-quality spectra and can achieve higher spectral resolution within the same acquisition time compared to systems with less sensitive detectors, such as indium gallium arsenide (InGaAs) detectors. The apodization function in FT-Raman spectroscopy is used to modify the interferogram in order to optimize spectral resolution and minimize artifacts. Since the optimal choice of apodization depends on various experimental parameters, it is advisable to follow the recommendations provided by the instrument manufacturers.
36. The exposure time refers to the duration over which the detector collects the Raman signal for each measurement, typically ranging from 5 to 60 s per acquisition, depending on sample and instrument properties. Since the optimal choice of exposure time depends on various experimental parameters (in particular, excitation laser frequency and power), it is advisable to follow the recommendations provided by the instrument manufacturers.
37. Some Raman systems for bulk measurements include automated plate-readers, enabling high-throughput analysis of well-plates and vials. The readers are referred to the following article for general procedures on high-throughput screening of biological samples [30].
38. Spectral preprocessing of Raman spectra typically involves removal of cosmic ray artifacts (particularly prevalent in dispersive spectrometers), baseline correction, and, if needed, conversion to derivative form. Baseline issues in Raman spectra arise due to several factors, including sample heating and fluorescence, stray light, and instrumental drift and artifacts, all of which can distort the baseline and complicate spectral preprocessing. A typical preprocessing procedure involves removal of cosmic ray artifacts (by applying algorithms that detect and eliminate sharp, isolated spikes), smoothing (or smoothing and transforming to the second derivative form) using the Savitzky–Golay algorithm, truncating the data to the 1900–700 cm^{-1} spectral region, and baseline correction. Baseline correction is typically performed using algorithms such as polynomial fitting, rubber band, and extended multiplicative scatter correction (EMSC) with linear and quadratic components. Raman bands usually have a relatively narrow range of values of the width of a spectral band (i.e., full width at half maximum, FWHM) and can therefore be easily separated from the high- and low-frequency interferent signals. However, if

spectral bands with a range of different bandwidths are present in the spectrum, this can complicate the selection of appropriate Savitzky-Golay parameters. The readers are referred to the following article for detailed guidance [17].

39. The specifics of QCL-based microspectroscopy approach is beyond the scope of this protocol, and readers are referred to the following study for detailed guidance [28].

40. For measurements on microscope slides, scattering of infrared radiation by pollen can be mitigated using analytical solutions for Mie-type scattering, which are implemented in spectral preprocessing and correction methods such as Mie extinction extended multiplicative signal correction (ME–EMSC), or by a scattering correction approach based on a deep convolutional neural network (DCNN). The readers are referred to the following study for detailed guidance [4, 24, 26]. Alternatively, scattering effects can be reduced by averaging data from multiple pollen grains, either through numerical averaging or by acquiring single-point measurements with a large microscope aperture. The readers are referred to the following studies for detailed guidance [5, 10].

41. For measurements of paraffin-embedded samples, Mie-type scattering effects are significantly suppressed. However, the embedding technique may result in nearly perfectly parallel sample surfaces, leading to another type of scattering to occur in the sample; i.e., multiple internal reflections. This causes interference fringes to be present in the absorbance spectra. Since fringe interferences may alter absorbance peak ratios, they often hamper the spectral analysis, and thus, they should be removed by fringe-correction preprocessing. The readers are referred to the following study for detailed guidance [4].

42. The available objectives in Raman microspectroscopy depend on the excitation laser frequency, but generally cover a magnification range from 10× to 100×.

43. The step size determines the spatial resolution of the measurement, and is typically set to match or be smaller than the laser spot size. Usually, step size is between 0.2 and 2.0 μm, depending on the objective magnification, excitation laser frequency, and desired mapping details.

Acknowledgments

This work was supported by the Research Council of Norway grants, project numbers 324670 and 351985.

References

1. Zimmermann B (2010) Characterization of pollen by vibrational spectroscopy. Appl Spectrosc 64(12):1364–1373
2. Kendel A, Zimmermann B (2020) Chemical analysis of pollen by FT-Raman and FTIR spectroscopies. Front Plant Sci 11:352. https://doi.org/10.3389/fpls.2020.00352
3. Zimmermann B, Bagcioglu M, Sandt C, Kohler A (2015) Vibrational microspectroscopy enables chemical characterization of single pollen grains as well as comparative analysis of plant species based on pollen ultrastructure. Planta 242(5):1237–1250. https://doi.org/10.1007/s00425-015-2380-7
4. Muthreich F, Magnussen EA, Solheim JH, Tafintseva V, Kohler A, Robin Seddon AW, Zimmermann B (2025) Analytical and experimental solutions for Fourier transform infrared microspectroscopy measurements of microparticles: a case study on Quercus pollen. Anal Chim Acta 1351:343879. https://doi.org/10.1016/j.aca.2025.343879
5. Bağcıoğlu M, Zimmermann B, Kohler A (2015) A multiscale vibrational spectroscopic approach for identification and biochemical characterization of pollen. PLoS One 10(9). https://doi.org/10.1371/journal.pone.0137899
6. Zimmermann B, Bagcioglu M, Tafinstseva V, Kohler A, Ohlson M, Fjellheim S (2017) A high-throughput FTIR spectroscopy approach to assess adaptive variation in the chemical composition of pollen. Ecol Evol 7(24):10839–10849. https://doi.org/10.1002/ece3.3619
7. Zimmermann B, Tkalcec Z, Mesic A, Kohler A (2015) Characterizing aeroallergens by infrared spectroscopy of fungal spores and pollen. PLoS One 10(4):e0124240. https://doi.org/10.1371/journal.pone.0124240
8. Muthreich F, Tafintseva V, Zimmermann B, Kohler A, Vila-Viçosa CM, Seddon AWR (2025) Evaluating the use of Fourier transform Raman spectroscopy for pollen chemical characterization. Appl Spectrosc 79(7):1142–1154. https://doi.org/10.1177/00037028251334405
9. Bağcıoğlu M, Kohler A, Seifert S, Kneipp J, Zimmermann B (2017) Monitoring of plant–environment interactions by high-throughput FTIR spectroscopy of pollen. Methods Ecol Evol 8(7):870–880. https://doi.org/10.1111/2041-210X.12697
10. Zimmermann B (2018) Chemical characterization and identification of Pinaceae pollen by infrared microspectroscopy. Planta 247(1):171–180. https://doi.org/10.1007/s00425-017-2774-9
11. Zimmermann B, Tafintseva V, Bagcioglu M, Hoegh Berdahl M, Kohler A (2016) Analysis of allergenic pollen by FTIR microspectroscopy. Anal Chem 88(1):803–811. https://doi.org/10.1021/acs.analchem.5b03208
12. Diehn S, Zimmermann B, Tafintseva V, Bagcioglu M, Kohler A, Ohlson M, Fjellheim S, Kneipp J (2020) Discrimination of grass pollen of different species by FTIR spectroscopy of individual pollen grains. Anal Bioanal Chem 412(24):6459–6474. https://doi.org/10.1007/s00216-020-02628-2
13. Seddon AWR, Festi D, Robson TM, Zimmermann B (2019) Fossil pollen and spores as a tool for reconstructing ancient solar-ultraviolet irradiance received by plants: an assessment of prospects and challenges using proxy-system modelling. Photochem Photobiol Sci 18(2):275–294. https://doi.org/10.1039/C8PP00490K
14. Zimmermann B, Kohler A (2014) Infrared spectroscopy of pollen identifies plant species and genus as well as environmental conditions. PLoS One 9(4):e95417. https://doi.org/10.1371/journal.pone.0095417
15. Muthreich F, Zimmermann B, Birks HJB, Vila-Viçosa CM, Seddon AWR (2020) Chemical variations in Quercus pollen as a tool for taxonomic identification: implications for long-term ecological and biogeographical research. J Biogeogr 47(6):1298–1309. https://doi.org/10.1111/jbi.13817
16. Diehn S, Zimmermann B, Tafintseva V, Seifert S, Bağcıoğlu M, Ohlson M, Weidner S, Fjellheim S, Kohler A, Kneipp J (2020) Combining chemical information from grass pollen in multimodal characterization. Front Plant Sci 10:1788. https://doi.org/10.3389/fpls.2019.01788
17. Kohler A, Solheim JH, Tafintseva V, Zimmermann B, Shapaval V (2020) 3.03 – model-based pre-processing in vibrational spectroscopy. In: Brown S, Tauler R, Walczak B (eds) Comprehensive chemometrics, 2nd edn. Elsevier, Oxford, pp 83–100. https://doi.org/10.1016/B978-0-12-409547-2.14607-4
18. Afseth NK, Kohler A (2012) Extended multiplicative signal correction in vibrational spectroscopy, a tutorial. Chemometr Intell Lab 117:92–99. https://doi.org/10.1016/j.chemolab.2012.03.004

19. Zimmermann B, Kohler A (2013) Optimizing Savitzky-Golay parameters for improving spectral resolution and quantification in infrared spectroscopy. Appl Spectrosc 67(8):892–902
20. Blumel R, Lukacs R, Zimmermann B, Bagcioglu M, Kohler A (2018) Observation of Mie ripples in the synchrotron Fourier transform infrared spectra of spheroidal pollen grains. J Opt Soc Am A Opt Image Sci Vis 35(10):1769–1779. https://doi.org/10.1364/JOSAA.35.001769
21. Lukacs R, Blumel R, Zimmermann B, Bagcoglu M, Kohler A (2015) Recovery of absorbance spectra of micrometer-sized biological and inanimate particles. Analyst 140. https://doi.org/10.1039/C5AN00401B
22. Bassan P, Kohler A, Martens H, Lee J, Byrne HJ, Dumas P, Gazi E, Brown M, Clarke N, Gardner P (2010) Resonant Mie Scattering (RMieS) correction of infrared spectra from highly scattering biological samples. Analyst 135(2):268–277. https://doi.org/10.1039/B921056c
23. Kohler A, Sule-Suso J, Sockalingum GD, Tobin M, Bahrami F, Yang Y, Pijanka J, Dumas P, Cotte M, van Pittius DG, Parkes G, Martens H (2008) Estimating and correcting Mie scattering in synchrotron-based microscopic Fourier transform infrared spectra by extended multiplicative signal correction. Appl Spectrosc 62(3):259–266
24. Magnussen EA, Zimmermann B, Blazhko U, Dzurendova S, Dupuy–Galet B, Byrtusova D, Muthreich F, Tafintseva V, Liland KH, Tøndel K, Shapaval V, Kohler A (2022) Deep learning-enabled inference of 3D molecular absorption distribution of biological cells from IR spectra. Commun Chem 5 (1): 175. doi:https://doi.org/10.1038/s42004-022-00792-3
25. Solheim JH, Borondics F, Zimmermann B, Sandt C, Muthreich F, Kohler A (2021) An automated approach for fringe frequency estimation and removal in infrared spectroscopy and hyperspectral imaging of biological samples. J Biophotonics 14(12):e202100148. https://doi.org/10.1002/jbio.202100148
26. Solheim JH, Gunko E, Petersen D, Grosseruschkamp F, Gerwert K, Kohler A (2019) An open-source code for Mie extinction extended multiplicative signal correction for infrared microscopy spectra of cells and tissues. J Biophotonics 12(8)
27. Blazhko U, Shapaval V, Kovalev V, Kohler A (2021) Comparison of augmentation and pre-processing for deep learning and chemometric classification of infrared spectra. Chemometr Intell Lab 215:104367. https://doi.org/10.1016/j.chemolab.2021.104367
28. Bird B, Baker MJ (2015) Quantum cascade lasers in biomedical infrared imaging. Trends Biotechnol 33(10):557–558. https://doi.org/10.1016/j.tibtech.2015.07.003
29. Langseter AM, Dzurendova S, Shapaval V, Kohler A, Ekeberg D, Zimmermann B (2021) Evaluation and optimisation of direct transesterification methods for the assessment of lipid accumulation in oleaginous filamentous fungi. Microb Cell Factories 20(1):59. https://doi.org/10.1186/s12934-021-01542-1
30. Rehman HU, Cord-Landwehr S, Shapaval V, Dzurendova S, Kohler A, Moerschbacher BM, Zimmermann B (2023) High-throughput vibrational spectroscopy methods for determination of degree of acetylation for chitin and chitosan. Carbohyd Polym 302:120428. https://doi.org/10.1016/j.carbpol.2022.120428

Chapter 9

Extracting and Quantifying Specific Pollen Proteins by Western Blot

Sara Parri, Chiara Piccini, Giampiero Cai, and Veronica Conti

Abstract

Western blotting is a useful technique to detect and quantify specific proteins in biological samples. Researchers can use it to investigate protein expression during key plant reproduction processes, such as pollen germination and tube growth. Protein quantification is critical for elucidating the molecular mechanisms governing pollen viability, polarity, and signaling. Western blotting has been used to examine the expression of cytoskeletal proteins, cell wall remodeling enzymes, and signaling molecules involved in pollen tube guidance and fertilization. Effective protein extraction from pollen and pollen tubes, which possess rigid cell walls, is necessary for reliable analysis. To maximize protein yield and maintain integrity, various extraction protocols are used. These protocols combine mechanical disruption, cell compartment fractionation, and chaotropic agents. Quantitative analysis requires careful normalization using total protein staining or reference proteins to ensure reproducibility across developmental stages or treatments. This technique complements molecular and imaging approaches by providing biochemical validation of gene expression data. This is relevant in studies of mutant pollen lines or stress responses, where protein levels may not directly correlate with transcript abundance. Western blotting is therefore essential for analyzing the dynamic protein composition of pollen cells and improving our understanding of processes, such as pollen tube growth and fertilization.

Key words Immunoblotting, Cell wall protein, Cytoplasmic protein, TCA/acetone, Differential extraction, Stain-free

1 Introduction

Protein extraction from pollen is essential for proteomic analysis. However, the unique properties of plant tissues, particularly pollen grains, present several challenges [1, 2]. Plant cells, such as pollen grains, have rigid cell walls and often contain substances that can interfere with analysis, including phenolic compounds, alkaloids, and other secondary metabolites [1, 2]. Effective homogenization is therefore required to release proteins for extraction [1, 2], as well as the inhibition of proteases, oxidative enzymes, and interfering metabolites (e.g., pigments, phenolic compounds, lipids, and

Giampiero Cai et al. (eds.), *Pollen and Pollen Tube Analysis: Methods and Protocols*, Methods in Molecular Biology, vol. 3042, https://doi.org/10.1007/978-1-0716-5308-1_9,

polysaccharides) [1]. Several methods have been developed and evaluated for extracting proteins from pollen for proteome analysis. Sheoran et al. [1] demonstrated the superiority of the trichloroacetic acid (TCA)-acetone and phenol methods for analyzing the tomato pollen proteome, due to improved separation on 2-DE gels and identification by mass spectrometry. The simplified TCA-acetone method consists of grinding the sample in liquid nitrogen, followed by extraction using an acetone solution containing TCA and DL-dithiothreitol (DTT) [1, 3]. Similar methods have been applied to *Arabidopsis* and *Picea meyeri* [3–5]. Using a Tris–phenol buffer for protein extraction produces a high yield, whereas direct extraction into an isoelectric focusing (IEF) buffer results in poorer resolution on a two-dimensional electrophoresis (2-DE) gel [1]. The Tris–HCl buffer method is incompatible with mass spectrometry (MS) analysis [1]. A rapid protocol for *Arabidopsis* uses a K-HEPES (4-(2-hydroxyethyl)-1-piperazineethanesulfonic acid, potassium salt) buffer [6]. Fíla et al. [2] compared various protocols, highlighting the effectiveness of the phenol-based method due to its simplicity. In contrast, the TCA-phenol-SDS method is more complex.

Western blotting is a robust method to detect and quantify specific proteins in pollen and plant cells. Based on antibody–antigen recognition, it provides specific information on a protein's presence, molecular weight, relative abundance, and possible post-translational modifications. This enables the differentiation of closely related isoforms and modified protein states, which is often not possible using transcript measurements. Therefore, Western blotting is essential for studying key processes such as pollen development, germination, and stress signaling pathways. Specialized software is essential for the relative quantification of Western blot signals, enabling the conversion of band images into objective and reproducible numerical data [7]. The software automatically performs background subtraction, integrates band intensity, and verifies linear range. Automated densitometry allows normalization to loading controls or total protein stains and supports calibration using known standards. Normalization proves particularly valuable when concentration analyses of extracts, like the Bradford assay, lack reliability. This situation frequently arises with plant material due to low protein extraction yields or interfering substances.

This chapter details a dual protein extraction protocol followed by Western blotting. The aim is to relatively quantify specific immunoreactive bands in pollen extracts. Protein extraction is based on either differential centrifugation and sequential extraction, which is used to extract cell wall-, membrane-, and cytosol-associated proteins, or a faster method involving TCA-acetone extraction (Fig. 1). The Stain-Free gel system was used in conjunction with fluorescently labeled secondary antibodies to optimize

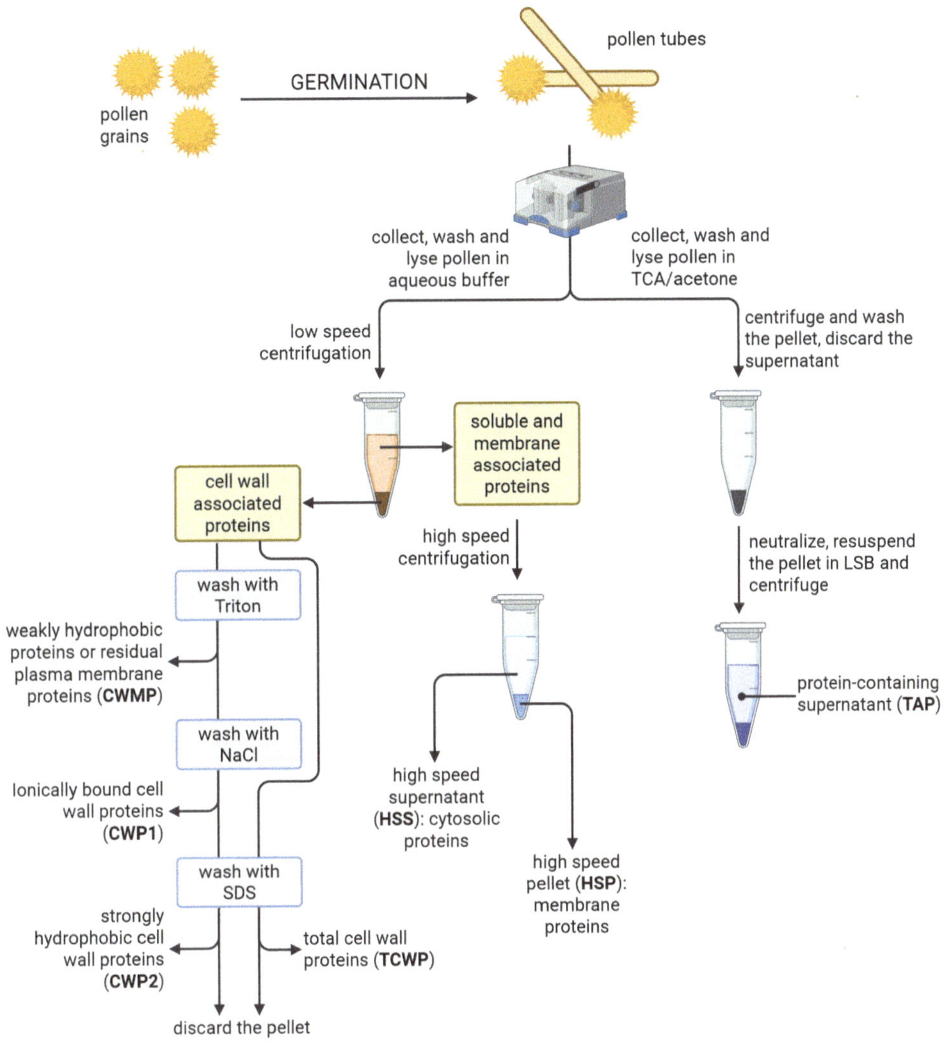

Fig. 1 Schematic diagram showing the two main pollen protein extraction methods detailed in this chapter. The right-hand side illustrates the faster method, which involves direct extraction with TCA/acetone. The left panel shows the more complex protocol involving sequential differential centrifugation and extraction with different buffers to selectively isolate proteins from the pollen tube cell wall

Western blotting, enhancing data quality and process efficiency. The Stain-Free system enables the immediate visualization and quantification of total proteins directly on the membrane, facilitating more accurate normalization and reducing dependence on reference proteins. Meanwhile, fluorescent antibodies enable highly sensitive and specific detection with the ability to analyze multiple protein targets simultaneously via multiplexing and linear quantification over a wide dynamic range. This, coupled with software-based analysis of immunoreactive bands, provides a fast and accurate method of relative quantitation of pollen proteins (Fig. 2).

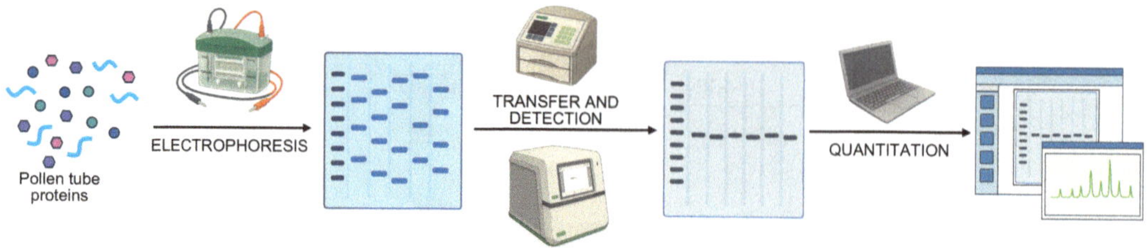

Fig. 2 Flow path of the methodological process after protein extraction. Proteins are separated by 1D (or 2D) electrophoresis on Stain-Free gels and then transferred to blotting membranes. All proteins and immunoreactive proteins are visualized using the Chemidoc apparatus. Following this, signals are quantified using specific software (Image Lab)

2 Materials

2.1 Pollen Germination Medium

1. BK medium: For a 100 mL 10× solution, dissolve 34.2 g of sucrose, 18.6 mg of H_3BO_3, 47.2 mg of $Ca(NO_3)_2{\cdot}4H_2O$, 10.1 mg of KNO_3 and 24.6 mg of $MgSO_4{\cdot}7H_2O$ in ~70 mL of distilled water with magnetic stirring. Bring the volume up to 100 mL with distilled water and mix thoroughly. If sterility is required, filter-sterilise through a 0.22 μm membrane into a sterile bottle. Store at −20 °C in aliquots. Before use, thaw the required 10× BK aliquot and add the necessary amount of sucrose (*see* **Note 1**) to a suitable container. Add water until the final volume is almost reached, stirring to dissolve the sucrose, before adjusting the volume.

2.2 Raw Extraction

1. Cold acetone.
2. 10% TCA in cold acetone: dissolve 10 g TCA in about 80 mL of pre-chilled acetone (kept at −20 °C) with magnetic stirring, ensure complete dissolution, bring to 100 mL with the same cold acetone, mix thoroughly, store at −20 °C until use.
3. 0.2 M NaOH: To prepare 10 mL, weigh 0.08 g of solid NaOH and carefully add it to approximately 8 mL of distilled water while stirring. Allow the solution to cool. Once fully dissolved, bring the final volume to 10 mL with distilled water. Mix thoroughly and store in a tightly sealed container at room temperature.
4. BME: β-Mercaptoethanol.
5. Laemmli Sample Buffer 1× (*see* **Note 2**).
6. Centrifuge.
7. Ice.
8. Heated block for microcentrifuge tubes.
9. Bead mill homogenizer.

2.3 Differential Extraction

1. 250 mM Hepes pH 7.5: to prepare 100 mL, weigh 5.96 g of Hepes powder and dissolve it in approximately 80 mL of distilled water. Measure the pH using a calibrated pH meter and adjust it to 7.5 by slowly adding drop by drop 5 M NaOH while stirring. Once the desired pH is reached, bring the final volume to 100 mL with distilled water. Mix thoroughly and store the solution at 4 °C for a few months.
2. 0.1 M EGTA: to prepare 100 L, weigh 3.8 g of EGTA disodium salt and add it to approximately 80 mL of distilled water. Stir the mixture thoroughly to help dissolve the EGTA. Measure the pH using a calibrated pH meter and adjust it to pH 8.0 by slowly adding drop by drop 5 M NaOH while stirring, as EGTA dissolves more efficiently at alkaline pH. Once fully dissolved, bring the final volume to 100 mL with distilled water. Mix well and store at 4 °C for a few months.
3. 0.1 M $MgCl_2$: to prepare 100 mL, weigh 0.95 g of anhydrous magnesium chloride or adjust the weight based on the hydrate form used. Add the salt to approximately 80 mL of distilled water and stir until fully dissolved. Once dissolved, bring the final volume to 100 mL with distilled water. Mix thoroughly and store the solution at 4 °C for a few months.
4. Sucrose.
5. 1 M DTT: to prepare 10 mL, weigh 1.54 g of DTT powder and add it to approximately 8 mL of distilled water. Stir until fully dissolved, then bring the final volume to 10 mL with distilled water. Mix thoroughly and store the solution in small aliquots at −20 °C to prevent oxidation and maintain stability.
6. Protease inhibitors (*see* **Note 3**).
7. Triton X-100.
8. NaCl.
9. Sodium Dodecyl Sulfate (SDS).
10. 10% TCA in cold acetone: *see* Subheading 2.2 for preparation instructions.
11. Cold acetone.
12. 0.2 M NaOH: *see* Subheading 2.2 for preparation instructions.
13. HEMS Pollen Washing Buffer: mix 5 mL of stock 250 mM Hepes pH 7.5, 1.5 mL of stock 0.1 M EGTA, and 0.1 mL of stock 0.1 M $MgCl_2$. Add sucrose to reach a final concentration of 12% (adjust depending on pollen type). Add distilled water to reach 40 mL, and dissolve completely. Adjust final volume to 50 mL. Store at room temperature.
14. Pollen Lysis Buffer: mix 1 mL of stock 250 mM Hepes pH 7.5, 0.3 mL of stock 0.1 M EGTA, and 0.2 mL of stock 0.1 M $MgCl_2$. Just before use, add 20 μL of 1 M DTT and protease

inhibitors (according to the manufacturer's instructions). Add sucrose to reach the required final concentration. Add distilled water to reach 8 mL, and dissolve completely. Adjust final volume to approximately 10 mL. Store on ice.

15. Solubilization Buffer with Triton X-100: mix 1 mL of stock 250 mM Hepes pH 7.5, 0.3 mL of stock 0.1 M EGTA, and 0.2 mL of stock 0.1 M $MgCl_2$. Just before use, add 20 μL of 1 M DTT. Add Triton X-100 to reach 0.1% final concentration. Add distilled water to reach 8 mL, and dissolve completely. Adjust final volume to approximately 10 mL. Store on ice.
16. Cell Wall Protein Extraction Buffer with NaCl: mix 1 mL of stock 250 mM Hepes, pH 7.5, and 0.3 mL of stock 0.1 M EGTA. Just before use, add 20 μL of 1 M DTT. Add NaCl to reach 0.5 M final concentration. Add distilled water to reach 8 mL, and dissolve completely. Adjust final volume to approximately 10 mL. Store on ice.
17. Cell Wall Protein Extraction Buffer with SDS: mix 1 mL of stock 250 mM Hepes, pH 7.5, and 0.3 mL of stock 0.1 M EGTA. Just before use, add 20 μL of 1 M DTT. Add SDS to reach 2% final concentration. Add distilled water to reach 8 mL, and dissolve completely. Adjust final volume to approximately 10 mL. Store on ice.
18. Centrifuge.
19. Ultracentrifuge.
20. Bead mill homogenizer.
21. Ice.
22. Vortex.

2.4 Electrophoresis

1. Tris-Glycine eXtended (TGX) Stain-Free FastCast Gel (Bio-Rad).
2. Mini-gel electrophoresis cell.
3. Short and spacer glass.
4. Electrophoresis power supply.
5. Tris-Glycine-SDS (TGS) running buffer (1×): 25 mM Tris, 192 mM Glycine, 0.1% SDS.
6. Clean 15 mL tubes.

2.5 Protein Concentration (Optional)

It is important to measure protein concentration before gel loading to ensure that equal amounts of protein are loaded in each lane, allowing for an accurate comparison between samples. However, when using stain-free gels, this step can sometimes be bypassed as the gel enables direct visualization and normalization of protein content. Nevertheless, various commercial kits are available for quantifying protein concentration prior to electrophoresis, such as the BCA, Bradford, or Lowry assays.

2.6 Western Blot

1. TRIS-buffered saline (TBS) (25 mM Tris; 150 mM NaCl; pH 7.4–7.6): Add approximately 70 mL of distilled water to a beaker and initiate magnetic stirring, add 2.42 g of Tris base and 8.77 g of NaCl, and stir until completely dissolved. Adjust the pH to 7.6 by slowly adding concentrated HCl. Bring the final volume to 100 mL with distilled water.
2. Tween-20, ideally in the form of a concentrated 1% solution.
3. Primary antibody (*see* **Note 4**).
4. Secondary antibody conjugated with StarBright Blue 700 (*see* **Note 5**).
5. Gel blot image analysis (*see* **Note 6**).
6. Transfer buffer: 5× Trans-Blot Turbo Transfer Buffer is compatible with Bio-Rad's TBT Ready-to-Assemble Transfer Kits. Follow the manufacturer's instructions to prepare it correctly.
7. A protein transfer apparatus such as the Trans-Blot Turbo Transfer System (Bio-Rad).

2.7 Band Quantification

Use integrated optical density quantification software such as Fiji/ImageJ, which is a free and widely used densitometry package. In this protocol, we used ImageLab software (Bio-Rad).

3 Methods

3.1 Pollen Germination

1. Weigh an appropriate amount of fresh or cold-stored pollen (*see* **Note 7**).
2. Hydrate pollen overnight in a humid chamber (*see* **Note 8**).
3. Place pollen in Petri dishes or Erlenmeyer flasks and germinate it by adding BK medium with appropriate sucrose concentration for 3–12 h at 25 °C with constant agitation, preferably in the dark.
4. Examine pollen germination efficiency by observing a suspension drop on a glass slide using a light microscope.

3.2 Raw Extraction of Proteins from Pollen

1. Collect the germinated pollen by centrifugation at 200 rcf for 5 min.
2. Discard the supernatant.
3. Transfer the pollen pellet (200 mg) into a mL microcentrifuge tube chilled on ice (*see* **Note 9**).
4. Add 1.8 mL 10% TCA acetone, 1.25 μL of BME, and resuspend by vortexing.
5. Place the samples at −20 °C for 20 min to precipitate the proteins.

6. Lyse germinated pollen using a bead mill homogenizer (*see* **Note 10**).
7. Centrifuge for 10 min at 10,000 rcf, 4 °C.
8. Discard the supernatant, resuspend the pellet in 1.8 mL of acetone and 1.25 μL of BME.
9. Vortex thoroughly.
10. Centrifuge for 10 min at 10,000 rcf at 4 °C and discard the supernatant.
11. Repeat **steps 7–9**.
12. Add 100 μL of 0.2 M NaOH to the pellet and incubate for 2 min.
13. Resuspend the pellet in 150 μL 1× LSB (*see* **Note 11**).
14. Heat at 95 °C for 5 min.
15. Centrifuge for 15 min at 10,000 rcf, room temperature.
16. Collect the supernatant (TAP) in a new microcentrifuge tube.

3.3 Differential Protein Extraction from Pollen

1. Collect the germinated pollen by centrifugation at 200 rcf for 5 min.
2. Wash pollen twice with 25 mL of washing buffer.
3. Lyse germinated pollen using a bead mill homogenizer (*see* **Note 10**) with 1–2 mL of cold lysis buffer.
4. Centrifuge the sample at a low speed of 500 rcf for 10 min at 4 °C using a microcentrifuge.
5. Place the pellets on ice temporarily.
6. Centrifuge supernatants from point 4 at 100,000 rcf for 60 min at 4 °C using an ultracentrifuge.
7. During ultracentrifugation, wash the pellet (point 5) with 0.5–1 mL of lysis buffer and centrifuge at 500 rcf for 10 min at 4 °C.
8. Repeat the washing process 2–3 times, discarding the resulting supernatants to remove any contaminating proteins (*see* **Note 12**).
9. Extract one of the pellets with 0.5–1 mL of solubilization buffer containing Triton X-100 using a plastic pestle.
10. Incubate for 10–15 min on ice.
11. Centrifuge at 500 rcf for 10 min at 4 °C.
12. Collect the supernatant (CWMP) for 1D electrophoresis and store it on ice. This will contain weakly hydrophobic or residual plasma membrane proteins.
13. Wash pellet 1–2 times with the Triton solubilization buffer, centrifuging as above to remove contaminants.

14. Resuspend pellet in 0.5–1 mL of NaCl extraction buffer.
15. Incubate for 15 min on ice.
16. Centrifuge at 12,000 rcf for 10 min at 4 °C.
17. Collect the supernatant (CWP1) for 1D electrophoresis and store it on ice. This contains proteins that are ionically bound to the cell wall.
18. Wash pellet 1–2 times with NaCl buffer, centrifuging as above to remove contaminants.
19. Resuspend the final pellet in 0.5–1 mL of SDS extraction buffer.
20. Incubate for 15 min on ice.
21. Centrifuge at 12,000 rcf for 10 min at 4 °C.
22. Collect the CWP2 supernatant for 1D electrophoresis and store it on ice. This sample contains highly hydrophobic cell wall proteins. Discard the pellets.
23. After the ultracentrifugation step (point 6), collect supernatant (HSS) for 1D electrophoresis and keep it on ice.
24. Resuspend pellet in LSB 1× (*see* **Note 11**) buffer for 1D electrophoresis (HSP).
25. Process previous supernatants (CWMP, CWP1, CWP2, HSS) by precipitating proteins with 10% TCA in cold acetone for 20 min.
26. Wash the pellets twice with cold acetone and centrifuge at 12,000 rcf for 15 min.
27. Neutralize the pH using approximately 20 μL of 0.2 M NaOH.
28. Resuspend the pellets in LSB for electrophoresis.

3.4 Electrophoresis of Extracted Proteins

To perform this stain-free method, use a TGX Stain-Free FastCast Gel (Bio-Rad) and follow the manufacturer's instructions for preparing the gel. Use Tris–Glycine–SDS (TGS) 1× as the running buffer. Perform the electrophoretic run at 200 V for about 40 min.

3.5 Western Blot Analysis

1. During electrophoresis, ensure that the blot membrane (*see* **Note 13**) and filter papers are wet with a transfer buffer.
2. After electrophoresis, remove the gel from the cassette and activate the proteins using the ChemiDoc, following the instructions for the instrument (Fig. 3a).
3. Assemble the blot sandwich in the Trans-Blot Turbo Transfer cassette by placing filter papers on the bottom, then place the gel over the filter papers and the blot membrane over the gel. Finally, cover with additional filter papers.
4. Place the transfer cassette in the Trans Blot Turbo apparatus.

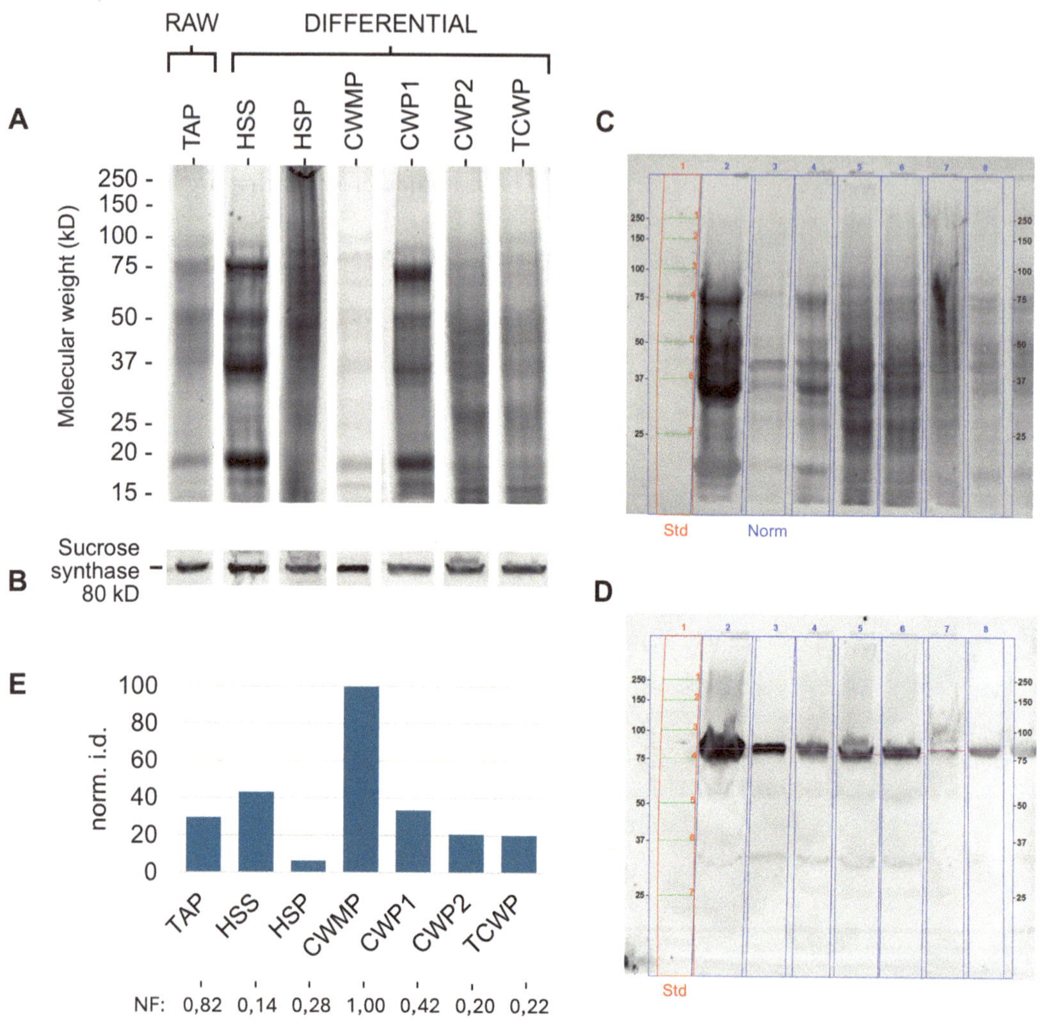

Fig. 3 Panel **a** shows an example of gel protein samples obtained using the two extraction methods described above: the TCA/acetone method (TAP: TCA/acetone proteins) and the differential extraction method (from HSS to TCWP). The abbreviations used to identify the samples are the same as those described in the method section above. On the left is the molecular weight scale in kilodaltons. Panel **b** shows an example of immunostaining with the anti-sucrose synthase antibody (Agrisera code AS152830) on the protein fractions shown in Panel **a**. Panel **c** shows the final adjusted frame within the analysis software, displaying the total protein signal from Stain-Free membrane results for channel 1. The third lane has been selected as the normalization lane (Norm). Panel **d**: blot band detection results as shown in channel 2. Blue vertical lines delineate lanes (1–8) for individual sample analysis. The first lane (Std) is aligned with the unstained standard (Bio-Rad). The green tick marks (1–7) on the y-axis indicate the calculated molecular weights in kilodaltons, providing a reference scale for protein sizing. Panel **b** has a thin purple line digitally overlaid on top of each detected band. Panel E shows the relative quantification of the sucrose synthase signal normalized to the protein content of each lane, as determined using Bio-Rad's Image Lab software. The integrated normalized density is shown on the y-axis. NF Normalization Factor

5. Run the protein transfer from gel to membrane according to the Trans Blot Turbo's preset transfer programs, which are dependent on the molecular weight of the target proteins to be investigated.
6. At the end of the transfer process, move the membrane to the ChemiDoc system and visualize the proteins to verify the quality of the transfer.
7. Place the membranes in the incubation trays, add 10 mL of EveryBlot Blocking Buffer (Bio-Rad) (*see* **Note 14**), and then block the membranes for five minutes with constant agitation at room temperature.
8. Add the primary antibody, diluted in blocking buffer, at the correct concentration (*see* **Note 15**). Incubate the membranes for 1 h at room temperature with constant agitation.
9. Wash the membranes twice in 20 mL of TBS containing 0.1% Tween-20. Leave them to soak for 5 min each time, agitating gently.
10. Replace the washing solution with 10 mL of a secondary antibody solution (*see* **Note 16**) directed toward the type of primary antibody, conjugated with StarBright Blue 700 and diluted as required in the blocking buffer.
11. Incubate the membranes for 1 h at room temperature with agitation.
12. Wash the membranes twice more with 20 mL of TBS containing Tween-20 for 5 min each time, with agitation throughout.
13. Place the membrane in the ChemiDoc system and select the appropriate channels for visualizing the immunoreactive bands (the Stain-Free channel to visualize all proteins in the membrane, and the StarBright Blue 700 channel to visualize only the immunoreactive bands) (Fig. 3b).
14. Export the saved images in their native format and analyze them for protein quantification using the Image Lab software (*see* **Note 17**).

3.6 Quantification Analysis of Immuno-Reactive Bands

This section details the steps for the densitometric quantification of fluorescent Western Blots using the Image Lab Software (Bio-Rad) and subsequent normalization against total protein loaded in each lane, as detected by the Stain-Free technology.

1. Open the target Western Blot file in the Image Lab software. Close the composite RGB image, retaining only the individual channels: Channel 1 (C1) for Stain-Free (total protein) and Channel 2 (C2) for the fluorescent secondary antibody (target protein).

2. Use the Image Tool to adjust display settings for optimal visualization. Invert the colors and remove the highlight saturated pixel feature, which is often preferred for densitometric analysis.
3. Display only the Stain-Free channel (C1). Click on "Lane and Bands" and define the lanes using either the manual or automatic detection tool. Carefully adjust the boundaries of each individual lane to ensure they encompass the entirety of the protein migration path, including the Protein Standard lane. If "resize frame" is unselected, individual adjustments can be made (Fig. 3c).
4. Click "Adjust Background" and modify the background subtraction parameters for the defined lanes. Set the disk size to 70 (recommended value for Stain-Free gels) to remove non-specific background signal from the total protein quantification.
5. Switch to the fluorescent antibody channel (C2); the lane definitions from the previous step should transfer automatically. Under "Lane and Bands," select "Bands," and click "Detect bands." Manually review, add, delete, or adjust the automatically detected bands to accurately enclose only the target protein signal in each lane.
6. Select the "Normalization" tab. Assign the STAIN-FREE channel (C1) as the Normalization Channel. Check the box for "Total Lane Protein" and select the lane that exhibits the highest overall protein signal as the reference lane.
7. Click on the "MW Protein Standard" tab. In the stain-free channel (C1) view, select the box corresponding to the Protein Standard lane. Review and adjust the automatically detected bands in the Standard lane (using "Lane and Bands", then "Bands") to ensure accurate molecular weight calibration. The final result should be similar to Fig. 3d.
8. Click on the "Analysis Table" tab to view the quantitative results. The key results for reporting are the Molecular Weights (kDa) and the Adjusted Volume (Adj Volume) of the target bands, which represent the final, loading-corrected densitometric signal (*see* **Note 18**) that you can use to create the graph (Fig. 3e).

4 Notes

1. The quantity of sucrose required for the BK medium depends on the pollen species used in the experiment. This is because each species of pollen requires a specific concentration in order

to germinate optimally. Please refer to the most up-to-date literature for the optimal sucrose concentration for your pollen of interest.

2. You can buy a pre-made solution, which is usually more concentrated (e.g., Bio-Rad 4× Laemmli Sample Buffer), or you can prepare it as follows: Mix 62.5 mM Tris HCl, 20% glycerol, 2% SDS, 5% beta-mercaptoethanol, and 0.032% bromophenol blue in distilled water.
3. Protease inhibitors can be purchased from several manufacturers. To counteract the negative effects of internal proteases, it is advisable to use a cocktail containing several protease inhibitors.
4. The choice of primary antibodies for Western blotting obviously depends on the protein of interest to be analyzed. There are several companies that provide antibodies for a wide range of proteins.
5. As we are using the Chemidoc system from Bio-Rad to detect immunoreactive bands, we need to use secondary antibodies whose fluorochrome can be recognized by the system. Several fluorochromes are suitable for use with the Chemidoc system. We prefer to use secondary antibodies conjugated to Starbright 700.
6. This chapter describes a method that makes use of the Chemidoc system from Bio-Rad. Consequently, other reagents are purchased from the same manufacturer due to their compatibility with the imaging system. However, it is worth noting that other imaging systems from different manufacturers are available worldwide.
7. Pollen can be collected as fresh from opening flowers, or alternatively, it can be stored after suitable dehydration and acclimatization in freezers at −20 °C. We have worked with both cases, and very little differences can be noted. Storing pollen in the freezer allows experiments to be performed even when fresh pollen is not available.
8. A humid chamber can be easily assembled using large Petri dishes covered with aluminum foil to protect them from light exposure. Place wet filter papers inside the dishes to create the chamber, then place a smaller Petri dish containing pollen inside.
9. To enhance precipitation and slow protease activity, all steps leading to protein precipitation should be performed on ice.
10. Due to the robust cell wall of the pollen, a harsh homogenization system may be required. There are several alternatives. One option is to use a small pestle adapted for use in Eppendorf tubes to lyse the pollen. For larger volumes, a motor-

driven Potter-Elvehjem homogenizer can be used. If several samples have to be processed and the lysing conditions standardized, a bead mill homogenizer such as the TissueLyser from Qiagen can be used.

11. Add the minimum amount of LSB 1× needed to completely wet the pellet and achieve the desired number of western blots. Adding a high volume of LSB 1× at this stage will dilute the precipitated protein. Do not add too much, or you will lose the Western blot signal.
12. During this step, a deviation from the standard sequential extraction protocol was implemented to obtain a reference sample of total cell wall proteins. The pellet obtained was divided into two fractions: one fraction followed the standard protocol, which uses a series of buffers with increasing polarity to collect proteins sequentially based on their binding strength (weakly, then locally, and finally strongly bound). The second optional fraction, however, bypassed these sequential steps and was immediately subjected to the final, high-detergent extraction buffer (corresponding to Step 19 of the full protocol, containing SDS). The proteins extracted from this parallel approach were designated "Total Cell Wall Proteins"(TCWP), intended to represent all cell wall-associated proteins extracted simultaneously. However, based on the subsequent Western Blot quantification, the amount of TCWP was unexpectedly similar to the quantity obtained from the last, most stringent sequential fraction (CWP2/strongly bound proteins). This observation suggests that under these conditions, the TCWP method primarily captured the highly hydrophobic or strongly bound proteins, rather than the full spectrum of cell wall proteins.
13. If nitrocellulose membranes are used, they can simply be wetted in the transfer buffer a few minutes before the transfer begins. If a PVDF membrane is used instead, it should be pre-wetted with pure methanol for a few seconds before wetting it with the transfer buffer.
14. Blocking the blot membranes is a mandatory step to prevent non-specific binding of the primary or secondary antibodies to the membrane. This step can be performed using BSA dissolved in suitable buffers, such as PBS or TBS. However, this can be expensive. A cheaper alternative is to use commercially available non-fat milk powder, although the quality may vary from batch to batch. We use the EveryBlot Blocking Buffer from Bio-Rad, which enables a very fast blocking step in just a few minutes.
15. Finding the most appropriate dilution of the primary antibody may be challenging. Luckily, information sheets of most

commercial antibodies report a suitable concentration range already tested under different conditions. We advise starting from this indicated concentration, and then making some attempts to find the best primary antibody dilution.

16. The type of secondary antibody used depends on the type of primary antibody. The conjugated molecule used to detect immunoreactive bands can range from classical horseradish peroxidase enzymes, which produce darker immunoreactive bands directly on membranes, to chemiluminescent systems and the use of fluorescently labeled secondary antibodies. The choice also depends on the type of imaging system available in the laboratory. Unlike primary antibodies, secondary antibodies have a more standardized concentration range, and the information sheet usually provides a very specific dilution range.
17. We use the Image Lab software from Bio-Rad in our lab because it is compatible with the file format produced by the Chemidoc system.
18. The analysis table provides additional information. It is interesting to note, however, that adj. volume is the density of the band normalized by the normalization factor. This factor is used to make bands on the same membrane comparable without loading the same amount of protein per lane.

References

1. Sheoran IS, Ross ARS, Olson DJH et al (2009) Compatibility of plant protein extraction methods with mass spectrometry for proteome analysis. Plant Sci 176:99–104
2. Fíla J, Čapková V, Feciková J et al (2011) Impact of homogenization and protein extraction conditions on the obtained tobacco pollen proteomic patterns. Biol Plant 55:499–506
3. Chen Y, Chen T, Shen S et al (2006) Differential display proteomic analysis of Picea meyeri pollen germination and pollen-tube growth after inhibition of actin polymerization by latrunculin B. Plant J 47:174–195
4. Zou J, Song L, Zhang W et al (2009) Comparative proteomic analysis of arabidopsis mature pollen and germinated pollen. J Integr Plant Biol 51:438–455
5. Chen T, Wu X, Chen Y et al (2009) Combined proteomic and cytological analysis of Ca2+-Calmodulin regulation in Picea meyeri pollen tube growth. Plant Physiol 149:1111–1126
6. Chang M, Huang S (2017) Rapid isolation of total Pprotein from *Arabidopsis* pollen. Bio-protocol 7
7. Mazet F, Dunster JL, Jones CI et al (2015) A high-density immunoblotting methodology for quantification of total protein levels and phosphorylation modifications. Sci Rep 5:16995

Chapter 10

Detecting ROS Production by Pollen Grains and Tubes

Maria Breygina, Oksana Luneva, and Anna Podobedova

Abstract

Reactive oxygen species (ROS) are signaling agents that actively participate in pollen germination. Pollen produces ROS both during the initial stages of germination and at later stages, including polar growth. ROS are distributed unevenly within the pollen tube, forming a gradient. Some ROS are released into the extracellular space, while others accumulate intracellularly, including in organelles. The main experimental difficulty lies in distinguishing between hydrogen peroxide, superoxide radicals, and other ROS. In this chapter, we describe methods for detecting ROS production in pollen: microscopic, spectrophotometric, and EPR-based. Fluorescence microscopy is accessible and easy to use, but it has limited accuracy and sensitivity. EPR spectroscopy is a relatively uncommon method, primarily used in physics and biophysics. However, it produces excellent results and allows for the quantification of total ROS production and superoxide radical production in liquid medium. For hydrogen peroxide detection, we recommend highly specific approaches: a non-oxidative dye PFBSF for pollen grains and tubes, and a spectrophotometric method FOX-1 for extracellular H_2O_2.

Key words ROS, Pollen grain, Pollen tube, Hydrogen peroxide, EPR, Spin trap, DCFH-DA, PFBSF, CAT1H, DEPMPO

1 Introduction

1.1 Evaluation of ROS Generation in Pollen Grains and Pollen Tubes Using Fluorescence Microscopy and Spectrophotometry

One of the most classic and accessible methods for assessing ROS production in cells is fluorescent staining. Here, we will present two options, one widely used and the other less popular but more specific. Both can be successfully used for pollen staining.

$DCFH_2$-DA—2′,7′-dichlorodihydrofluorescein diacetate is one of the earliest developed probes used to date to study total ROS production in living cells, including pollen [1–3]. ROS-mediated oxidation produces the fluorescent product 2′,7′-dichlorofluorescein, which fluoresces brightly in the green region of the spectrum [4]. It has been found that oxidation of DCFH to dichlorofluorescein is induced by H_2O_2, hydroperoxides, nitric oxide, and peroxynitrite [2]. Therefore, the fluorescent response based on the oxidation of DCFH provides information

Giampiero Cai et al. (eds.), *Pollen and Pollen Tube Analysis: Methods and Protocols*, Methods in Molecular Biology, vol. 3042,
https://doi.org/10.1007/978-1-0716-5308-1_10,

about the total level of oxidants present in pollen grains or tubes. Measuring total ROS/NOS and other oxidizing components is an important task in research and practice, so the dye continues to be used [5]. It is also convenient that a commercially available dye can be easily converted into a de-esterified form. This allows for the assessment of total extracellular ROS production by pollen grains [6].

There are more specific dyes than DCFH. Since hydrogen peroxide is one of the key components produced by plant tissues, one can use pentafluorobenzenesulfonyl fluorescein (PFBSF), which is highly specific for this ROS [7]. This dye is intracellular; to assess extracellular ROS production, the OxyBURST Green H_2HFF BSA can be used [8]. Despite the convenience and availability of dyes, it should be noted that the sensitivity of visual methods is limited, and if it is necessary to detect very low amounts of ROS or measure their level with high precision, one should consider EPR spectroscopy [9–11].

1.2 Evaluation of ROS Generation in Pollen Grains by EPR Spectroscopy

Electron paramagnetic resonance (EPR) spectroscopy is an analytical method based on the phenomenon of resonant absorption of high-frequency electromagnetic radiation in the radio range by unpaired electrons in an external magnetic field [12]. Inorganic cellular radicals, including ROS, do not produce their own EPR signal due to their short lifetime. However, EPR spectroscopy allows for the direct detection and identification of short-lived free radicals, including those produced in biological systems due to the use of *spin traps* and *spin probes*. A spin trap, a substance that is non-paramagnetic (i.e., has no unpaired electrons), is introduced into the experimental system and, upon interaction with a short-lived inorganic radical, forms a stable radical (spin adduct), which has a lifetime sufficient for recording an EPR spectrum. Parameters of the EPR spectrum allow one to determine the type of inorganic radical, and the intensity of the corresponding signal is proportional to its concentration [12].

Currently, a wide range of spin traps has been synthesized to solve various experimental problems. Among them are highly selective and low-selective ones. For example, commercially available DEPMPO (5-Diethoxyphosphoryl-5-methyl-1-pyrroline-n-oxide) spin trap belongs to the first group and, upon interaction with superoxide anion radical ($O^{\bullet}{}_2{}^-$), forms a stable spin adduct with a characteristic EPR spectrum. The second group includes DMPO (5,5-Dimethyl-1-pyrroline N-oxide), a spin trap which forms the same spin adduct with both $O^{\bullet}{}_2{}^-$ and hydroxyl radical (OH•) [13].

Spin probes from the cyclic hydroxylamine group are also used to study inorganic radicals. These probes bind non-selectively to inorganic radicals, but, unlike spin traps, have a high affinity for them. The reaction results in the formation of nitroxyl radicals, which produce a characteristic EPR spectrum. An example of such a spin probe is CAT1H (1-Hydroxy-2,2,6,6-tetramethyl-4-(trimethylammonio)-piperidinium dichloride), which has a low distribution coefficient between the aqueous medium and the membrane, and therefore practically does not penetrate the cell, which allows one to evaluate the production of inorganic radicals into the extracellular environment [14]. Cyclic hydroxylamines are convenient for rapid assessment of the total number of inorganic radicals and the oxidative status of the system [15]. Thus, the experimental approaches described above using EPR spectroscopy enable the qualitative and quantitative analysis of short-lived radicals in experimental systems. These methods are considered the "gold standard" in the study of inorganic radicals, including ROS [10, 11].

2 Materials

2.1 Pollen Germination Medium

The optimal medium for pollen germination varies for different plant species. General guidelines for selecting a medium for a given plant can be found in the compendium [16] (*see* **Note 1**).

We conducted the experiments described below with three plant species, and here we present the composition of the optimal medium that we used for them:

1. *Nicotiana tabacum*: 1.6 mM H_3BO_3, 3 mM $Ca(NO_3)_2$, 0.8 mM $MgSO_4$, 1 mM KNO_3, 0.3 M sucrose in 25 mM MES–Tris buffer, pH 5.8 [17].
2. *Streptocarpus* hybr.: 0.81 mM H_3BO_3, 0.31 mM Ca $(NO_3)_2{\cdot}4H_2O$, 0.3 M sucrose [18].
3. *Picea pungens*: 1 mM H_3BO_3, 1 mM $CaCl_2$, 0.3 M sucrose in 25 mM MES–Tris buffer, pH 5.5 [19] (*see* **Note 2**).

2.2 Dyes, Spin Probe, and Spin Trap

1. $DCFH_2$-DA stock solution of 5 mM in EtOH (can be stored at −20 °C).
2. PFBSF stock solution of 1 mM in DMSO (can be stored at −20 °C).
3. OxyBURST Green H_2HFF BSA stock solution in phosphate-buffered saline 50 μg/ml. should be stored at 4 °C (do not freeze!).
4. CAT1H stock aqueous solution of 20 mM can be best stored at −80 °C (at −20 °C also possible).
5. DEPMPO stock solution of 100 mM in DMSO can be best stored at −80 °C.

3 Methods

3.1 Pollen Suspension

For experiments, use freshly collected pollen from plants and a germination medium containing all the necessary components. For plants whose pollen remains viable, pre-collected pollen stored at −20 °C can be used. For angiosperms, we recommend moistening the pollen before germination, i.e., incubating it in a humid atmosphere (*see* **Note 3**). To study pollen tubes growing flat, we used the agarose pad method (*see* **Note 4**).

3.2 Pollen Humidification

1. Collect/defrost the required amount of pollen and place it in an Eppendorf tube.
2. Close the tube and shake it to evenly distribute the pollen.
3. Assemble a sealed humidity chamber using a container with a lid that will comfortably accommodate the rack holding the tube with pollen. Line the inside of the container and/or lid with filter paper and moisten it with distilled water (*see* **Note 5**).
4. Place the open tube inside the humidity chamber. Humidify tobacco or *Streptocarpus* pollen for 2 h at 25 °C.

3.3 Pollen Germination in Liquid Medium

1. Add germination medium at room temperature to the tube containing the humidified pollen, leaving at least 1/5 of the tube empty, as air is required for pollen germination.
2. Close the tube and shake well to distribute the pollen evenly throughout the medium. Place the closed tube horizontally on a shaker set on gentle agitation at 25 °C. Leave the sample on the shaker for the required time: 1 h for *Nicotiana tabacum*, 4 h for *Streptocarpus*, 12 h for *Picea* (*see* **Note 6**).

3.4 Pollen Germination in Agarose Pads

1. To prepare an agarose suspension, put some humidified pollen onto an 18 × 18 or 24 × 24 cover slip, depending on the intended purpose, with a spatula, distributed evenly.
2. Pour a low-melting agarose solution on pollen, keeping the temperature no higher than 35 °C. If the agarose is too hot, it is best to wait (*see* **Note 7**).
3. Cover with a second cover slip and let it solidify for 5 min.
4. Then remove one or both coverslips and place the pad in the small Petri dish.
5. Pour 1–2 mL of germination medium on top. Place in a thermostat for tubes to grow.

3.5 Evaluation of ROS Generation in Pollen Grains and Pollen Tubes Using Fluorescence Microscopy

Dyes are added to the suspension from stock solutions, after which the suspension (or agarose pads) should be stored in the dark (*see* **Note 8**). An example of a pollen tube stained with PFBSF can be seen in Fig. 1.

1. Add the dye to the sample containing germinated pollen to a final concentration of 25 μM ($DCFH_2$-DA) or 5 μM (PFBSF).

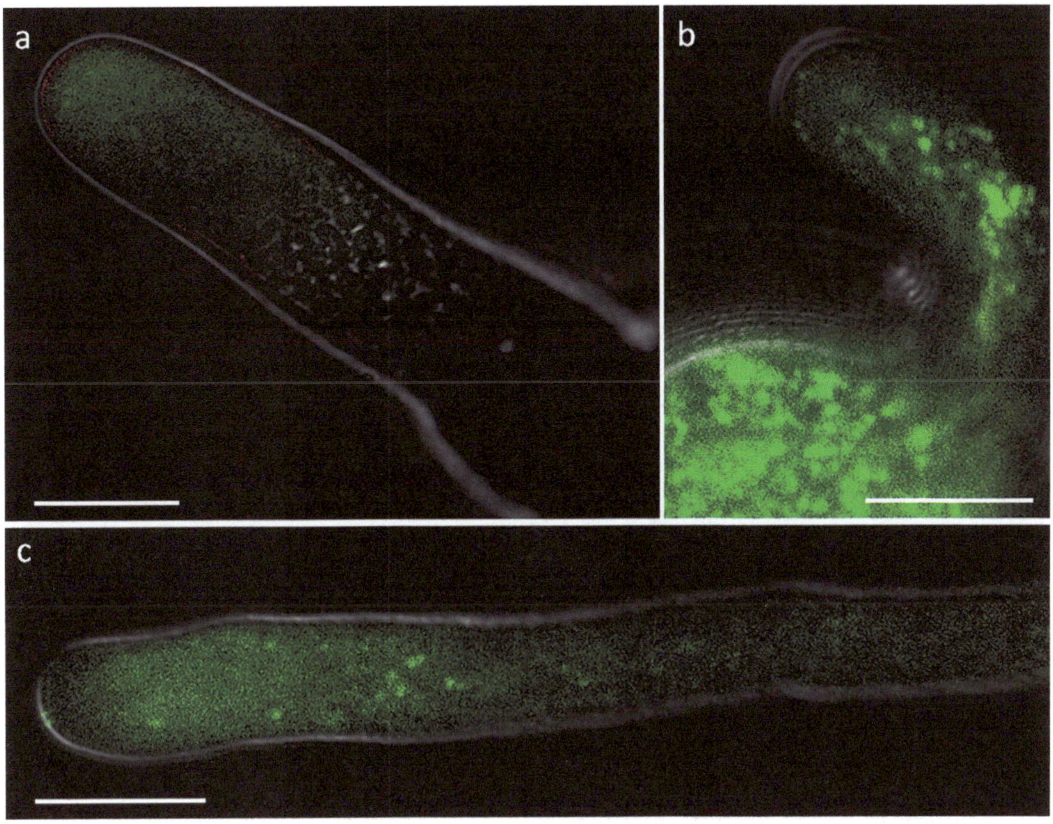

Fig. 1 Pollen tubes of *Lilium longiflorum* (**a**) and *Nicotiana tabacum* (**b**, **c**) stained with PFSBF (5 μM). Fluorescence was excited with a 488 nm laser and combined with DIC images. Imaging was performed with a confocal Nikon A1 microscope at the White Sea Biological Station of Moscow State University. Scale bar: 10 μm

2. Mix the sample and incubate for 15 min at 25 °C on a shaker at medium speed.
3. For angiosperm pollen, centrifuge the suspension at low speed and take a sample from the sediment for microscopy (*see* **Note 9**).
4. Place the suspension sample (approximately 20 μL) on a glass slide and cover with a coverslip.
5. If staining agarose pads, remove the pad from the dye solution and place it on a glass slide, and cover with a coverslip if needed.
6. Excite fluorescence with light at a wavelength of 485 ± 20 nm.
7. Record the results immediately after fluorescence excitation, as fluorescence intensity may decrease (or increase) with time (*see* **Note 10**).

3.6 Spectrophotometric Method for Determining the Amount of ROS Released by Pollen

1. Prepare the working medium (WM) for staining. To do this, mix the germination medium with OxyBURST Green H_2HFF BSA stock solution (100:1).
2. Use the resulting medium as pollen germination medium (*see* Subheading 3.3): mix the sample with an aliquot of the WM (for example, 3.9 mg of spruce pollen +650 μL) and incubate for the required time.
3. To test the involvement of NADPH oxidase in ROS generation, add an inhibitor (diphenylene iodonium, DPI) to the test samples before incubation. You can also check the specificity of the dye by adding ascorbic acid, Mn-5,10,15,20-tetrakis (1-methyl-4-pyridyl)21H, 23H-porphin (MnTMPP), and other ROS quenchers.
4. Remove the pollen from the suspension by filtration using a syringe and a Swinnex filter holder (nylon filter with a mesh size 10–40 μm) or by centrifugation (not recommended for saccate gymnosperm pollen).
5. Transfer the resulting solution to a spectrofluorimeter cuvette and record the spectrum at 488 nm excitation and emission wavelengths 500–600 nm. As the dye peaks at 530 nm, this wavelength can be used to obtain final fluorescence values (Fig. 2).

3.7 EPR-Based Determination of ROS Released by Pollen

1. Prepare the sample according to the following scheme: 290 μL germination medium +2 μL EDTA solution ($C = 20$ mM) + 8 μL CAT1H stock solution ($C = 20$ mM) or 190 μL germination medium +10 μL DEPMPO stock solution ($C = 100$ mM).
2. Incubate the pollen as usual, but with a spin trap or spin probe. It can be added either at the beginning of incubation or 30 min (CAT1H) to 1 h (DEPMPO) before the end (*see* **Note 11**).
3. Remove pollen from the suspension after incubation by filtration using a syringe and a Swinnex filter holder (nylon filter with a mesh size of 10–40 μm) or mild centrifugation.
4. Fill the measuring cuvette with the resulting solution.
5. Record EPR spectra at a temperature of 21–22 °C, microwave power of 20 mW, and a time constant of 0.1 s (*see* **Note 12**).
6. Assess the oxidative status of the medium or superoxide concentration based on the intensity of the central component (I) of the CAT1H or DEPMPO EPR spectra, respectively (Fig. 3).

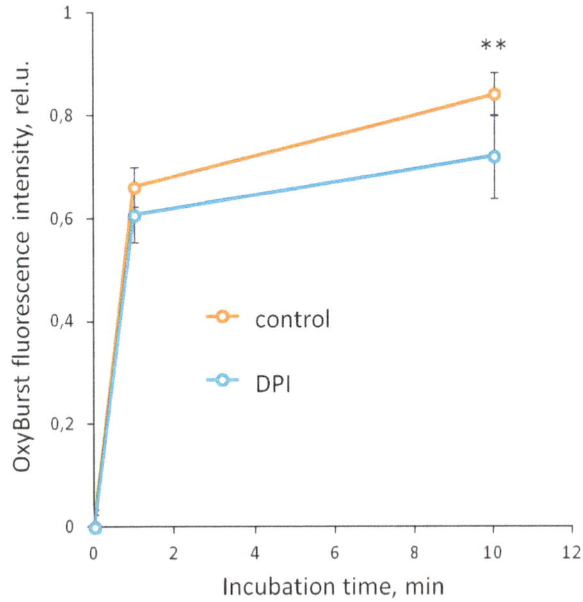

Fig. 2 An example of the use of a BSA-linked dye OxyBURST Green, to detect extracellular ROS. A significant increase in fluorescence was demonstrated in the 0–1 min interval, with this increase continuing for 1–10 min in the presence of *Picea pungens* pollen. In the presence of an NADPH oxidase inhibitor DPI, we observed an initial increase in ROS, apparently associated with release from the cell wall; however, the 1- and 10-min points do not differ significantly. Apparently, at this stage, NADPH oxidase is already involved in the process

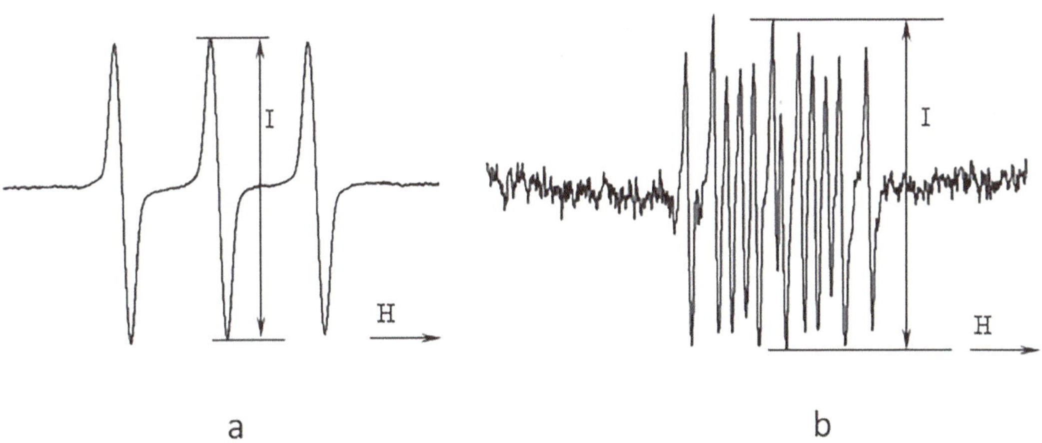

Fig. 3 Typical EPR spectrum of CAT1H (**a**) and DEPMPO (**b**) in an aqueous solution in the presence of ROS and the I measurement scheme. H shows the direction of the magnetic field

4 Notes

1. Medium containing sucrose can be stored at −20 °C for a long time, while at 4 °C it can be stored for about a week. Medium without sucrose can be stored at 4 °C for months, with occasional pH checks.
2. Low-melting agarose is prepared using germination medium as the base for pollen tube growth. We used 1.5% agarose; it requires heating to dissolve the powder, and then, before each use, it should be heated again until it liquefies.
3. For angiosperms, moistening the pollen before germination will increase the percentage of germinated pollen and improve the staining results. For gymnosperms, this preparation does not affect pollen germination success.
4. The agarose pad method is suitable for both standard and inverted microscopes, as agarose pads can be positioned upside down or up, depending on the position of the objective lens.
5. Ensure the entire surface of the filter paper is wet, but avoid excess water, as this could cause droplets to fall into the test tube.
6. The best results of germination success were found at 2 mg pollen/ml; this concentration can be used as a guideline, but there is a broad optimum.
7. We used a skin test (a drop on the skin of the hand) to avoid overheating the pollen, which would prevent germination.
8. To keep the suspension (or agarose pad) in the dark, one can wrap the test tube/Petri dish in foil.
9. For saccate conifer pollen, even gentle centrifugation damages pollen tubes and is ineffective, so it is best to let the tube stand for 1–5 min and then take a sample of floating pollen from the surface. It is important to use a wide-ended pipette tip.
10. It is better to use the shutter to avoid excessive illumination of the preparation area.
11. The incubation time with spin reagents was selected differently, since the spin probe (CAT1H) is more sensitive to low concentrations of ROS and gives a good signal even at short times. To see the DEPMPO signal, we had to incubate for an hour, since the trap is selective for superoxide radical.
12. To obtain a well-resolved DEPMPO EPR spectrum, signal accumulation is necessary. In our published results, the presented DEPMPO spectra are the result of averaging approximately 20 primary spectra.

References

1. Hiscock SJ, Bright J, McInnis SM, Desikan R, Hancock JT (2007) Signaling on the stigma. Potential new roles for ROS and NO in plant cell signaling. Plant Signal Behav 2:23–24. https://doi.org/10.1111/j.1469-8137.2006.01875.x
2. Kolbert Z, Petô A, Lehotai N, Feigl G, Ördög A, Erdei L (2012) In vivo and in vitro studies on fluorophore-specificity. Acta Biol Szegediensis 56:37–41
3. Smirnova AV, Timofeyev KN, Breygina MA, Matveyeva NP, Yermakov IP (2012) Antioxidant properties of the pollen exine polymer matrix. Biophysics (Oxf) 57:174–178. https://doi.org/10.1134/S0006350912020224
4. Keston AS, Brandt R (1965) The fluorometric analysis of ultramicro quantities of hydrogen peroxide. Anal Biochem 11:1–5. https://doi.org/10.1016/0003-2697(65)90034-5
5. Breygina M, Luneva O, Babushkina K, Kushunina M, Lazareva N, Klimenko E (2025) Which reactive oxygen species are produced on wet stigmas of different monocot plants? Plant Growth Regul. https://doi.org/10.1007/s10725-025-01330-6
6. Smirnova AV, Matveyeva NP, Polesskaya OG, Yermakov IP (2009) Generation of reactive oxygen species during pollen grain germination. Russ J Dev Biol 40:345–353. https://doi.org/10.1134/S1062360409060034
7. Maeda H, Fukuyasu Y, Yoshida S, Fukuda M, Saeki K, Matsuno H, Yamauchi Y, Yoshida K, Hirata K, Miyamoto K (2004) Fluorescent probes for hydrogen peroxide based on a non-oxidative mechanism. Angew Chem Int Ed 43:2389–2391. https://doi.org/10.1002/anie.200452381
8. Maksimov N, Evmenyeva A, Breygina M, Yermakov I (2018) The role of reactive oxygen species in pollen germination in *Picea pungens* (blue spruce). Plant Reprod 18:761–767. https://doi.org/10.1007/s00497-018-0335-4
9. Steffen-Heins A, Steffens B (2015) EPR spectroscopy and its use in planta-a promising technique to disentangle the origin of specific ROS. Front Environ Sci 3:1–6. https://doi.org/10.3389/fenvs.2015.00015
10. Suzen S, Gurer-Orhan H, Saso L (2017) Detection of reactive oxygen and nitrogen species by electron paramagnetic resonance (EPR) technique. Molecules 22. https://doi.org/10.3390/MOLECULES22010181
11. Menezes FAF, Oliveira JG, Guimarães AO (2024) Electron paramagnetic resonance applied to free radicals and reactive oxygen species detection in plant systems. Appl Magn Reson 55:335–355. https://doi.org/10.1007/S00723-023-01625-9
12. Hawkins CL, Davies MJ (2014) Detection and characterisation of radicals in biological materials using EPR methodology. Biochim Biophys Acta 1840:708–721. https://doi.org/10.1016/j.bbagen.2013.03.034
13. Zhu H (2016) Assays for detecting biological superoxide. React Oxygen Species 1:65–80. https://doi.org/10.20455/ros.2016.813
14. Dikalov SI, Kirilyuk IA, Voinov M, Grigor'Ev IA (2011) EPR detection of cellular and mitochondrial superoxide using cyclic hydroxylamines. Free Radic Res 45:417–430. https://doi.org/10.3109/10715762.2010.540242
15. Dikalov SI, Polienko YF, Kirilyuk I (2018) Electron paramagnetic resonance measurements of reactive oxygen species by cyclic hydroxylamine spin probes. Antioxid Redox Signal 28:1433–1443. https://doi.org/10.1089/ars.2017.7396
16. Tushabe D, Rosbakh S (2021) A compendium of in vitro germination media for pollen research. Front Plant Sci 12. https://doi.org/10.3389/fpls.2021.709945
17. Brewbaker JL, Kwack BH (1963) The essential role of calcium ion in pollen germination and pollen tube growth. Am J Bot 50:859. https://doi.org/10.2307/2439772
18. Baptiste FJ, Fang JY (2023) Study on pollen viability and stigma receptivity throughout the flowering period in the selected taxa of the Gesneriaceae family. Folia Hortic 35:123–133. https://doi.org/10.2478/fhort-2023-0009
19. Breygina M, Maksimov N, Polevova S, Evmenyeva A (2019) Bipolar pollen germination in blue spruce (*Picea pungens*). Protoplasma 256:941–949. https://doi.org/10.1007/s00709-018-01333-3

Chapter 11

Quantitative Phenotyping of Ion Fluxes in Tip-Growing Cells Under Varying Growth Regimes: A Data Analysis Protocol

Maria Teresa Portes, Daniel Santa Cruz Damineli, and José A. Feijó

Abstract

Tip-growing cells exhibit complex growth regimes in vitro, alternating between growing and non-growing intervals, oscillatory or more steady behavior. In *Arabidopsis thaliana*, pollen tube growth arrest is often accompanied by spiking behavior in intracellular ion concentrations and extracellular ion fluxes. Thus, selecting comparable growing regimes is critical for quantifying ion dynamics across cells and genotypes. Defining non-growing regimes is a fundamental step to filter out their associated data points. Here, we provide computational and statistical procedures for the quantitative phenotyping of ion fluxes associated with growth dynamics in tip-growing cells. The goal is to provide reliable estimates of ion fluxes given pairwise time-series comparisons (ion fluxes *vs.* growth rate), focused on growth-associated intervals and avoiding data stemming from regimes associated with growth arrest. We consider extracellular ion fluxes in growing tubes, but the analysis is applicable to other quantitative variables and tip-growing cells. After visualizing both series in a common timeframe, we extract the growth rate baseline and determine the non-growing regime threshold with a Gaussian Mixture Model, then predict the growth state at sampled flux times, filtering, and finally quantification. The protocol is presented in R but is of general use, since multiple software routines can yield similar results.

Key words Tip growth, Pollen tubes, Ion fluxes, Oscillations, Growth rate, Time-series analysis, R scripts, Quantitative phenotyping

1 Introduction

Pollen tubes are tip-growing cells that highly depend on ion dynamics, which are associated with other cellular processes underlying tube elongation and the chemotropic response, culminating in sperm cell delivery for double fertilization. Combined with such dynamics, pollen tube tip growth in vitro shows a variety of growth regimes, including oscillations in growth rate, followed by oscillations in extracellular ion fluxes and intracellular ion concentrations [1, 2]. Pollen tubes are also robust and highly adequate cellular

Maria Teresa Portes and Daniel Santa Cruz Damineli contributed equally to the chapter.

Giampiero Cai et al. (eds.), *Pollen and Pollen Tube Analysis: Methods and Protocols*, Methods in Molecular Biology, vol. 3042, https://doi.org/10.1007/978-1-0716-5308-1_11,

models for studying oscillatory growth dynamics driven by ions and other cellular processes. They generate long nonlinear time-series, which must be properly analyzed to allow phenotypic identification and/or pharmacological characterization of responses. Thus, time-series analysis requires quantitative frameworks for comparisons across species, genotypes, and conditions, and enables further quantifications of oscillatory parameters such as frequency, amplitude, and phase, as we have demonstrated previously in Damineli et al. [2].

Just like other tip-growing cells, pollen tubes show complex growth regimes, including growing *vs.* non-growing; oscillatory *vs.* steady, and complex *vs.* simple oscillations. Pollen tubes under growth arrest often show spiking behavior, seen in intracellular ion concentrations and extracellular ion fluxes [2], but these may also disappear. Thus, to yield comparable results, one must avoid these time points where the growth rate falls below a given threshold. Here, we offer computational and statistical procedures to analyze each pair of time-series (ion flux *vs.* growth rate) with a comparable approach, focused on growth-associated intervals. The goal is to improve the estimates of extracellular ion fluxes, which provides a sensitive and quantitative phenotyping tool, capable of distinguishing genotypes lacking evident morphological phenotypes [3, 4]. However, the same approach works for other quantitative variables that depend on the growth regime.

Often, there are two main technical issues with pairs of ion flux and image data: (1) assigning a common timeframe; (2) filtering data by growth regime. Aiming to solve these problems, we developed the present protocol for quantitative analysis of these processes. Despite being of general use, here the protocol will consider time-series of extracellular ion fluxes (e.g., Ca^{2+}, H^{+}, Cl^{-}) acquired by ion-selective electrodes, simultaneously recorded with tip growth rate (estimated from tip position or length) in pollen tubes or root hairs. Given its popularity and widespread usage, we use the programming language R [5] for data manipulation, visualization, and analysis. However, multiple other software options could yield similar results.

2 Materials

2.1 Biological Samples & Instrumentation

Germinate the collected pollen grains in vitro using an optimized liquid medium with reduced ion concentrations. This will decrease the ionic background when measuring extracellular ion fluxes (recorded by scanning ion-selective electrode technique—SIET) or when estimating intracellular ion concentrations using genetically encoded probes simultaneously with tip growth [3, 4]. Perform quantitative measurements and imaging acquisitions in pollen tubes that are at least 250–300 mm long. Given the variability in growth dynamics, the sampling size should be considerably high to ensure a comparable set of time-series per dataset, which is also relevant for statistical analysis.

Here, we are going to use as an example a pollen tube from *Arabidopsis thaliana* growing in vitro, used for measuring extracellular ion fluxes using SIET and simultaneous bright-field imaging acquisition, following the protocol described by Portes & Feijó [3].

2.2 Software

R (> = 4.3) and RStudio, with the mix tools package installed [6].

3 Methods

3.1 Acquisition and Synchronization

1. Record extracellular ion fluxes and pollen tube growth with a shared clock, so that their time values correspond to a common frame in the same time units (*see* **Note 1**).
2. Open the flux data. Here, extracellular H^+ fluxes were recorded at approximately 4 s intervals with the probe position aligned with the growing tip (*see* **Note 2**).

```
# Read a csv file with time in the first and flux in the second column

flux <- read.csv("~/Documents/MyData/FluxData.csv")


# Look at the beginning of the file

head(flux)

##        Time      Flux

## 1  0.000000 1.4311445

## 2  3.996000 4.7141375

## 3  7.956996 5.8250800

## 4 11.989998 3.4029830

## 5 15.957000 2.3832419

## 6 19.998996 0.8571556
```

3. Open the growth rate data. Here, bright-field images were recorded at 2 s intervals and growth rate estimated with MetaMorph's template-matching algorithm. However, this data may have been acquired by a variety of methods. Bright-field imaging tracking can be performed using a template-matching algorithm on ImageJ or commercial tools, such as the extinct MetaMorph [7], or using AI-assisted tools such as DeepLabCut [8]. For fluorescence imaging, when a fluorescent reporter allows to determine the tip location, one can also use the same tracking tools or estimate the tip location from kymographs. In kymographs, fluorescence threshold methods are commonly used, but one can resort to higher resolution methods such as the subpixel tip detection algorithm in CHUKNORRIS [2].

```
# Read a csv file with time in the first and growth rate in the second column

growth <- read.csv("~/Documents/MyData/GrowthData.csv", header = TRUE)


# Look at the beginning of the file

head(growth)

## Time Growth

## 1    2 0.3059926

## 2    4 0.1343753

## 3    6 0.2022292

## 4    8 0.2442763

## 5   10 0.1171459

## 6   12 0.2550594
```

3.2 Preprocessing and Quality Control

1. Plot the data and ensure that the acquisition times of the ion flux and growth rate match and are suitable for analysis (Fig. 1; *see* **Note 3**).

```
# Get the maximum x-axis range considering both files
time.range <- range(c(flux[, 1], growth[, 1]))

# Make room for axis label on the right
par(mar = c(5, 4, 4, 5) + 0.1)

# Plot flux data
plot(flux,
  type = "o", pch = 16, cex = 0.35, col = "#CC79A7",
  xlim = time.range)

# Color left axis tick labels and title
axis(2, col = "#CC79A7", col.axis = "#CC79A7")
mtext("Flux", side = 2, line = 3, col = "#CC79A7")

# Plot growth data on the same plot with the same x-axis
par(new = TRUE)
plot(growth,
  type = "o", pch = 16, cex = 0.35, col = "#009E73",
  xlim = time.range,
  xaxt = "n", yaxt = "n", ylab = "")

# Right axis with matching color
axis(4, col = "#009E73", col.axis = "#009E73")
mtext("Growth", side = 4, line = 3, col = "#009E73")
```

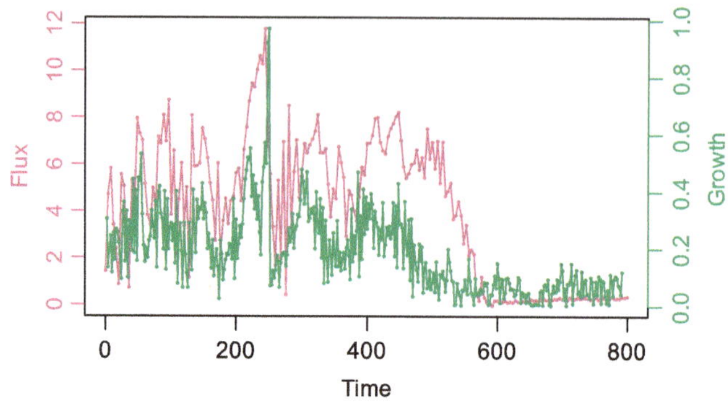

Fig. 1 Overlay of ion flux (red) and growth rate (green) time series

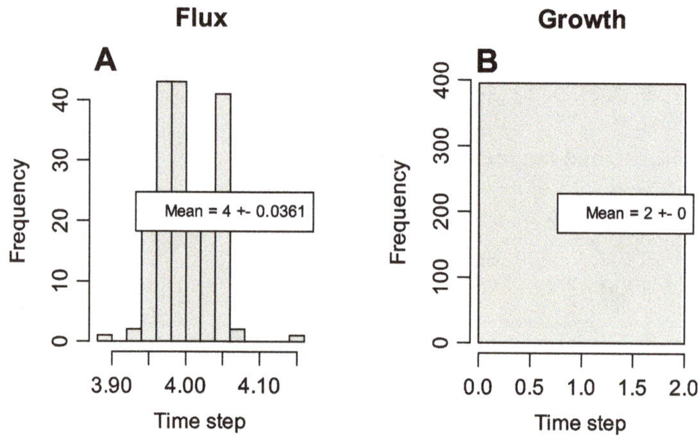

Fig. 2 Distribution of time steps for flux (**a**) and growth (**b**) series

2. Check both series for gross outliers and consider whether the quality of the data is sufficient for the analysis.
3. Check the time intervals between samples for each variable (Fig. 2). Make sure the precision digits shown in the legend make sense for your particular case. In general, extracellular flux series are not sampled at perfectly regular intervals. This usually requires some form of interpolation to continue the analysis. Analyzing oscillations requires regularly sampled data; however, the objective here is to obtain the most reliable estimate of average flux values. Thus, we will proceed differently and use the actual times when flux was measured as a reference (assigning it to the variable "sampled.times").

```
# Get flux time steps

tstep.flux <- diff(flux[, 1])

# Get growth time steps

tstep.growth <- diff(growth[, 1])


# Plot the distribution of flux time steps

par(mfrow = c(1, 2))

hist(tstep.flux, main = "Flux", xlab = "Time step")

legend("right", legend = paste("Mean =", round(mean(tstep.flux), digits = 4), "+-",

round(sd(tstep.flux), digits = 4)), cex = 0.75, bg = "white")


# Plot the distribution of growth time steps

hist(tstep.growth, main = "Growth", xlab = "Time step")

legend("right", legend = paste("Mean =", round(mean(tstep.growth), digits = 4), "+-",

round(sd(tstep.growth), digits = 4)), cex = 0.75, bg = "white")

# Establish reference times for later, here when flux was measured

sampled.times <- flux[, 1]
```

3.3 Determining Growth Regime

1. Smooth the growth series to establish a baseline (Fig. 3). This will enable you to choose an adequate threshold for determining growing and non-growing time points. This may be challenging since the growth rate is noisy, so adjust the amount of smoothing as necessary.

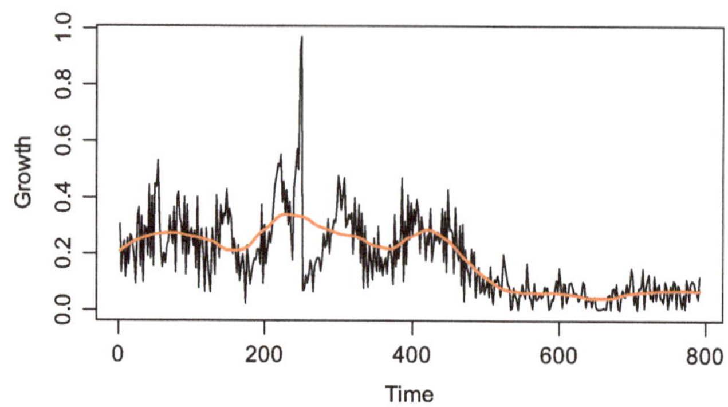

Fig. 3 Smoothed growth rate series (coral) over original data (black), yielding the growth rate baseline

```
# Smooth growth series with loess, choosing span between 0 and 1
growth.smooth <- loess(growth[, 2] ~ growth[, 1],
                degree = 2,
                span = 0.25) # span is the fraction of the points used
                        # in each window of the local polynomial
                        # fit, so larger values makes it smoother

# Plot original series and smoothed series
plot(growth, type="l")
lines(growth.smooth$x, growth.smooth$fitted, col = "coral", lwd = 2)
```

2. Set the growth threshold using a Gaussian Mixture Model of time points (Fig. 4). Pooled data from all smooth growth series would allow for a more precise approach (*see* **Note 4**).

```
# Get smoothed growth values to analyze the baseline distribution
growth.baseline <- c(growth.smooth$fitted)


# Perform a Gaussian Mixture Model (GMM) of two components
# to separate growing vs non-growing time points
library(mixtools)

## mixtools package, version 2.0.0.1, Released 2022-12-04
## This package is based upon work supported by the National Science Foundation under
Grant No. SES-0518772 and the Chan Zuckerberg Initiative: Essential Open Source Software
for Science (Grant No. 2020-255193).

set.seed(123)
gr.gmm <- normalmixEM(growth.baseline)

## number of iterations= 23

# Identify the two main distributions (lowest and highest means)
i = sort(gr.gmm$mu, index.return = TRUE)$ix[1]
j = sort(gr.gmm$mu, index.return = TRUE)$ix[2]
```

```
# Use intersect to define boundaries between the two distributions

intrsct <- uniroot(function(x) gr.gmm$lambda[i]*dnorm(x, mean = gr.gmm$mu[i],
                                  sd = gr.gmm$sigma[i],
                                  log = FALSE) -
                        gr.gmm$lambda[j]*dnorm(x, mean = gr.gmm$mu[j],
                                  sd = gr.gmm$sigma[j],
                                  log = FALSE),
            c(gr.gmm$mu[i], gr.gmm$mu[j]))

# Get the threshold

thrsh <- intrsct$root

# Plot the GMM

plot(gr.gmm, whichplots = 2)

abline(v = thrsh, col = "magenta", lwd = 3, lty = "dotdash")
```

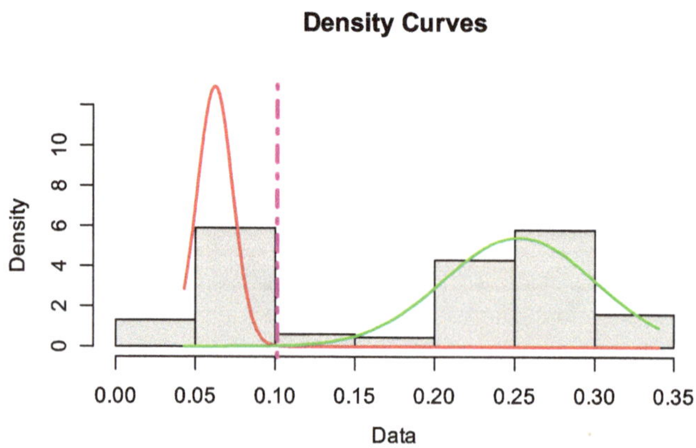

Fig. 4 Gaussian Mixture Model of the growth rate baseline with the calculated threshold (magenta dashed line)

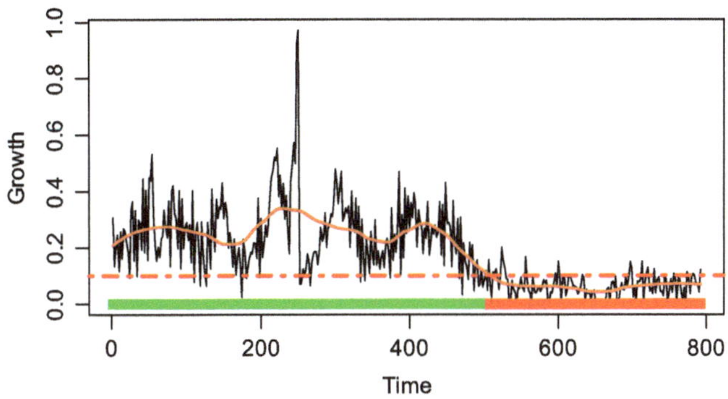

Fig. 5 Smoothed growth series with growth threshold (red dashed line). Green points indicate growing time points and red points indicate non-growing time points

3. Visualize the threshold on the smoothed growth series (Fig. 5). Here, coincidentally all the points with low growth occur at the end of the series. However, this protocol is sufficiently general to handle intermittent growth arrest.

```
plot(growth, type="l")

lines(growth.smooth$x, growth.smooth$fitted, col = "coral", lwd = 2)

abline(h = thrsh, col = "red", lty = "dotdash", lwd = 2.5)


# Visualize included points

growing <- which(growth.baseline >= thrsh)

points(growth.smooth$x[growing], rep(0, length(growing)),

    pch = 15, col = "green")


# Visualize excluded points

stopped <- which(growth.baseline < thrsh)

points(growth.smooth$x[stopped], rep(0, length(stopped)),

    pch = 15, col = "red")
```

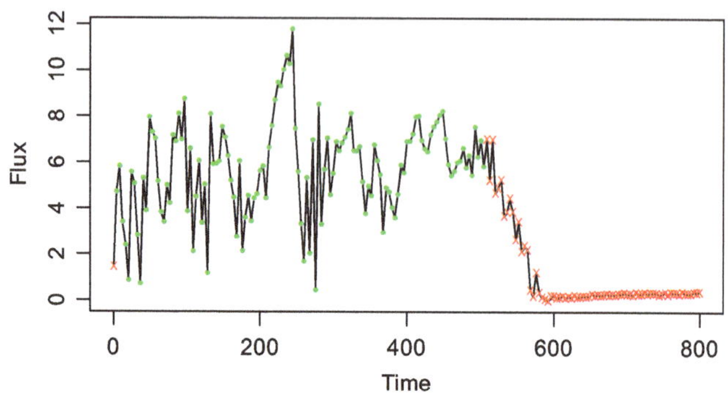

Fig. 6 Flux series with growing time points indicated in green circles and non-growing time points indicated in red Xs

4. Use the reference times at which flux was directly measured ("sampled.times") to estimate growth at those precise time points (Fig. 6). Since the threshold was calculated in the growth series, we now need to determine which flux time points correspond to growing intervals.

```
# Predict growth baseline (loess model) at flux sampled times

growth.at.flux <- predict(growth.smooth, sampled.times)

# Determine which predicted growth baseline is above the threshold

growing.times <- which(growth.at.flux >= thrsh)


# Visualize growing time points on flux series

plot(flux, type="l")

points(flux[growing.times, 1], flux[growing.times, 2],

    pch = 16, col = "green", cex = 0.5)

points(flux[-growing.times, 1], flux[-growing.times, 2],

    pch = "X", col = "red", cex = 0.5)
```

5. Finally, filter the flux series to retain only the relevant time points (Fig. 7). As we can see, filtering greatly increases the mean flux value and decreases the standard deviation.

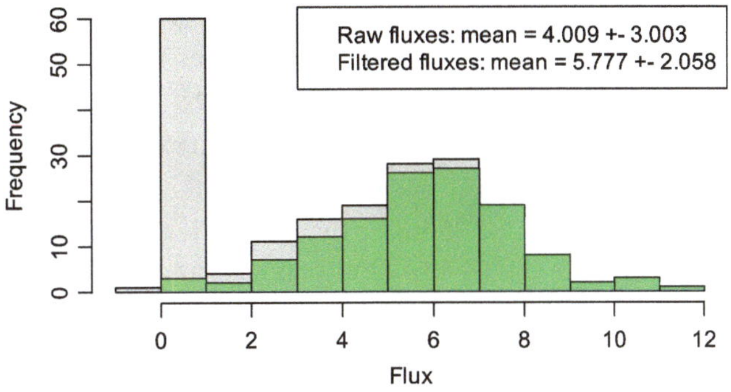

Fig. 7 Histogram of raw fluxes (black) and filtered fluxes (green)

```
flux.with.growth <- flux[growing.times, 2]


h0 <- hist(flux[, 2], main = "", xlab = "Flux")
hist(flux.with.growth, add = T, col = adjustcolor("limegreen", alpha.f = 0.5), breaks =
h0$breaks)
legend("topright",
   legend = c(paste("Raw fluxes: mean =",
              round(mean(flux[, 2]), digits = 3),
              "+-",
              round(sd(flux[, 2]), digits = 3)),
          paste("Filtered fluxes: mean =",
              round(mean(flux.with.growth), digits = 3),
              "+-",
              round(sd(flux.with.growth), digits = 3))
          ))
```

3.4 Genotype-Level Statistics

1. Apply the above steps to all cells in each genotype, storing per-cell summaries (e.g., mean flux during growth) in a data frame (*see* **Note 5**).
2. Perform an adequate statistical test to compare means between groups (e.g., t-test, ANOVA, etc.);
3. Filter by growth regime. This is especially useful since there is spiking behavior and intermittent growth arrest.

3.5 Reporting Standards

1. Include the following in your report: the estimated growth threshold, sampling intervals, the Loess span and degree, excluded fractions, per-cell and per-genotype tables and code for reproducibility.

4 Notes

1. We refrain from using units here, since they will be specific to each experimental setup. However, make sure to use consistent units throughout the analysis.
2. Here we assume clean data, without NAs, adequate units, and stored in comma-separated values (CSV).
3. If time steps differ between flux and growth series, resampling both to a common time step will be essential. Beware that if the sampling rate is too discrepant, it may not make sense to perform a combined analysis. Other requirements in terms of sampling rate will apply if the goal is to analyze oscillations.
4. In this particular example, the distribution of growing vs. growth-arrested time points worked beautifully, but this depends on the behavior of each cell. Thus, the best approach is to pool the data of the smoothed growth series (baseline) of the entire population.
5. Remember that the growth threshold defined using the smoothed growth series from the entire population is more robust than using per-cell thresholds. Per-cell thresholds assume your cell has both growing and non-growing intervals, and during sufficient time points, which may not be the case.

Acknowledgments

MTP was supported by FAPESP (Fundação de Apoio à Pesquisa do Estado de São Paulo 2019/26129-6; 2021/09025-2); JAF was supported by the National Science Foundation (MCB-1930165) and the National Institute of Health (1R35GM13781524).

References

1. Portes MT, Damineli DSC, Moreno N, Colaço R, Costa S, Feijó JA (2015) The pollen tube oscillator: integrating biophysics and biochemistry into cellular growth and morphogenesis. In: Mancuso S, Shabala S (eds) Rhythms in plants: dynamic responses in a dynamic environment. Springer, Cham, pp 121–156
2. Damineli DSC, Portes MT, Feijó JA (2017) Oscillatory signatures underlie growth regimes in Arabidopsis pollen tubes: computational methods to estimate tip location, periodicity, and synchronization in growing cells. J Exp Bot 68(12):3267–3281
3. Portes MT, Feijó JA (2021) Measuring extracellular proton and anionic fluxes in *Arabidopsis* pollen tubes. Bio-protocol 11(3):e3908
4. Portes MT, Damineli DSC, Feijó JA (2021) Spatiotemporal quantification of cytosolic pH in *Arabidopsis* pollen tubes. Bio-protocol 11(14):e4084
5. R Core Team (2024) R: a language and environment for statistical computing. R Foundation for Statistical Computing, Vienna. https://www.R-project.org/
6. Benaglia T, Chauveau D, Hunter DR, Young D (2009) Mixtools: an R package for analyzing finite mixture models. J Stat Softw 32(6):1–29
7. Molecular Devices (2020) MetaMorph microscopy automation and image analysis software [Software manual]. Molecular Devices, LLC
8. Mathis A, Mamidanna P, Cury KM, Abe T, Murthy VN, Weygandt MM, Bethge M (2018) DeepLabCut: markerless pose estimation of user-defined body parts with deep learning. Nat Neurosci 21(9):1281–1289

Chapter 12

Quantitatively Analyzing the Structural Organization at the Tip of Growing Tobacco Pollen Tubes

Carolin Fritz and Benedikt Kost

Abstract

Rapid, unidirectional pollen tube tip growth is a widely employed model for polar cell expansion, which is essential for plant morphogenesis. At the pollen tube tip, various proteins and lipids playing pivotal roles in regulating polar cell expansion are associated with distinct domains of the plasma membrane (PM) or accumulate within different cytoplasmic regions. These PM domains and cytoplasmic regions must be dynamically maintained during tip growth, which is contingent on massive secretory and endocytic membrane trafficking. The molecular and cellular mechanisms responsible for PM and cytoplasmic compartmentalization at the pollen tube tip are not well understood. To further characterize these mechanisms, standardized methods are required, which enable quantitative analysis of the structural organization of normally growing *Nicotiana tabacum* (tobacco) pollen tube tips, and enhance the comparability of data generated in independent studies. The application of stably transformed marker lines greatly facilitates quantitative structural analyses and enhances the reliability of results obtained. This chapter describes methods required for the generation of transgenic tobacco lines and for the expression level-independent quantitative characterization of PM domains and cytoplasmic regions at the pollen tube tip. Furthermore, a FM4-64 and brefeldin A-based method allowing quantitative analysis of endocytic lipid and protein recycling is discussed.

Key words Pollen tube, *Nicotiana tabacum*, Plasma membrane domains, Membrane recycling, Generation of transgenic tobacco lines, YFP tagging, Confocal microscopy, Fiji, Relative fluorescence intensity, BFA compartment

1 Introduction

Pollen tubes elongate rapidly and strictly unidirectionally by tip growth, an extreme form of polarized cell expansion [1], which is also displayed by root hairs [2] and apical cells of moss protonemata [3]. Pollen tube tip growth depends on massive secretion of cell wall material, which must be compensated by endocytic recycling of excess material delivered to the plasma membrane (PM) as a result of this process [4–6]. Consequently, pollen tube growth depends on precisely coordinated interplay between exo- and endocytosis

Giampiero Cai et al. (eds.), *Pollen and Pollen Tube Analysis: Methods and Protocols*, Methods in Molecular Biology, vol. 3042,
https://doi.org/10.1007/978-1-0716-5308-1_12,

[7, 8]. These two processes appear to occur within distinct PM domains at the pollen tube tip, between which PM components are subjected to drift [7, 8]. During tip growth, different proteins or lipids with important regulatory functions remain associated with distinct, clearly defined PM domains. Furthermore, the polarized positioning of cytoplasmic structures or regions displaying elevated ion concentrations needs to be maintained, which play key roles in tip growth and are in direct contact with different regions of the PM [8–14]. The remarkably pronounced cytoplasmic and PM compartmentalization of growing pollen tube tips clearly needs to be maintained dynamically, despite rapid PM turnover and drift of PM components.

Very little is currently known about the cellular and molecular mechanisms underlying the establishment and maintenance of different PM domains within the highly dynamic pollen tube PM. To support the identification and characterization of these mechanisms, a quantitative model of the structural organization of the tobacco pollen tube tip has recently been developed [14]. The model provides a map of the exact positions of different PM domains, which are enriched in key signaling proteins or lipids, or are in direct contact with cytoplasmic structures or regions playing important roles in apical membrane traffic. To generate this map, the intracellular distributions of eYFP (enhanced yellow fluorescent protein)-tagged tobacco pollen tube RAC/ROP signaling proteins, and of fluorescent fusion proteins serving as makers for lipids, Ca^{2+}, organelles, or F-actin, were quantitatively analyzed in transgenic tobacco pollen tubes, which elongated at normal rates (> 3.0 μm/min) and displayed no morphological defects [14].

This chapter describes methods required for the generation of transgenic lines expressing fluorescent marker proteins, which greatly facilitated the structural analyses summarized in the previous paragraph and helped generating reproducible data. Furthermore, methods enabling the determination of pollen tube growth rates are presented, which were systematically applied as a fundamental basis for all other quantitative analyses. Pollen tubes elongating at rates below 3.0 μm/min were excluded from these analyses (except for the investigation of brefeldin A [BFA]-treated pollen tubes) to ensure that data generated reflected structural organization during normal tip growth. In addition, methods used to quantify the start and endpoints of analyzed PM domains and to determine fluorescence intensity distributions along these domains are presented. These methods include background subtraction procedures, the application of an intensity-based LUT (lookup table) for image segmentation, the normalization of fluorescence intensity raw data, as well as distance measurements along the curved perimeter of pollen tube tips. To complement this toolbox, a method allowing the quantification of endocytic lipid and protein recycling is presented. This method is based on combining BFA treatment, which blocks TGN (trans Golgi network)exit [15, 16], with the visualization of lipid and protein dynamics using the lipid dye FM4-64 and different eYFP-fusion proteins. Digital image analyses discussed in this chapter rely on algorithms provided by the ImageJ version Fiji [14, 17].

The standardized procedures described in detail in this chapter enable comparative quantitative analysis of distinct PM domains and cytoplasmic regions at the tip of normally growing tobacco pollen tubes. These procedures were employed to develop a quantitative model of the structural organization at the tobacco pollen tube tip [14], and can be used to extend this model in future studies in other laboratories, which characterize additional lipid, protein (including markers for structures and organelles), or ion distributions within the pollen tube PM or cytoplasm.

2 Materials

2.1 Plants for Pollen Collection and for Agrobacterium-Mediated Transformation

1. Plants for pollen collection: Wild-type (WT) or transgenic tobacco (*Nicotiana tabacum* Petit Havana SR1) plants grown in soil (Hawita) from seedlings, which have emerged from soil-germinated seeds and were transferred to larger pots 2–3 weeks later. Seedlings and plants are maintained under long-day conditions (16 h light at 22 °C; 8 h darkness at 18 or 22 °C) in a growth chamber or in a greenhouse, which are set to a constant relative humidity of 60% (*see* **Notes 1** and **2**). Pollen harvested from mature flowers is best used to establish pollen tube cultures immediately after collection. Alternatively, pollen can be stored for later use. For this purpose, pollen-containing anthers are transferred into gelatin capsules (https://kapselwelt.de; *see* **Note 3**), frozen in liquid nitrogen, and stored at −80 °C.
2. Plants for Agrobacterium-mediated transformation: axenic tobacco plants grown from sterilized seeds in sterile glass jars (Weck; Fig. 1A) or in similar growth containers on half-strength Murashige Skoog (MS; Duchefa) medium [18] containing 2% (w/v) sucrose (Roth; *see* Subheading 3.1). Cultures are kept under long-day conditions in a growth chamber (16 h light at 22 °C; 8 h darkness at 18 or 22 °C) (*see* **Note 4**).

2.2 Materials for Nicotiana Tabacum Transformation

1. *Nicotiana tabacum* Petit Havana SR1 seeds.
2. 50-mL plastic tubes.
3. 500-mL glass beaker with 100-mL water.
4. Microwave oven.
5. Sterile glass jars (Weck; Fig. 1A) or similar growth containers suitable for plant tissue culture.
6. Sterile hood.

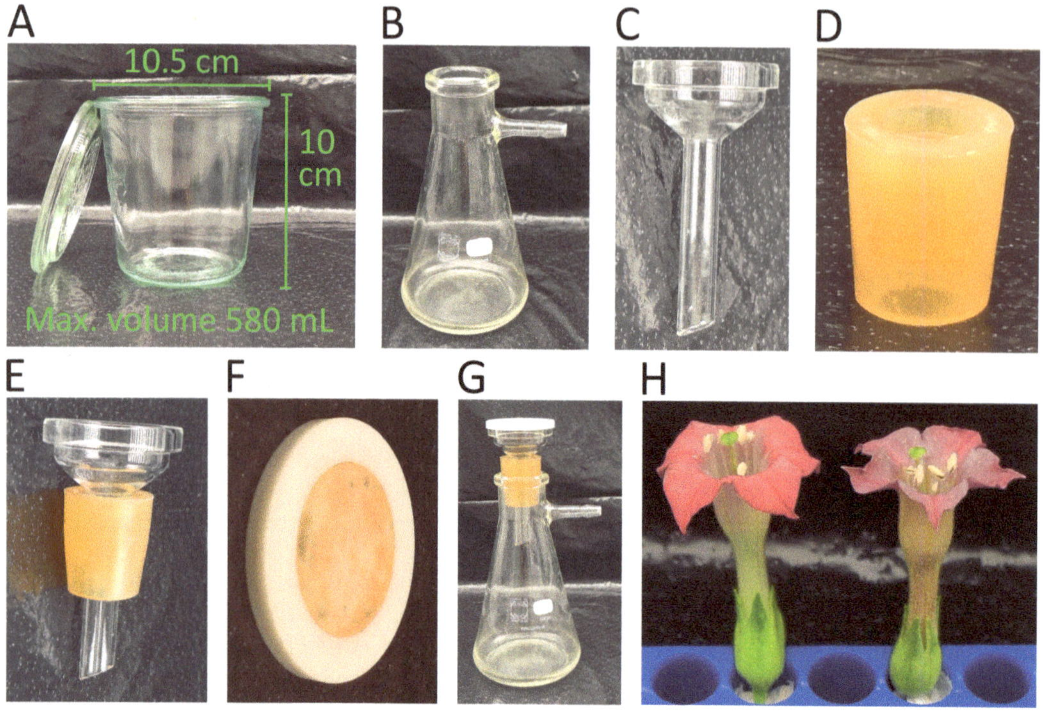

Fig. 1 Materials used for the cultivation of axenic tobacco plants and the preparation of pollen tube cultures. (**A**) Glass jar employed for axenic plant cultivation (size parameters indicated in green). The lid (shown on the left side) can be placed on top of the jar and sealed with micropore tape. (**B**) Conical Büchner flask. (**C**) GV050/0/02 glass frit with lower glass base. (**D**) Conical Büchner flask adapter. (**E**) Assembly of components shown in (**C**) and (**D**). (**F**) PTFE-centering ring GV 050/0/05 with ceramic filter in the center. (**G**) Complete assembly of components shown in (**B**), (**E**) and (**F**). (**H**) Mature *Nicotiana tabacum* flowers at the optimal stage for pollen collection (left) and after the onset of senescence (right)

7. Sterile scalpel.
8. Sterile tweezers.
9. Erlenmeyer flask containing transformed Agrobacterium in LB medium (Lysogeny Broth [19]; *see* **Note 5**).
10. Timer.
11. Petri dishes: different sizes (*see* **Note 6**).
12. Phyto agar (0.8% [w/v] for medium solidification; Duchefa).
13. Half-strength Murashige Skoog (MS) medium containing 2% (w/v) sucrose.
14. Aluminum foil.
15. Sugar-free full-strength MS medium.
16. MG I medium (full-strength MS medium containing 1.6% [w/v] glucose [Roth], 1 mg/L BA [6-Benzylamino-Purine; Sigma], 0.2 mg/L NAA [Alpha-Naphaleneacetic acid; Sigma], and 500 mg/L Ticarcillin [Duchefa]; *see* **Note 7**)

supplemented with the appropriate selective agent (corresponding to the resistance conferred by the selective marker gene used: e.g., 100 mg/L Kanamycin [Roth or Serva]).

17. Full-strength MS medium containing 2% (w/v) sucrose and 250 mg/L Ticarcillin.
18. Micropore tape.

2.3 Materials for the Preparation of PTNT Medium and Plates

Preparation (*see* Subheading 3.2) was performed essentially as previously described [20].

1. 50-mL plastic tubes.
2. 5.5-cm plastic Petri dish.
3. 5-mL serological pipettes.
4. Pipette boy.
5. Sterile hood.
6. pH meter.
7. KOH (Sigma) for pH adjustment.
8. 10× salt stock solution: 1.0 mM $CaCl_2$ (Roth), 1.0 mM KCl (Roth), 0.8 mM $MgSO_4$ (Roth), 1.6 mM H_3BO_4 (Sigma), 30 μM CuSO4 (Roth). Store at −20 °C (*see* **Note 8**).
9. 1.5% (w/v) casein acid hydrolysate (Merck; "amicase") stock solution. Store at −20 °C (*see* **Note 9**).
10. 1-L glass beaker.
11. Sucrose.
12. MES hydrate (Sigma).
13. PEG-6000 (Roth).
14. 20 mg/mL rifampicin (Duchefa) in methanol (Roth). Store at −20 °C.
15. 0.5% (w/v) phytagel (Sigma).
16. Microwave oven.
17. 300-mL glass beaker filled with water.
18. 2× PTNT medium: 10% (w/v) sucrose, 0.6% (w/v) MES hydrate, 0.06% (v/v) casein acid hydrolysate (1.5% stock solution), 10× salt stock solution (*see* **Step 8**): 5× diluted, 20 mg/mL rifampicin, 25% (w/v) PEG-6000. Adjust pH with KOH to 5.9 prior to adding PEG-6000. Filter sterilize and store at 4 °C (*see* **Note 10**).

2.4 Materials for Pollen Tube Cultures

Cultures were prepared essentially as previously described [20] with some modifications (*see* Subheading 3.2).

1. 50-mL plastic tubes.
2. PTNT plates (prepared as described below: *see* Subheading 3.2). Store at 4 °C.
3. 2× PTNT medium (prepared as described above: *see* Subheading 2.3). Store at 4 °C.
4. Sterile water (H_2O_{bidest}).
5. Pollen-containing anthers collected from mature tobacco flowers: fresh or stored within gelatin capsules at −80 °C (*see* Subheading 2.1).
6. Tweezers.
7. Filter sieve.
8. Vortex mixer.
9. Polyamide filter membrane (pore size 0.45 μm, diameter 47 mm; Sartorius).
10. Vacuum pump.
11. 5.5-cm plastic Petri dishes.
12. 500-mL Büchner flask (vacuum resistant; Fig. 1B+G).
13. GV 050/0/02 glass frit filter with lower glass base (Fig. 1C+E +G).
14. Conical Büchner flask adapter to seal the flask (Fig. 1D+E+G).
15. Whatman PTFE-centering ring GV 050/0/05 with ceramic filter in the center (Fig. 1F+G).
16. Parafilm.
17. Incubator (*see* **Note 11**).

2.5 Materials for FM4-64 Application and BFA Treatment

1. Pollen tube cultures on plates (prepared as described below: *see* Subheading 3.2).
2. Micropipettes and tips.
3. Liquid 1× PTNT medium prepared by diluting 2× PTNT medium (*see* Subheading 2.3) 1:1 (v/v) with sterile water (-H_2O_{bidest}). Store at 4 °C.
4. Dimethyl sulfoxide (DMSO; Roth).
5. 10 mM FM4-64 (ThermoFisher) in DMSO stock solution. Store at −20 °C.
6. 10 mM BFA (ThermoFisher) in DMSO stock solution. Store at −20 °C.

2.6 Imaging and Image Analysis

1. Slides and cover slips (22 × 22 mm; VWR; *see* **Note 12**).
2. Scalpel.
3. Timer.

4. Confocal microscope equipped with a 20× multi-immersion objective, a 63× oil- or water-immersion objective, and an argon laser allowing fluorophore excitation at different wavelengths: i.e., 515 nm (YFP [yellow fluorescent protein] and simultaneous YFP/FM4-64 imaging), and 488 nm (GFP [green fluorescent protein] and simultaneous GFP/FM4-64 imaging). The microscope also needs to be set up for the selective detection of fluorescence emission in different wavelength ranges (emission filters or spectral sensors with adjustable detection windows): i.e., 497–525 nm (GFP imaging), 525–575 nm (YFP imaging), and 650–790 nm (FM4-64 imaging).
5. Image acquisition software (i.e., LAS X software; Leica).
6. Image analysis software: Fiji/ImageJ [17].

3 Methods

3.1 *Generation of Transgenic* Nicotiana Tabacum *Plants*

For considerations and recommendations regarding expression vector construction, *see* [20]. Transgenic plants were generated essentially as previously described [14, 21]. A schematic illustration of the procedure is shown in Fig. 2. To minimize the risk of fungal or bacterial contamination, axenic material needs to be handled in a sterile hood using disinfected tweezers and scalpels.

1. Surface sterilize WT tobacco seeds in a microwave oven (*see* **Note 13**) and subsequently transfer them onto solid (0.8% Phyto agar) half-strength MS medium containing 2% sucrose in a sterile glass jar or similar growth container.
2. Seal glass jars or growth containers with micropore tape (also: all other glass jars, Petri dishes, or growth containers used throughout this procedure).
3. Grow plants for 4–5 weeks under long-day growth conditions (*see* Subheading 2.1).
4. Cut leaves into 1 cm^2 explants (*see* **Note 14**).
5. Submerge leaf explants in Agrobacterium suspension in a 5.5-cm Petri dish and incubate for 10–15 min at RT (*see* **Note 6**).
6. Place leaf explants bottom-side-down onto solid sugar-free full-strength MS medium in a Petri dish (*see* **Note 15**).
7. Incubate leaf explants in the dark for 2 days at 21 °C.
8. Transfer leaf explants onto solid MG I medium containing the appropriate selective agent (corresponding to the resistance conferred by the selective marker gene used) in a Petri dish and maintain cultures under long-day growth conditions (*see* Subheading 2.1).

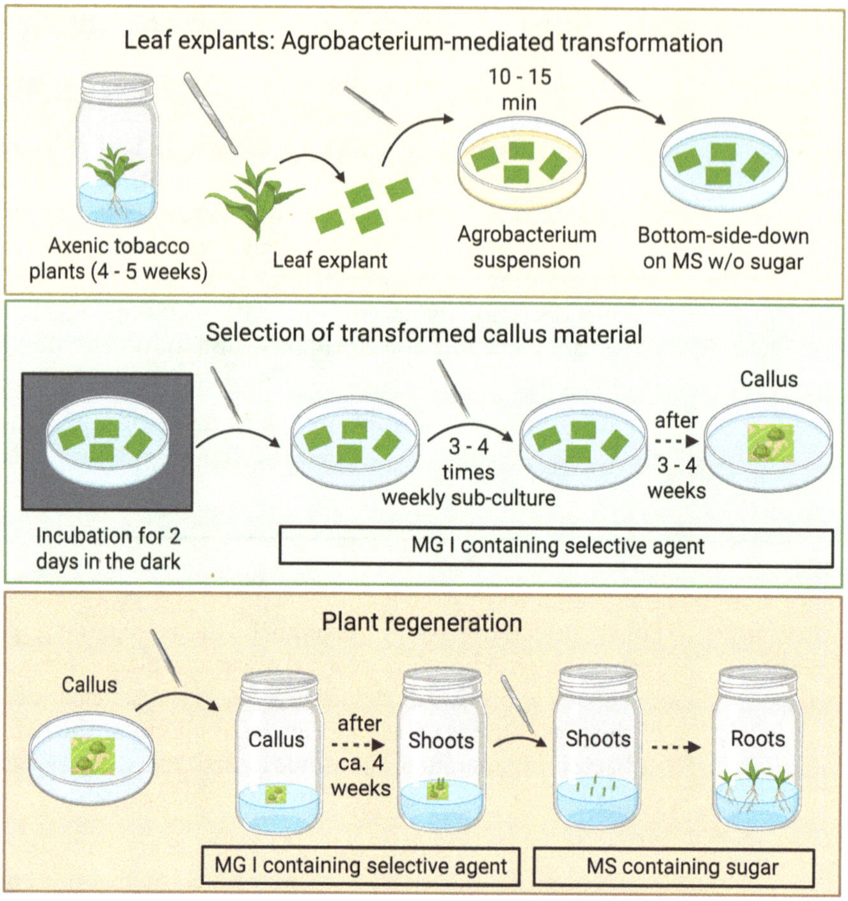

Fig. 2 Schematic illustration of the procedure used for tobacco transformation. *Top panel*: Leaf explants (without mid rip) are excised from axenic tobacco plants, incubated in a suspension of Agrobacterium harboring a binary T-DNA construct, and subsequently placed bottom-side-down on a solid, sugar-free MS medium. *Central panel*: Leaf explants are incubated in the dark for 2 days, and then transferred onto solid MG I medium containing the appropriate selective agent (corresponding to the resistance conferred by the selective marker gene used). Explants are sub-cultured weekly on this medium until callus formation is observed, typically after 3–4 weeks. *Bottom panel*: Leaf explants with developing calli are transferred to glass jars (Fig. 1A) containing MG I medium supplemented with the appropriate selective agent, and maintained under these conditions until shoots are emerging from callus tissues, typically after 3–4 weeks. Developing shoots are cut off and transferred to glass jars containing MS medium containing 2% sucrose to promote root formation. Once rooted, plants can be transferred to soil

9. Sub-culture leaf explants weekly onto fresh solid MG I medium containing the appropriate selective agent.
10. After 3–4 weeks, transfer leaf explants forming callus onto MG I medium containing the appropriate selective agent in a glass jar or similar culture container (*see* **Note 6**).
11. After ca. 4 weeks, cut off shoots developing from callus tissue and transfer them to full-strength MS medium containing 2%

sucrose and 250 mg/L Ticarcillin in a glass jar or similar culture container (*see* **Note 16**).

12. Transfer shoots with well-developed roots to soil (*see* **Note 17**).
13. After 6–8 weeks (depending on growth conditions; *see* **Note 1**), germinate pollen collected from primary regenerants in vitro and screen growing pollen tubes for fluorescence emission indicating transgene expression (*see* **Note 18**).

3.2 Preparation of PTNT Plates and Plating of Pollen Grains

Procedures were performed essentially as previously described [20].

1. To prepare PTNT plates, dilute 2× PTNT medium 1:1 with preheated 0.5% phytagel (*see* **Note 19**).
2. Pipette 3.5 mL of liquid (heated) 1× PTNT medium containing 0.25% phytagel into a 5.5-cm Petri dish under a sterile hood (*see* **Note 20**). Wait until the medium has solidified, and store plates at 4 °C (*see* **Note 21**).
3. Collect mature flowers (cut from plants) to harvest anthers and pollen. To minimize loss of pollen, remove the corolla with tweezers after cutting it open from the bottom to the top on 2 opposite sides [20]. Harvest the anthers using tweezers and transfer them into a 50-mL tube containing the appropriate volume (*see* **Step 4**) of 1× PTNT medium (2× PTNT diluted 1:1 with sterile H_2O_{bidest}; *see* **Note 22**). Instead of freshly harvested anthers, pollen within gelatin capsules, which were stored at −80 °C (*see* Subheading 2.1), can be transferred into a 50-mL tube containing 1× PTNT medium.
4. Suspend 5 anthers (1 flower), which were either freshly collected or stored in gelatin capsules at −80 °C (*see* **Step 3**), in 10–15 mL 1× PTNT medium (*see* **Note 23**). To enhance pollen germination rate, anthers retrieved from storage at −80 °C should be incubated in 1× PTNT medium for 30 min at 4 °C (*see* **Note 3**) before proceeding to **step 5**.
5. Vortex 50-mL tubes containing anthers in liquid 1× PTNT medium for 3 min to release pollen grains.
6. Remove anthers and gelatin capsules (only if anthers retrieved from storage at −80 °C are used) by filtering the suspension through a sieve into another tube. Alternatively, it is possible to remove anthers and gelatin capsules from the suspension using ethanol-sterilized tweezers.
7. To plate pollen grains on solid medium, vacuum filtrate 5 mL pollen suspension through a pre-wetted polyamide filter membrane (pollen grains are retained on the surface of the membrane), which has been positioned on top of an assembled filter system (Fig. 1G).

8. Place the membrane up-side-down on 3.5 mL solid PTNT medium in a 5.5-cm Petri dish and remove the filter immediately (pollen grains remain on the surface of the culture medium).
9. Seal plates containing plated pollen grains with Parafilm and incubate them at 21 °C in the dark.

3.3 Confocal Imaging

1. Switch on the confocal microscope and activate the 488 nm (for GFP [green fluorescent protein] and simultaneous GFP/FM4-64 imaging) or 514 nm (for YFP [yellow fluorescent protein] and simultaneous YFP/FM4-64 imaging) laser lines. If gas lasers (e.g., argon laser) are used, start them at least 15 min prior to imaging (*see* **Note 24**). When spectral detectors are available, select the following detection windows to image GFP, YFP, and FM4-64 emission, respectively: 497–525 nm, 525–575 nm, and 650–790 nm. Use emission filters with similar characteristics when working with a filter-based microscope. Also, activate the transmitted light channel, enabling the acquisition of bright-field reference images, if the microscope is set up with the necessary equipment.
2. Using a scalpel, cut out a section (ca. 1.5 × 1.5 cm) of solid culture medium covered with growing pollen tubes from a culture plate and transfer it onto a microscope slide (up-side-up). Carefully place a coverslip directly on top of the pollen tubes with the help of tweezers.
3. Place pollen tube cultures mounted between slide and coverslip on the microscope stage (coverslip facing downwards when working with an inverted microscope) and use a high magnification objective (e.g., 63×) to identify and image fluorescent pollen tube tips that are evenly growing in a plane parallel to the coverslip (*see* **Note 25**).
4. Keep key imaging settings (i.e., image resolution, zoom factor, pixel size, averaging, scanning speed, pinhole size) constant to enable quantitative comparative analysis of data collected in different experiments (*see* **Note 26**).
5. Adjust laser intensity, detector gain, and tunable transmission filters to maximize image brightness without saturating detectors (*see* **Note 27**).
6. After acquiring images for subsequent quantitative analyses, record a second image of each investigated pollen tube 2 min later to enable growth rate determination (*see* Subheading 3.4; *see* **Note 28**).
7. To preserve metadata, save acquired images in the original format determined by the imaging software that runs the microscope used. In addition, apply a grey-scale lookup table (LUT) and export images as.tif files, which can be imported into Fiji for quantitative analyses.

3.4 Growth Rate Determination

To confirm that pollen tubes are growing normally at the time images used for quantitative analyses are acquired, determine pollen tube growth rates immediately after recording these images. Procedures that can be employed for this purpose are described in detail below and have, in essence, been reported previously [20]. They rely on tools provided by the Fiji version of ImageJ, and can easily be adapted for use with other image analysis software offering comparable functionality.

1. After applying a grey-scale LUT, export the originally acquired images as.tif files, and import these files into Fiji using the "Drag and Drop" function.
2. In Fiji, open the 2 fluorescence images acquired for each analyzed pollen tube: "image 1" (t_0) used for quantitative image analysis and "image 2" (t_{2min}) recorded 2 min later for growth rate determination (Fig. 3; *see* Subheading 3.3: **Step 6**; *see* **Note 29**).
3. Start the Fiji tool "Image → Color → Merge Channel," define "image 1" as C1 (red) and "image 2" as C2 (green), and click "OK". This generates a merged image showing an overlay of "image 1" and "image 2" (Fig. 3).
4. Apply the Fiji "straight-line" tool (Fig. 3A), or the "segmented-line" tool (Fig. 3B) to determine the distance in μm between the positions of the extreme pollen tube apex in the two overlaid images (*see* **Note 30**). Dividing this distance by 2 yields the growth rate of the analyzed pollen tubes in μm/min. Zooming in to enlarge the image display on the computer screen helps drawing accurate straight or segmented lines.
5. Growth rate data sets are best displayed as boxplots or violin plots, which can be generated using freely accessible web-based software tools (*see* **Note 31**). Such tools are also available for the statistical analysis of multiple growth rate data sets based on ANOVA testing (*see* **Note 32**).

3.5 Determination of PM Domain Position and Extension

To enable comparative quantitative analysis of the position and extension of different PM domains, standardized procedures were developed to determine the meridional distance (MD; along the pollen tube perimeter) from the apex (MD = 0) of the proximal (closer to the apex) and distal (further away from the apex) endpoints of these domains (Fig. 4A). To determine MDs of domain endpoints, the Fiji version of ImageJ can be employed to draw segmented lines delineating either regions of the PM (PM-associated markers; Fig. 4A: red line) or of the cortical cytoplasm directly underneath (markers associated with cytoplasmic structures or regions in direct contact with the PM; Fig. 4A: blue line). Fiji-based procedures used for this purpose have, in essence, been previously reported [14] and are described in detail below.

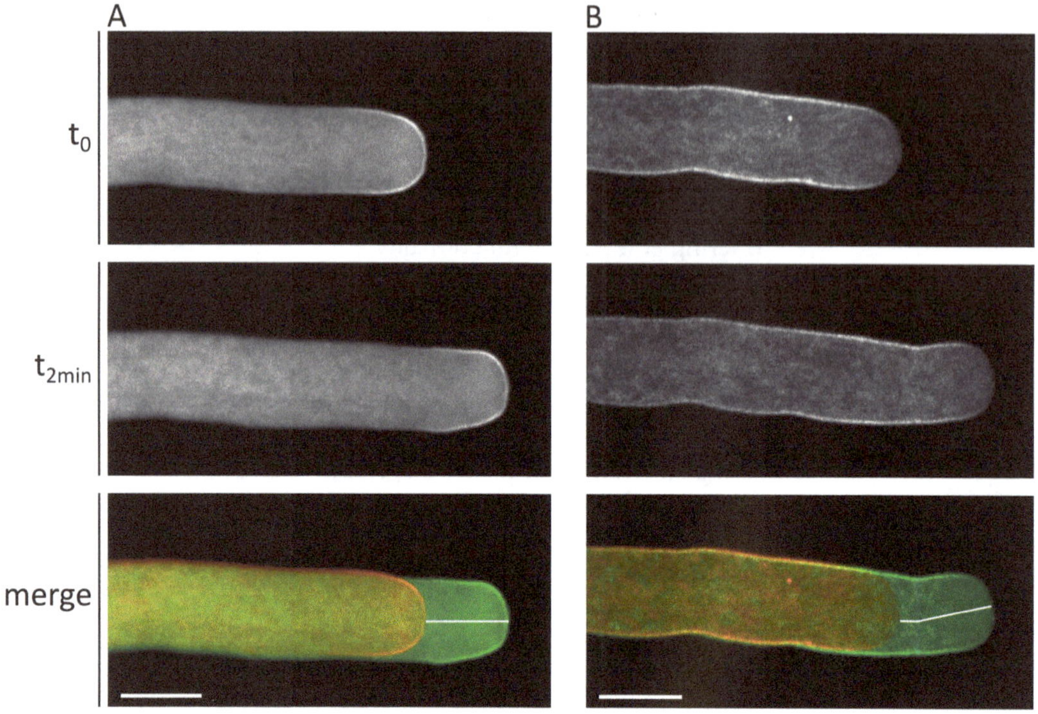

Fig. 3 Determination of the rate of straight and curved pollen tube tip growth. *Top row*: Medial confocal optical sections through tobacco pollen tubes expressing different fluorescent fusion proteins associated with distinct PM domains. These images were acquired for comparative quantitative analysis of the labeled PM domains (t_0). *Central row*: Images of the same pollen tubes acquired 2 min later to enable growth rate determination (t_{2min}). *Bottom row*: Merged images showing an overlay of the two images acquired of each analyzed pollen tube (t_0: red; t_{2min}: green). The distance between the positions of the extreme pollen tube apex in the overlaid images can be determined in Fiji by drawing straight (**A**) or segmented (**B**) lines (white) connecting these positions. Segmented lines are only drawn if the analyzed pollen tube has changed its growth direction between the acquisition of the 2 images. Scale bars: 10 μm

They can easily be adapted for use with other image analysis software offering comparable functionality.

1. After applying a grey-scale LUT, export the originally acquired images as.tif files, and import these files into Fiji using the "Drag and Drop" function. Convert all images to the same dynamic range (i.e., 8 bit/256 grey levels) (*see* **Note 33**).
2. Apply the Fiji "rolling ball" background subtraction tool: "Process → Subtract background. . . → choose a radius". Select a radius small enough to avoid the generation of artifacts (Fig. 4B; *see* **Note 34**).
3. Use the Fiji tool "Image → Lookup Table → 16 colors" to apply a 16-color LUT. This facilitates the identification of domain endpoints (Fig. 4C; *see* **Note 35**).

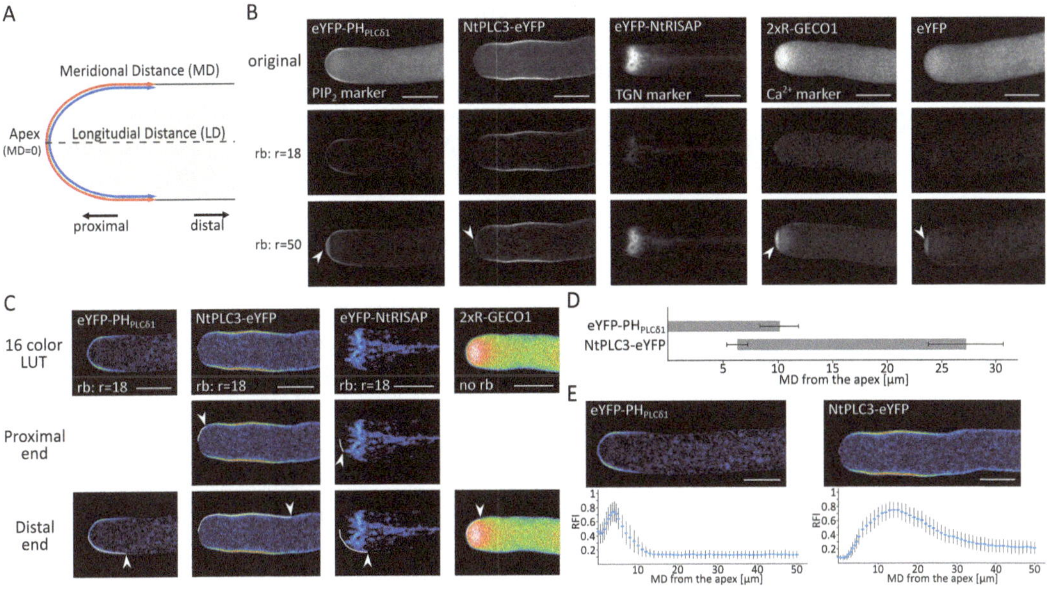

Fig. 4 Determination of PM domain endpoints and of relative fluorescence intensity (RFI) distributions along defined PM regions. (**A**) Schematic representation of a pollen tube tip illustrating the determination of the meridional distance (MD) from the apex (MD = 0) to the endpoints of PM domains. For this purpose, segmented lines can be drawn in Fiji, which delineate either regions of the PM (PM-associated markers; red line) or of the cortical cytoplasm directly underneath the PM (markers associated with cytoplasmic structures or regions in direct contact with the PM; blue line). MDs from the apex of both the proximal (closer to the apex) and distal (further away from the apex) domain endpoints can be measured using this approach. The MD from the apex of the proximal endpoint of apical domains is defined as 0. (**B**) Application of Fiji's "rolling ball" (rb) background subtraction to medial confocal optical sections through pollen tubes expressing the indicated fluorescent proteins. *Top row*: Original images before background subtraction. *Central row*: The images shown in the top row after background subtraction using a radius of 18 pixels ($r = 18$). *Bottom row*: The images shown in the top row after background subtraction using a radius of 50 pixels ($r = 50$). White arrowheads indicate artifacts generated when background subtraction was performed using this larger radius. (**C**) Application of Fiji's 16-color lookup table (LUT) to facilitate the identification of domain endpoints. *Top row*: Confocal optical sections through pollen tubes expressing the indicated fluorescent proteins. Images were either subjected in Fiji to "rolling ball" background subtraction using a radius of 18 pixels (rb; $r = 18$), or left unprocessed (no rb). They are displayed using Fiji's 16-color LUT to highlight domain endpoints. Immediately adjacent to these endpoints, pixel intensity (as reflected by the LUT color) should match that of cytoplasmic background regions. *Central row*: Segmented lines (white) drawn in Fiji to determine the MD from the apex of proximal domain endpoints (arrowheads) highlighted by the 16-color LUT. The proximal endpoints of apical PM domains associated with eYFP-PH$_{PLC\delta1}$ [22], which labels phosphatidylinositol-4,5-bisphosphate (PIP$_2$), or in direct contact with the Ca^{2+}-enriched cytoplasmic region visualized by 2xR-GECO1 [23], are defined as 0. *Bottom row*: Segmented lines (white) drawn in Fiji to determine the MD from the apex of distal domain endpoints (arrow heads) highlighted by the 16-color LUT. (**D**) Gantt chart displaying means and standard deviations of the MDs from the apex of the endpoints of the PM domains labeled by eYFP-PH$_{PLC\delta1}$ and NtPLC3-eYFP [10], which were determined as shown in (**C**). (**E**) Analysis of relative fluorescence intensity (RFI) distributions along segmented lines drawn in Fiji, which start at the extreme pollen tube apex (MD from the apex = 0) and extend beyond the distal endpoint of the PM domains associated with eYFP-PH$_{PLC\delta1}$ or NtPLC3-eYFP. *Upper row*: Representative individual images analyzed using this procedure, which are also shown in (**C**). PM domain endpoints are highlighted by the application of Fiji's 16-color LUT. *Lower row*: RFI plots

4. Apply the Fiji "segmented line" tool to draw lines along the perimeter of analyzed pollen tubes, which connect the extreme apex to the proximal or distal endpoints of investigated PM domains (Fig. 4C). As discussed above, these lines delineate either regions of the PM (markers for PM-associated lipids or proteins: e.g., eYFP-PH$_{PLC\delta1}$ [phosphatidylinositol-4,5-bisphosphate; PIP_2] [22] or NtPLC3-eYFP [10]), or of the cortical cytoplasm directly underneath the PM (markers associated with cytoplasmic structures or regions in direct contact with the PM: e.g., eYFP-NtRISAP [TGN] [12], or 2xR-GECO1 [Ca^{2+}-enriched cytoplasmic region] [23]) (Fig. 4A+C). Zooming in to enlarge the image display on the computer screen helps drawing accurate segmented lines. The length of segmented lines in μm corresponds to the MD from the apex of domain endpoints (*see* **Note 36**).
5. Means and standard deviations of MDs from the apex of different domain endpoints are best displayed as Gantt charts [14] (Fig. 4D), which can be generated using Microsoft Excel. Freely accessible web-based software tools are available for the statistical analysis of multiple domain endpoint data sets based on ANOVA testing (*see* **Note 32**). Note that statistical analysis of different data sets defining the 2 endpoints of the individual investigated domains needs to be independently performed.

3.6 Determination of Relative Fluorescence Intensity Distributions Along PM Domains

To enable comparative quantitative analysis of fluorescence intensity distributions along different PM domains, standardized procedures were developed to plot relative fluorescence intensities (RFIs) along segmented lines drawn using the Fiji version of ImageJ. These lines delineate either regions of the PM (PM-associated markers; Fig. 4A: red line) or of the cortical cytoplasm directly underneath the PM (markers associated with cytoplasmic structures or regions in direct contact with the PM; Fig. 4A: blue line). Fiji-based procedures that can be employed for this purpose have, in essence, been previously reported [14] and are described in detail below. They can easily be adapted for use with other image analysis software offering comparable functionality.

Fig. 4 (continued) displaying mean relative fluorescence intensities and their standard deviations along the analyzed PM regions at different MDs from the apex (0–50 μm; distance between data points: 1 μm). The analyzed region completely covers the 2 investigated PM domains and extends beyond their distal endpoints. The RFI plots represent averaged data collected by analyzing 30–50 images similar to those shown in the upper panel, which were acquired from different pollen tubes expressing the same fluorescent proteins. Scale bars: 10 μm

1. After applying a grey-scale LUT, export the originally acquired images as.tif files, and import these files into Fiji using the "Drag and Drop" function.
2. Apply the Fiji "rolling ball" background subtraction tool: "Process → Subtract background... → choose a radius". Select a radius small enough to avoid the generation of artifacts (Fig. 4B; *see* **Note 34**).
3. Using the Fiji "segmented line" tool, draw a 1 pixel-wide line along the perimeter of each analyzed pollen tube tip, which starts at the extreme apex (MD from the apex: 0) and extends beyond the distal endpoint of the investigated PM domain (Fig. 4C; *see* **Note 37**). As discussed above, these lines delineate either regions of the PM (markers for PM-associated lipids or proteins) or of the cortical cytoplasm directly underneath the PM (markers associated with cytoplasmic structures or regions in direct contact with the PM) (Fig. 4A+C). Zooming in to enlarge the image display on the computer screen helps drawing accurate segmented lines.
4. Apply the Fiji "profile plotting" tool "Analyze → Plot Profile" (shortcut: Ctrl + k) to plot fluorescence intensities along segmented lines drawn as outlined above (*see* **Step 3**). Click on "list" in the plot window to display all intensity values along the analyzed segmented line, which can be copied and pasted into Microsoft Excel or another spreadsheet software with similar functionality for further analysis.
5. Using Microsoft Excel (or a similar spreadsheet software), normalize the intensity values of each individual plot by setting the maximal value within this plot to 1 (i.e., divide all values of each plot by the maximal value) (*see* **Note 36**) [24]. RFI (relative fluorescence intensity) plots generated using the normalized intensity values are directly comparable, regardless of the proportion of the detector's dynamic range used during image acquisition (provided that detector saturation was avoided; *see* Subheading 3.3: **Step 5**).
6. Average all RFI plots obtained for each investigated PM domain to generate composite plots, which display mean RFI values at different MDs from the apex, along with their standard deviations (Fig. 4E) [14]. Optionally, maximal as well as half-maximal mean RFI values can be indicated in the composite RFI plots to highlight differences between similar plots obtained for overlapping PM domains [14].

3.7 Quantitative Analysis of Endocytic Lipid and Protein Recycling

To enable comparative quantitative analysis of the endocytic recycling of different PM lipids and proteins, standardized procedures were developed to determine relative levels of the accumulation of fluorescent lipid or protein markers within the cytoplasm and within brefeldin A (BFA)-induced "BFA compartments". These procedures have been employed to investigate possible endocytic recycling of lipids and proteins associated with PM regions, which overlap with a lateral endocytic PM domain identified in tobacco pollen tubes [8, 14].

BFA effectively blocks pollen tube tip growth by disrupting apical membrane traffic [8]. More precisely, BFA inhibits the formation of exocytic vesicles at the TGN [15, 16], but does not interfere with the endocytic uptake of PM material [25, 26]. Consequently, BFA blocks the recycling of endocytosed material back to the PM and causes this material to accumulate within BFA-induced aberrant TGN compartments, which have been designated "BFA compartments" [16, 27, 28].

Procedures described in detail below can be used (1) to visualize the BFA compartment in tobacco pollen tubes based on endocytic internalization of FM4-64-labeled PM material [29], and (2) to quantify the relative accumulation of fluorescent markers for PM lipids or proteins in the cytoplasm and in the FM4-64-labeled BFA compartment. Simultaneous imaging in separate channels of the red-fluorescent styryl dye FM4-64 together with yellow-fluorescent lipid or protein markers is required for this purpose. The described procedures rely on different functions of the Fiji version of ImageJ, which can easily be adapted for use with other image analysis software offering comparable functionality.

1. Grow transgenic pollen tubes expressing fluorescent lipid or protein markers for 2.5–3 h on solid PTNT medium in a 5.5-cm Petri dish at 21 °C in the dark. Carefully drop 200 μL 50 μM FM4-64 in 1× PTNT medium onto the center of culture plates and shake gently to evenly distribute the dye solution over a small central region (*see* **Note 38**). After FM4-64 application, continue to grow pollen tubes for another 30 min at 21 °C in the dark.
2. Carefully drop 200 μL 70 μM BFA in 1× PTNT medium onto the FM4-64-treated region in the center of culture plates and incubate for 30 min at 21 °C in the dark (*see* **Note 39**).
3. Cut out sections of the pollen tube cultures and mount them between a slide and a coverslip as described above (*see* Subheading 3.3: **Step 2**; *see* **Note 40**). Using a confocal microscope, simultaneously image red-fluorescent FM4-64 together with a yellow-fluorescent lipid or protein marker in separate channels (Fig. 5; *see* Subheading 3.3).
4. After applying a grey-scale LUT, export the images acquired in both channels as.tif files, and import these files into Fiji using the "Drag and Drop" function.

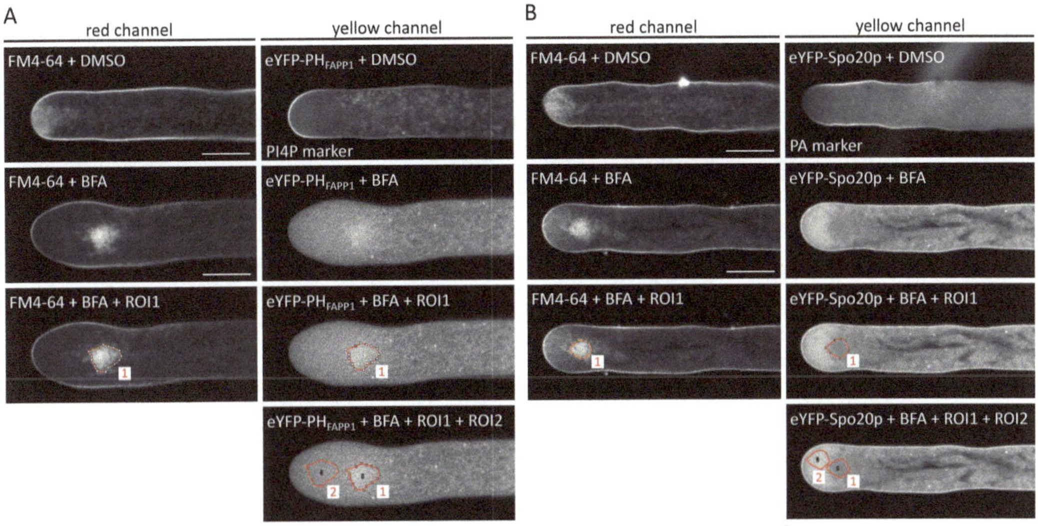

Fig. 5 Quantitative analysis of lipid accumulation in BFA compartments. Data showing that some, but not all, PM lipids are endocytically recycled and trapped within BFA compartments. (**A**) Phosphatidylinositol-4-phosphate (PI4P) labeled by eYFP-PH$_{FAPP1}$ [33, 34] strongly accumulates in these compartments, whereas (**B**) Phosphatidic acid (PA) labeled by eYFP-Spo20p [35] does not. *First row*: Confocal optical sections through DMSO-treated pollen tubes (solvent controls), which have been labeled for 30 min with the dye FM4-46, and expressed the indicated fluorescent proteins. FM4-64 labeling (red) and fluorescent proteins serving as lipid markers (yellow) were simultaneously imaged in separate channels. *Second row*: Confocal optical sections through BFA-treated pollen tubes, which were labeled with FM4-64 the same way, expressed the same fluorescent proteins serving as lipid markers, and were simultaneously imaged in 2 channels like the pollen tubes shown in the first row. *Third row*: The same images as shown in the second row, with ROI1 outlining the BFA compartment. ROI1 was defined using Fiji's "polygon selection" tool in the image showing FM4-64 labeling (red channel), and then transferred using the Fiji's "ROI manager" tool to the exact same position in the corresponding image showing lipid distribution in the same pollen tube (yellow channel). *Fourth row*: The same images as shown in the third row (yellow channel), with a copy of ROI1 positioned in a cytoplasmic reference region (ROI2) using Fiji's "ROI manager" tool. Scale bars: 10 μm

5. Employ the Fiji "polygon selection" tool to outline the FM4-64-labeled BFA compartment in the image recorded in the red channel (Fig. 5: red channel, third row).
6. Import the selected polygon as a new region of interest (ROI) into the Fiji "ROI manager" tool "Analyze → Tools → ROI manager" by clicking on "Add[t]".
7. Using the Fiji "ROI manager" tool, select the image recorded in the yellow channel and click on the previously imported ROI (*see* **Step 6**). Consequently, this ROI will be applied to the image recorded in the yellow channel at the exact same position as in the red channel image (Fig. 5: yellow channel, third row).
8. Apply the Fiji tool "Analyze → Measure" (shortcut: Ctrl + m) to determine the mean fluorescence intensity within the ROI applied to the image recorded in the yellow channel (*see* **Note 41**).

9. Import the same ROI a second time into the Fiji "ROI manager" tool as described above (*see* **Step 7**) and select it. Move the selected second ROI in the image recorded in the yellow channel to a cytoplasmic region not overlapping with the BFA compartment.
10. Apply the Fiji tool "Analyze → Measure" (shortcut: Ctrl + m) to determine the mean fluorescence intensity within the second ROI applied to the image recorded in the yellow channel, which overlaps with a cytoplasm reference region (*see* **Note 41**).
11. The ratio of the mean fluorescence intensities within the 2 ROIs applied to the image recorded in the yellow channel ("relative fluorescence intensity [RFI] BFA compartment/cytoplasm") corresponds to the level of accumulation of the investigated lipid or protein within the BFA compartment. Accumulation within the BFA compartment indicates endocytic recycling of this lipid or protein (*see* **Note 42**) [14].
12. Means and standard deviations of all "RFI BFA compartment/cytoplasm" values determined for individual pollen tubes expressing a specific marker are best displayed as boxplots or violin plots, which can be generated using freely accessible web-based software tools (*see* **Note 31**). Optionally, to facilitate data interpretation, plots can be complemented with a horizontal line indicating RFI = 1 [14]. Free web-based software tools are also available for the statistical analysis and comparison of multiple "RFI BFA compartment/cytoplasm" data sets based on ANOVA testing (*see* **Note 32**).

4 Notes

1. *Nicotiana tabacum* plants grown in soil under constant standard conditions (*see* Subheading 2.1) typically flower after about 2 months. However, tobacco plant development and the timing of flowering induction may vary considerably when plants are cultivated in a greenhouse, where environmental factors such as temperature, light intensity, and light quality fluctuate.
2. The lifespan and flowering period of tobacco plants can be substantially prolonged by regularly removing senescent flowers and maturing capsules (fruits), as well as by applying fertilizer to the soil once per week.
3. Place gelatin capsules containing anthers with pollen into a 15-mL tube before freezing them in liquid nitrogen. This prevents bursting of the capsules and facilitates storage. Pollen can be stored in gelatin capsules at −80 °C for at least 4 years.

However, as pollen germination rate declines with prolonged storage, the use of fresh pollen is recommended whenever possible. To enhance the germination rate after freezing, pollen can be incubated for 30 min at 4 °C in liquid 1× PTNT medium prior to plating on solid medium. To this end, suspend pollen-containing frozen gelatin capsules in liquid 1× PTNT. After about 1 min, when the capsules have softened, open them using tweezers or by vortexing.

4. In general, transgenic pollen tubes stably expressing lipid or protein markers are the preferred material for all quantitative analyses described here, although it typically takes about 6 months to regenerate flowering plants from transformed tobacco leaf explants. Unlike transiently transformed pollen tubes, stably transformed pollen tubes derived from the same transgenic line generally express marker proteins at similar levels. This substantially reduces the time required to obtain sufficiently large data sets for quantitative analyses (in general: $n > 30$) and greatly improves the reproducibility of these analyses.
5. The Agrobacterium strain AGL1 [30] can be used effectively for *Nicotiana tabacum* transformation.
6. Tobacco leaf explants can be incubated in Agrobacterium suspension in small Petri dishes (diameter: 5.5 cm). Subsequently, leaf explants should be cultured on solid medium in standard Petri dishes (diameter: 9.4 cm, height: 1.6 cm) until callus formation is observed. Once callus formation is visible, the explants are generally directly transferred onto solid medium in sterile glass jars (Figs. 1A+2) to promote further growth and shoot formation. Alternatively, it is possible to temporarily maintain the explants on solid medium in large Petri dishes (diameter: 10 cm, height: 2 cm) before they are transferred onto solid medium in sterile glass jars.
7. Pre-dissolve BA and NAA in ethanol.
8. Considering the relatively low salt concentrations, even in the 10× stock solution, it is advisable to prepare a large volume of this stock solution (e.g., 1 L). Prepare and freeze 50-mL aliquots. To prepare 500 mL 2× PTNT medium, 100 mL 10× salt stock solution is required.
9. Prepare and freeze 10-mL aliquots of the 1.5% casein acid hydrolysate stock solution. To prepare 500 mL 2× PTNT medium, 20 mL of this stock solution is required.
10. Weigh sucrose and MES hydrate into a 1-L beaker. Thaw 1.5% casein acid hydrolysate and 10× salt stock solutions in a water bath (or a beaker filled with water heated in a microwave oven), and add the required volumes of both stock solutions to the 1-L beaker containing sucrose and MES hydrate. Fill the

beaker to half the final volume with H_2O_{bidest} and add rifampicin as required. Adjust the pH to 5.9 with KOH and add PEG-6000. Once the PEG-6000 is completely dissolved (takes a while; gentle heating to ca. 60 °C speeds the process up), transfer the solution to a measuring cylinder and adjust to the final volume with H_2O_{bidest}. Filter sterilize and store aliquots at 4 °C.

11. If no incubator is available, pollen tubes can be grown in an air-conditioned laboratory at RT (20–23 °C) in the dark (e.g., wrapped in aluminum foil or in a box). However, even small temperature fluctuations substantially affect pollen tube growth rates.
12. To achieve the highest image quality, cover slips with a thickness matching objective design specifications (generally 0.17 mm) should be used for confocal microscopy.
13. Transfer seeds to be sterilized into a 50-mL tube. Place the tube on its side with the lid loosely attached (so that it cannot fall off but still allows gas exchange) in a microwave oven together with a 500-mL glass beaker containing 100-mL sterile water. Ensure that all seeds are distributed in a single layer and do not lie on top of one another. Switch on the microwave oven for 6 min at maximum power to generate steam for surface-sterilization. Do not open the microwave oven and allow its contents to cool for 2 min. Repeat the steam-based surface-sterilization for another 6 min. Sterilized seeds may be stored in the same 50-mL tube. Note that this procedure effectively surface-sterilizes tobacco seeds, but may damage smaller seeds (e.g., Arabidopsis seeds). Alternative chemical surface-sterilization methods may be used if necessary.
14. Remove the midrib before excising leaf explants and handle them gently to avoid damage.
15. For each T-DNA construct used, prepare 8–10 plates, each containing several leaf explants. In most cases, this will yield several independent transgenic lines per T-DNA construct.
16. To generate independent transgenic lines, transfer only one shoot per callus. In most cases, transferring about 70 shoots per T-DNA constructs yields several independent lines.
17. In case shoot cultures become contaminated with fungi or bacteria before rooting, unrooted shoots can be transferred to soil, where they have a better chance of survival under these conditions.
18. Because of differences in transgene copy number, genomic integration site, and silencing, transgene expression levels often vary considerably among independent lines and individual plants of the same line. Furthermore, transgene expression

levels may change between generations. Therefore, it is advisable to establish multiple independent transgenic lines and to collect as many pollen-containing anthers (to be stored in gelatin capsules at −80 °C) as possible from mature transgenic plants displaying suitable expression levels.

19. Boil water in a 300-mL beaker using a microwave oven and place a 50-mL tube containing 2× PTNT medium into the hot water to reduce the viscosity of the medium. Preheat the 0.5% phytagel solution by gently heating it in a microwave oven until the first bubbles appear. Add an equal volume of preheated 0.5% phytagel solution to the 2× PTNT medium and mix gently by inverting the 50-mL tube several times until the resulting 1x PTNT medium containing 0.25% phytagel is homogeneous.
20. Place 50-mL tubes containing liquid 1× PTNT medium with 0.25% phytagel prepared as described above (*see* **Note 19**) back into hot water to prevent premature solidification. Distribute 3.5 mL of this medium into 5.5-cm Petri dishes. Equilibrate the pipette by pipetting up and down several times and release the medium slowly into the Petri dishes to avoid bubble formation. Gently shake the Petri dishes to evenly distribute the medium.
21. PTNT plates can be stored for up to 4 weeks at 4 °C.
22. Handle plants and harvested flowers carefully to avoid loss of pollen. Collect pollen only from fully open, mature flowers with white and pink corolla, not from flowers turning brownish and purple (Fig. 1H), which indicates beginning senescence.
23. Suspensions containing fresh pollen can be stored at 4 °C for several hours. Plating pollen at too high densities reduces pollen tube growth rate and hinders the identification of pollen tube tips evenly growing in a plane parallel to the cover slip, which is suitable for high-quality imaging.
24. The intensity of gas (e.g., argon) lasers typically stabilizes about 15 min after switching them on.
25. Optionally, specimens can be pre-screened at low magnification (e.g., 20× objective) for pollen tubes that may be suitable for imaging. However, because it is often difficult at low magnification to determine whether fluorescent pollen tubes are growing in a plane parallel to the coverslip, pre-screening is generally only beneficial when transiently transformed pollen tubes are investigated. To facilitate switching between 2 magnifications, which is required for pre-screening, objectives designed for the same immersion medium (oil or water) should be used. Water immersion objectives reduce spherical aberrations resulting from mismatches in the refractive indices of the immersion medium and pollen tube cultures.

26. The following settings are generally suitable for standardized high-quality imaging: pinhole size 1 Airy unit, zoom factor 2.5 (63× objective), pixel size 72 × 72 nm, scanning speed 600–700 Hz, and 3× line averaging. Laser intensity, detector gain and tunable transmission filters should be adjusted to optimally utilize the detectors' dynamic range. These imaging parameters do not need to be kept constant when determining relative fluorescence intensities (RFIs). Detailed information about standardized imaging settings suitable for specific transgenic tobacco marker lines has previously been reported ([14]: Supplemental Table S4).
27. It is essential to avoid overexposing images used for quantitative analyses. This can be greatly facilitated by lookup tables (LUTs), which assign a bright color to the maximal pixel intensity of the detector. Confocal microscopes generally allow online application of such LUTs during live imaging.
28. Do not acquire more than 2 images of the same pollen tube, as prolonged time-lapse imaging causes phototoxicity and reduces pollen tube growth rates. At an interval of 2 min, capture 2 images of the same rectangular sample area containing the tip of analyzed pollen tubes. Position the sample area so that the pollen tube tip remains within the field of view for at least 2 min. Align the longer edges of the rectangular sample area parallel to the longitudinal pollen tube axis to facilitate subsequent image analysis and presentation.
29. Transmitted light bright-field images can alternatively be used for growth rate determination. However, the apex position is less clearly defined in bright-field images, which reduces measurement accuracy.
30. Draw a segmented line only if the analyzed pollen tube has changed its growth direction between the acquisition of the 2 consecutive images (Fig. 3B). Only images of pollen tubes growing at a normal rate of at least 3.0 μm/min should be used for further analysis, as the intracellular distribution of the analyzed markers may be altered when tip growth is inhibited and occurs at lower rates.
31. A free R-based web tool for generating boxplots and violin plots is available at http://shiny.chemgrid.org/boxplotr/ [31]. This tool allows easy data upload and modification of blot parameters (color, size, scale, axes, etc.). Plots can be exported as.svg or.pdf files.
32. A free web tool for one-way ANOVA testing is available at https://houssein-assaad.shinyapps.io/TableReport/ [32]. Using this tool, data can be uploaded as Microsoft Excel (.xlsx) files. The tool allows selecting different significance levels and post-hoc tests.

33. The dynamic range of images can be adjusted using the Fiji tool "Image → Type → select the desired depth". To standardize the dynamic range across groups of comparatively analyzed images, adjust all images to match the lowest dynamic range within each group. Increasing the dynamic range of images requires interpolation and is therefore not recommended.
34. When a large radius is selected (e.g., 50 pixels), application of the Fiji "rolling ball" background subtraction tool can generate severe artifacts in images showing a fluorescent signal at the PM or the cytoplasm at the extreme pollen tube apex (Fig. 4B, row 3). No such artifacts are observed, and background subtraction is effective, when a smaller radius (e.g., 18 pixels) is selected and the "rolling ball" tool is applied to images showing labeling of PM domains (Fig. 4B, row 2: eYFP-$PH_{PLC\delta1}$ [phosphatidylinositol-4,5-bisphosphate; PIP_2] [22] and NtPLC3-eYFP [10]) or cytoplasmic structures (Fig. 4B, row 2: eYFP-NtRISAP [TGN] [12]). For comparative quantitative analysis of different PM domains, the same radius should be selected consistently for background subtraction. Additionally, it should be confirmed that background subtraction using this radius does not generate artifacts when applied to images of any of the investigated PM domains. Note that "rolling ball" background subtraction, even with a small radius (e.g., 18 pixels), cannot be effectively applied to images showing diffuse fluorescent labeling of cytoplasmic regions: although no artifacts are generated, image brightness is reduced to near background levels (Fig. 4B, row 2: 2xR-GECO1 [Ca^{2+}-enriched cytoplasmic region] [23] and eYFP).
35. Around PM domain endpoints, pixel intensity as reflected by the LUT color is expected to match that of cytoplasmic background regions (Fig. 4C).
36. It is generally sufficient to analyze about $n > 30$ pollen tubes expressing a fluorescent marker protein, to generate reproducible data and detect small, statistically significant differences between distinct sets of such data. In most cases, the domain extension and position can be determined on both sides of analyzed pollen tubes. Since the 2 resulting measurements are not independent, they should be averaged and treated as a single data point.
37. If the distal domain endpoints are out of focus in individual analyzed pollen tubes, shorter segmented lines covering only the proximal region of analyzed domains (which is in focus) may be included into the analysis. However, the majority of the analyzed segmented lines should fully cover the investigated domain and extend at least to a MD of 30–50 μm from the apex (Fig. 4E).

38. If the FM4-64 solution is distributed too widely over the culture plate, the effective dye concentration becomes too low for consistent PM labeling. Increasing the FM4-64 concentration is not feasible, as the DMSO solvent would reach levels that inhibit pollen tube tip growth. Furthermore, increasing the volume to the FM4-64 solution applied to pollen tube cultures, or plating pollen at too high densities, promotes uneven pollen tube growth, making it difficult to identify pollen tube tips suitable for imaging, which are growing in a plane parallel to the cover slip.
39. Include a DMSO (solvent)-only control in all experiments (Fig. 5, row 1), and determine the growth rate of both DMSO- and BFA-treated pollen tubes. BFA compartments only form if pollen tube growth is completely inhibited. It is essential to apply FM4-64 30 min prior to the BFA treatment to allow the intracellular dye distribution to equilibrate before tip growth is arrested.
40. FM4-64 and BFA application softens the solid PTNT medium within culture plates. Consequently, cutting out sections of FM4-64-labeled and/or BFA-treated pollen tube cultures and transferring them onto a microscope slide must be done with particular care to avoid damage.
41. Application of this tool generates an output table containing a column labeled "mean" (if not, adjust the Fiji settings to enable the display of this column). The values in this column represent the mean fluorescence intensity within the selected ROI.
42. "RFI BFA compartment/cytoplasm" values equal or greater than 1 indicate accumulation in BFA compartments and endocytic recycling. Lower values suggest that endocytic recycling does not occur [14].

Acknowledgment/Funding

This work was supported by the "German Research Foundation (DFG)" within the framework of the "Research Training Group (RTG) 1962" (project 7; BK) and via 2 "Major Equipment Grants" (INST90/1074-1FUGG [SP8 confocal microscope]; BK and INST90/1025-1FUGG [plant growth chamber facility]; BK). CF received an equal opportunities stipend (Bavarian Equal Opportunities Sponsorship—Realisierung von Chancengleichheit von Frauen in Forschung und Lehre (FFL)—Realization Equal Opportunities for Women in Research and Teaching).

References

1. Steer MW, Steer JM (1989) Pollen tube tip growth. New Phytol 111:323–358. https://doi.org/10.1111/j.1469-8137.1989.tb00697.x
2. Carol RJ, Dolan L (2002) Building a hair: tip growth in *Arabidopsis thaliana* root hairs. Philos Trans R Soc London Ser B Biol Sci 357:815–821. https://doi.org/10.1098/rstb.2002.1092
3. Menand B, Calder G, Dolan L (2007) Both chloronemal and caulonemal cells expand by tip growth in the moss *Physcomitrella patens*. J Exp Bot 58:1843–1849. https://doi.org/10.1093/jxb/erm047
4. Picton JM, Steer MW (1983) Membrane recycling and the control of secretory activity in pollen tubes. J Cell Sci 63:303–310. https://doi.org/10.1242/jcs.63.1.303
5. Derksen J, Rutten T, Lichtscheidl IK et al (1995) Quantitative analysis of the distribution of organelles in tobacco pollen tubes: implications for exocytosis and endocytosis. Protoplasma 188:267–276. https://doi.org/10.1007/BF01280379
6. Ketelaar T, Galway ME, Mulder BM, Emons AMC (2008) Rates of exocytosis and endocytosis in Arabidopsis root hairs and pollen tubes. J Microsc 231:265–273. https://doi.org/10.1111/j.1365-2818.2008.02031.x
7. Grebnev G, Ntefidou M, Kost B (2017) Secretion and endocytosis in pollen tubes: models of tip growth in the spot light. Front Plant Sci 8:154. https://doi.org/10.3389/fpls.2017.00154
8. Grebnev G, Cvitkovic M, Fritz C et al (2020) Quantitative structural organization of bulk apical membrane traffic in pollen tubes. Plant Physiol 183:1559–1585. https://doi.org/10.1104/pp.20.00380
9. Potocký M, Pleskot R, Pejchar P et al (2014) Live-cell imaging of phosphatidic acid dynamics in pollen tubes visualized by Spo20p-derived biosensor. New Phytol 203:483–494. https://doi.org/10.1111/nph.12814
10. Helling D, Possart A, Cottier S et al (2006) Pollen tube tip growth depends on plasma membrane polarization mediated by tobacco PLC3 activity and endocytic membrane recycling. Plant Cell 18:3519–3534. https://doi.org/10.1105/tpc.106.047373
11. Klahre U, Kost B (2006) Tobacco RhoGTPase ACTIVATING PROTEIN1 spatially restricts signaling of RAC/Rop to the apex of pollen tubes. Plant Cell 18:3033–3046. https://doi.org/10.1105/tpc.106.045336
12. Stephan O, Cottier S, Fahlén S et al (2014) RISAP is a TGN-associated RAC5 effector regulating membrane traffic during polar cell growth in tobacco. Plant Cell 26:4426–4447. https://doi.org/10.1105/tpc.114.131078
13. Fratini M, Krishnamoorthy P, Stenzel I et al (2021) Plasma membrane nano-organization specifies phosphoinositide effects on Rho-GTPases and actin dynamics in tobacco pollen tubes. Plant Cell 33:642–670. https://doi.org/10.1093/plcell/koaa035
14. Fritz C, Reimann TM, Adler J et al (2024) Plasma membrane and cytoplasmic compartmentalization: a dynamic structural framework required for pollen tube tip growth. Plant Physiol 197:558. https://doi.org/10.1093/PLPHYS/KIAE558
15. Geldner N, Friml J, Stierhof YD et al (2001) Auxin transport inhibitors block PIN1 cycling and vesicle trafficking. Underw Nat 413:425–428. https://doi.org/10.1038/35096571
16. Nebenführ A, Ritzenthaler C, Robinson DG (2002) Brefeldin A: deciphering an enigmatic inhibitor of secretion. Plant Physiol 130:1102–1108. https://doi.org/10.1104/pp.011569
17. Schindelin J, Arganda-Carreras I, Frise E et al (2012) Fiji: an open-source platform for biological-image analysis. Nat Methods 9:676–682. https://doi.org/10.1038/nmeth.2019
18. Murashige T, Skoog F (1962) A revised medium for rapid growth and bio assays with tobacco tissue cultures. Physiol Plant 15:473–497. https://doi.org/10.1111/j.1399-3054.1962.tb08052.x
19. Bertani G (1951) Studies on lysogenesis I: the mode of phage liberation by lysogenic *Escherichia coli*. J Bacteriol 62:293–300. https://doi.org/10.1128/jb.62.3.293-300.1951
20. Fritz C, Kost B (2020) Testing pollen tube proteins for in vivo binding to phosphatidic acid by n-butanol treatment and confocal microscopy. In: Methods in Molecular Biology. Humana Press Inc, pp 307–325
21. Horsch RB, Fry JE, Hoffmann NL et al (1985) A simple and general method for transferring genes into plants. Science 227:1229–1231. https://doi.org/10.1126/science.227.4691.1229
22. Stauffer TP, Ahn S, Meyer T (1998) Receptor-induced transient reduction in plasma membrane PtdIns(4,5)P_2 concentration monitored in living cells. Curr Biol 8:343–346. https://doi.org/10.1016/s0960-9822(98)70135-6

23. Zhao Y, Araki S, Wu J et al (2011) An expanded palette of genetically encoded Ca^{2+} indicators. Science 333:1888–1891. https://doi.org/10.1126/science.1208592
24. Chebli Y, Kaneda M, Zerzour R, Geitmann A (2012) The cell wall of the Arabidopsis pollen tube—spatial distribution, recycling, and network formation of polysaccharides. Plant Physiol 160:1940–1955. https://doi.org/10.1104/pp.112.199729
25. Baluška F, Hlavacka A, Šamaj J et al (2002) F-actin-dependent endocytosis of cell wall pectins in meristematic root cells. Insights from Brefeldin A-induced compartments. Plant Physiol 130:422. https://doi.org/10.1104/PP.007526
26. Emans N, Zimmermann S, Fischer R (2002) Uptake of a fluorescent marker in plant cells is sensitive to Brefeldin A and Wortmannin. Plant Cell 14:71–86. https://doi.org/10.1105/TPC.010339
27. Lippincott-Schwartz J, Yuan L, Tipper C et al (1991) Brefeldin A's effects on endosomes, lysosomes, and the TGN suggest a general mechanism for regulating organelle structure and membrane traffic. Cell 67:601–616. https://doi.org/10.1016/0092-8674(91)90534-6
28. Tse YC, Lo SW, Hillmer S et al (2006) Dynamic response of prevacuolar compartments to brefeldin A in plant cells. Plant Physiol 142:1442–1459. https://doi.org/10.1104/pp.106.090423
29. Parton RM, Fischer-Parton S, Watahiki MK, Trewavas AJ (2001) Dynamics of the apical vesicle accumulation and the rate of growth are related in individual pollen tubes. J Cell Sci 114:2685–2695. https://doi.org/10.1242/JCS.114.14.2685
30. Lazo GR, Stein PA, Ludwig RA (1991) A DNA transformation–competent Arabidopsis genomic library in Agrobacterium. Nat Biotechnol 9:963–967. https://doi.org/10.1038/nbt1091-963
31. Spitzer M, Wildenhain J, Rappsilber J, Tyers M (2014) BoxPlotR: a web tool for generation of box plots. Nat Methods 11:121–122. https://doi.org/10.1038/nmeth.2811
32. Assaad HI, Zhou L, Carroll RJ, Wu G (2014) Rapid publication-ready MS-Word tables for one-way ANOVA. Springerplus 3:474. https://doi.org/10.1186/2193-1801-3-474
33. Godi A, Di Campli A, Konstantakopoulos A et al (2004) FAPPs control Golgi-to-cell-surface membrane traffic by binding to ARF and PtdIns(4)P. Nat Cell Biol 6:393–404. https://doi.org/10.1038/ncb1119
34. Dowler S, Currie RA, Campbell DG et al (2000) Identification of pleckstrin-homology-domain-containing proteins with novel phosphoinositide-binding specificities. Biochem J 351:19–31. https://doi.org/10.1042/BJ3510019
35. Nakanishi H, de los Santos P, Neiman AM (2004) Positive and negative regulation of a SNARE protein by control of intracellular localization. Mol Biol Cell 15:1802–1815. https://doi.org/10.1091/mbc.E03-11-0798

Chapter 13

Implementing In Vitro and In Vivo Approaches to Analyze Lipid–Protein Interactions in Growing Tobacco Pollen Tubes

Eliška Škrabálková, Ondřej Novotný, Přemysl Pejchar, and Martin Potocký

Abstract

The spatial organization of anionic phospholipids such as PI4P, PI(4,5)P_2, phosphatidic acid, and phosphatidylserine is a defining feature of plasma membrane domains that support the extreme polarity of pollen tube growth. Many key protein regulators—including exocyst subunits, ROP pathway components, and cytoskeletal effectors—associate with the membrane through lipid-dependent mechanisms. This chapter provides comprehensive protocols for analyzing these interactions using complementary in vitro and in vivo approaches. We describe the production of tagged plant proteins in *Escherichia coli* and wheat-germ extracts, followed by assessment of lipid-binding specificity using protein–lipid overlays and quantitative liposome co-sedimentation assays. To test the physiological relevance of these interactions, we detail methods for transient transformation of tobacco pollen via biolistics, followed by microscopic colocalization of proteins of interest with genetically encoded fluorescent lipid reporters, and changes in their membrane recruitment upon overexpression of lipid-modifying enzymes. The integrated workflow enables multifaceted characterization of lipid–protein interactions in their native context and provides a versatile platform for studying membrane identity, signaling, and vesicle trafficking in plant tip-growing cells.

Key words Pollen tube, Lipid–protein interactions, Anionic phospholipids, Lipid biosensors, Transient transformation, Biolistics, Liposome co-sedimentation, Protein–lipid overlay, In vitro translation, Membrane recruitment

1 Introduction

Pollen tubes represent one of the most rapidly growing and highly polarized plant cell types. Their elongation occurs exclusively at the apex, where localized secretion delivers cell-wall materials and membrane components required for sustained tip growth and, ultimately, for sperm cell delivery during fertilization. This extreme polarity depends on the precise spatial organization of the plasma membrane (PM) into discrete exocytic and endocytic domains, as well as on rapid, targeted vesicle trafficking that supports continuous growth [1]. These processes are coordinated by a network of

Giampiero Cai et al. (eds.), *Pollen and Pollen Tube Analysis: Methods and Protocols*, Methods in Molecular Biology, vol. 3042,
https://doi.org/10.1007/978-1-0716-5308-1_13,

Rho-family GTPases (ROPs), protein kinases, phospholipases, cytoskeletal regulators, and a sharply patterned distribution of anionic phospholipids—including phosphatidylinositol 4-phosphate (PI4P), phosphatidylinositol 4,5-bisphosphate [PI $(4,5)P_2$], phosphatidic acid (PA), and phosphatidylserine (PS). The segregation of these lipids contributes to the electrostatic identity of specific membrane regions and mediates the recruitment of peripheral proteins that participate in membrane trafficking and signaling [1, 2]. Lipid–protein interactions at the PM, therefore, constitute a fundamental mechanism through which polarity is established and reinforced.

In tobacco and Arabidopsis pollen tubes, distinct phospholipid pools demarcate the apical and lateral PM domains, thereby defining the sites of polarized secretion and endocytosis. PI4P and PI $(4,5)P_2$ are strongly enriched in the extreme apex, where they function as key determinants of tip identity and participate in the regulation of ROP GTPases, calcium gradients, and vesicle fusion events [2]. In contrast, PA accumulates in a broader subapical region along the flanks of the tube, segregated from the apical PI $(4,5)P_2$-rich domain [3, 4]. This PA-enriched region supports endocytic membrane retrieval and is associated with processes that require membrane curvature or remodeling [3, 5]. Recent findings indicate that PS is likewise enriched at the apex, a localization established by P4-ATPase lipid flippases that mediate its delivery to the luminal leaflet of the PM [6]. The accumulation of PS contributes additional negative charge to the apical PM, which facilitates the recruitment of PS-binding regulatory proteins. Disrupting these lipid patterns, either enzymatically or pharmacologically, causes rapid depolarization of pollen tubes, manifesting as growth arrest and mislocalization of polarity-associated proteins [7]. Such perturbations underscore the importance of maintaining the native lipid composition of the apex for sustained tip growth.

Many pollen tube regulators—including ROPs and their GEFs/GAPs, actin-associated proteins, kinases, and components of the exocytic and endocytic machineries—associate with the PM in a lipid-dependent manner [8]. A clear example of such dependence is found in the exocyst complex, an evolutionarily conserved heterooctamer responsible for tethering secretory vesicles to the PM. In plants, the exocyst contains numerous EXO70 subunit paralogs, many of which display distinct localization patterns in tip-growing cells and bind lipids with different specificities. In tobacco pollen tubes, EXO70A1a localizes to the extreme apex, corresponding to the zone of active exocytosis, whereas EXO70B1 occupies a subapical collar associated with clathrin-mediated endocytosis [9]. These isoforms exhibit minimal spatial overlap and instead colocalize with different phospholipid reporters: EXO70A1a with $PI(4,5)P_2$ at the tip and EXO70B1 with PA in the subapical domain. EXO70 proteins are known to bind PI4P, PI $(4,5)P_2$, PA, and PS; indeed, the C-terminal domain of EXO70A1

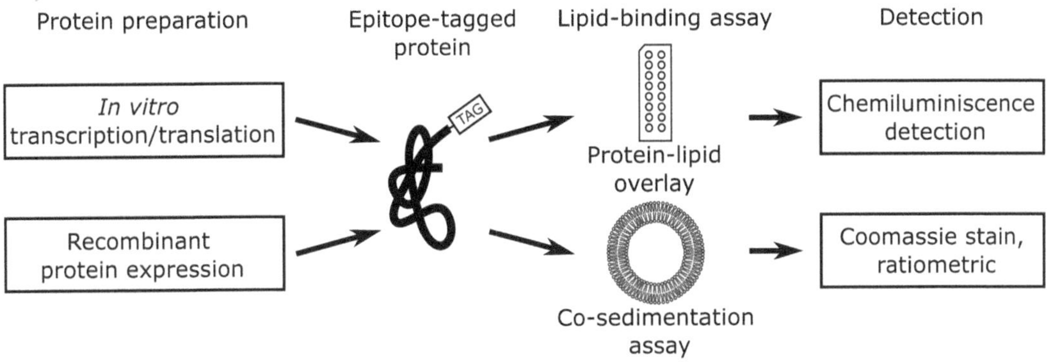

Fig. 1 Simplified pipeline for various protein preparation and protein-lipid binding techniques described in the text

has recently been shown to mediate PM association via multiple anionic phospholipids, underscoring the role of a specific lipid–protein code in exocyst recruitment [10]. Lipid–protein interactions of this type highlight the pivotal role of anionic phospholipids in specifying membrane identity and modulating the localization of regulatory complexes.

Dissecting the biochemical basis of such lipid-mediated protein localization requires methods capable of determining lipid selectivity and binding affinity under controlled conditions. Two in vitro assays—protein–lipid overlay (PLO, commonly known as "lipid strip" binding) and liposome co-sedimentation—are widely used for this purpose. PLO assays provide a straightforward qualitative readout of binding preferences across a panel of lipids, while liposome co-sedimentation allows a more quantitative analysis of binding strength and specificity using defined synthetic liposomes [11–13]. These techniques are technically accessible, low-cost, and adaptable for high-throughput testing of multiple protein variants, including EXO70 paralogs or engineered mutants of other pollen tube regulators. Proteins can be expressed recombinantly in *E. coli* or synthesized in cell-free wheat germ systems, which often alleviate solubility problems encountered with membrane-associated plant proteins [14, 15] (Fig. 1). Because lipid-binding properties are not easily predicted from amino acid sequence alone, such biochemical assays remain essential for confirming candidate lipid interactions [13]. However, in vitro assays necessarily rely on artificial membrane systems that lack the organization, curvature, compositional complexity, and electrostatic gradients characteristic of the pollen tube PM. Consequently, in vitro binding results may not accurately reflect whether a given protein requires a particular lipid for membrane recruitment in vivo.

To address this limitation, microscopy-based assays have become indispensable for testing the lipid dependence of protein localization in living cells. Genetically encoded fluorescent lipid biosensors—such as the Spo20p-PABD for PA, PH domains for PI $(4,5)P_2$, the P4M domain for PI4P, and the LactC2 domain for PS—allow live-cell imaging of lipid dynamics at high spatial and temporal resolution [16]. When transiently expressed in pollen tubes, these sensors reveal strongly polarized lipid distributions, such as the PA-enriched subapical annulus and the $PI(4,5)P_2$-enriched tip, and they enable real-time monitoring of lipid remodeling in response to experimental perturbations. Molecular tools that manipulate lipid composition in vivo further strengthen these approaches. Overexpression or mislocalization of lipid-modifying enzymes, including PI4P 5-kinases, PI phosphatases, diacylglycerol kinases, and phospholipase D isoforms, can locally elevate or deplete specific lipid species. Such perturbations provide a means to assess whether the PM targeting a protein of interest is sensitive to changes in the abundance of a particular lipid [17].

Several microscopy-based readouts can thus be used to probe lipid–protein interactions in vivo. Colocalization assays determine whether the spatial distribution of a protein overlaps with that of a lipid sensor, providing initial evidence of co-enrichment within specific membrane nanodomains. Relocalization assays test how protein distribution changes when lipid composition is experimentally modified. Elevating $PI(4,5)P_2$ levels through overexpression of PI4P 5-kinases, for example, can broaden or increase the PM association of $PI(4,5)P_2$-binding proteins, including EXO70 isoforms whose recruitment depends on this lipid [10]. Conversely, depletion of PI4P or $PI(4,5)P_2$ using targeted phosphatases can cause candidate lipid-binding proteins to redistribute from the PM to the cytosol [18]. Pharmacological treatments provide additional means of perturbation. Inhibiting PA synthesis using 1-butanol (which blocks phospholipase D activity) or inhibitors targeting diacylglycerol kinases induces rapid loss of the characteristic apical vesicle-rich "clear zone," accompanied by dissociation of PA-binding proteins from the PM. Similarly, inhibition of PI4P production by phenylarsine oxide (PAO) leads to rapid detachment of PI4P-binding proteins from the PM [19]. Quantitative analysis of fluorescence signals—such as PM-to-cytoplasm intensity ratios or tip-to-shank distribution profiles—enhances the precision of these in vivo assays and enables rigorous comparisons across treatments and protein variants.

The combined application of in vitro lipid-binding assays and in vivo microscopy, therefore, constitutes a comprehensive and mutually reinforcing workflow for elucidating protein–lipid interactions in pollen tubes. Biochemical assays identify candidate lipid interactions and allow mechanistic hypotheses to be formulated under well-defined conditions, whereas imaging approaches confirm their functional relevance within the polarized physiological environment of the growing pollen tube. This integrated strategy is particularly effective for investigating the membrane recruitment of key peripheral regulators such as exocyst subunits, ROP GTPase components, and cytoskeletal effectors, whose spatially confined localization determines the robustness and directionality of tip growth.

In this chapter, we integrate these complementary approaches and present protocols optimized for pollen tubes that analyze biochemical binding profiles obtained from PLO and liposome co-sedimentation assays of proteins generated by both recombinant protein expression and plant cell-free wheat germ systems and supplement them with in vivo methods like colocalization of proteins with fluorescent phospholipid biosensors and evaluation of lipid-dependent membrane recruitment by overexpressing or misdirecting lipid-modifying enzymes. By linking established biochemical methodologies with state-of-the-art live-cell imaging, this chapter provides a comprehensive methodological framework for dissecting the lipid–protein interfaces that underlie pollen tube polarity and growth.

2 Materials

All solutions are prepared with ultrapure water produced by a Milli-Q ultrapure filtration system or similar device. Water resistivity was approximately 18.2 M-Ohms.

2.1 Expression and Purification of Recombinant Proteins in E. coli

1. *E. coli* strain allowing expression of recombinant eukaryotic proteins. For example, we use ArcticExpress DE3 cells (Agilent Technologies).
2. Luria–Bertani medium (LB): Mix 1.51 g Luria Broth base and 0.95 g NaCl in 100 mL of water, autoclave sterilize, and store at room temperature.
3. Solid LB medium plates: Mix 1.51 g Luria Broth base, 0.95 g NaCl, and 1.5 g of Micro agar (e.g., Duchefa Biochemie, cat. no. M1002) in 100 mL water, autoclave sterilize, cool to room temperature, add antibiotics, store at 4 °C.
4. Gentamicin sulfate (stock solution 25 mg/mL in water, filter sterilized).
5. Ampicillin (stock solution 100 mg/mL in water, filter sterilized).
6. Lactose monohydrate.
7. Purification buffer: Dissolve 0.61 g Tris in 50 mL of water, adjust the pH with HCl to 7.5. Add 1.17 g NaCl, 0.1 μL β-mercaptoethanol, 5 μL glycerol, and 100 μL protease inhibitor mixture (e.g., Sigma-Aldrich, cat. no. T9599). Add water to 100 mL and store at 4 °C.
8. Glutathione Sepharose™ 4B (Cytiva).

9. Elution buffer: Dissolve 77 mg of reduced L-glutathione (e.g., Sigma-Aldrich, cat. no. G4251) in 10 mL of purification buffer. Store at 4 °C.
10. Probe sonicator.

2.2 In Vitro Coupled Transcription/Translation

1. Wheat germ in vitro protein expression system (e.g., TNT® SP6 High-Yield Wheat Germ Protein Expression System; Promega). Store in 30 μL aliquots at −80 °C.
2. Plasmid vector allowing expression of recombinant protein by T7 RNA polymerase (e.g., pTNT from Promega).

2.3 Protein-Lipid Overlay Assay

1. Lipids of interest applied on the membrane (e.g., Membrane lipid strips, Echelon Biosciences, cat. no. P-6002).
2. 10× Phosphate-buffered saline (10 × PBS): Dissolve 8 g NaCl, 0.2 g KCl, 2.4 g $Na_2HPO_4 \times 7H_2O$ and 0.2 g KH_2PO_4 in 100 mL of water. Adjust pH to 7.4 and store at 4 °C.
3. 1× PBS containing 0.1% Tween 20 (PBST): Dilute 10 × PBS in 1:9 water and add 1/1000th volume of Tween 20. Store at 4 °C.
4. Blocking solution: 3% bovine serum albumin in PBST. Store at 4 °C.
5. Antibody recognizing affinity tag (e.g., anti-glutathione S-transferase [GST] antibody or anti-HA (ThermoFisher, cat. no. 26183).
6. Secondary antibody coupled to horse radish peroxidase (HRP).
7. Enhanced chemiluminescence (ECL) reagent (e.g., Cytiva, Amersham™ ECL™ Prime, cat. no. RPN2236).
8. Chemiluminescence imager (e.g., BioRad, ChemiDoc XRS).

2.4 Liposome Co-Sedimentation Assay

1. Mini-extruder set with holder (e.g., Avanti Polar Lipids, cat. no. 610000)
2. Phospholipid stock solutions: 4 mM 1,2-dioleoyl-sn-glycero-3-phosphocholine (DOPC), 4 mM 1,2-dioleoyl-sn-glycero-3-phosphoethanolamine (DOPE), 4 mM 1,2-dioleoyl-sn-glycero-3-phosphate (DOPA), 4 mM 1,2-dioleoyl-sn-glycero-3-phosphoserine (DOPS) dissolved in chloroform.
3. Phosphoinositide stock solutions: 500 μM 1,2-dioleoyl-sn-glycero-3-phospho-(1′-myo-inositol-4′-phosphate) (DOPI4P) and 500 μM 1,2-dioleoyl-sn-glycero-3-phospho-(1′-myo-inositol-4′,5′-bisphosphate) ($DOPIP_2$) are dissolved in a mixture of chloroform:methanol:water in a ratio of 20:9:1. We obtain phospholipid and polyphosphoinositol mixtures from Avanti Polar lipids. Products from other suppliers may

also be appropriate. Store lipid solutions in sealed glass vials at −20 °C.

4. Extrusion buffer: Dissolve 3.72 g raffinose pentahydrate in 20 mL of warm water. After cooling, add 75.70 mg Tris–HCl pH 7.5, and 3.85 mg dithiothreitol (DTT). Add water to obtain a final volume of 25 mL (*see* **Note 1**).
5. Binding buffer: Mix 0.56 g KCl, 0.18 g Tris–HCl pH 7.5, 9.24 mg DTT, and 17.5 mg EDTA in 60 mL water (*see* **Note 1**).
6. 6× sodium dodecyl sulfate (SDS) loading buffer: Mix 1.2 mL of 0.5 M Tris–HCl, pH 6.8, 4.7 mL glycerol, 1.2 g SDS, 0.93 g DTT, and 6 mg of bromophenol blue, and add water to 10 mL. Store in 1.0 mL aliquots at −20 °C.

2.5 Tobacco Pollen Cultivation

1. Tobacco plants (Nicotiana tabacum): cv. Petit Havana SR1 or Samsun are suitable. Plants should be grown for approximately 2 months under long-day conditions (16 h light/8 h dark) at approximately 20 °C in a growth chamber or greenhouse.
2. Harvested pollen: Fresh pollen is collected from fully opened flowers showing pink corolla and opened anthers (avoid senescent, purple flowers). For applications that need frequent repetition, working with freeze-stored dry pollen is more convenient. For that, harvest unopened flowers, extract the anthers onto filter paper, and leave them to dehisce overnight. The next day, sift the dried pollen into a tube and freeze at −20 °C. Pollen prepared this way can be stored for several years without significant loss of germination efficiency.
3. 1.5% casein acid hydrolysate: Dilute 1.5 g casein acid hydrolysate (e.g., Millipore cat. no. A2427) in 100 mL water. Store in 10 mL aliquots at −20 °C.
4. 10× salts solution (8 mM $MgSO_4$, 10 mM $CaCl_2$, 16 mM H_3BO_3, 10 mM KCl, and 0.3 mM $CuSO_4$): Dilute 986 mg $MgSO_4 \times 7H_2O$, 735 mg $CaCl_2 \times 2H_2O$, 493 mg H_3BO_3, 373 mg KCl and 37.5 mg $CuSO_4 \times 5H_2O$ in 500 mL of water. Store in 50 mL aliquots at −20 °C.
5. 2× liquid pollen germination medium (2 × PTNT) stock, that contains 10% (w/v) sucrose, 25% (w/v) PEG-6000, 2 mM $CaCl_2$, 2 mM KCl, 1.6 mM $MgSO_4$, 3.2 mM H_3BO_3, 60 μM $CuSO_4$, 0.06% (w/v) casein acid hydrolysate and 0.6% (w/v) MES. For 500 mL of 2 × PTNT, mix 50 g of sucrose, 3 g MES (e.g., Duchefa, cat. no. M1503), 20 mL 1.5% casein acid hydrolysate, and 100 mL 10× salts solution. Adjust pH to 5.9 with KOH and add 125 g PEG-6000 (e.g., Sigma-Aldrich, cat. no. 81255). Fill up with water to obtain a final volume of 500 mL. Store in 50 mL aliquots at −20 °C.

6. Petri dishes, 5.5 cm diameter.
7. Solid PTNT medium plates: For approximately 20 plates, heat up 50 mL aliquot of 2 × PTNT. Mix 0.25 g Phytagel (e.g., Sigma-Aldrich, cat. no. P8169) in 50 mL of water and heat until dissolved. Mix equal volumes of hot 2 × PTNT and hot Phytagel solution to obtain 1 × PTNT containing 0.25% Phytagel. Pour approximately 4 mL per 5.5-cm Petri dish and allow it to solidify in the laminar box. Store upside down at 4 °C (*see* **Note 2**).
8. Tweezers, scalpel, filter sieve, 50 mL tubes, vortex mixer.

2.6 Biolistic Pollen Transformation

1. PDS-1000/He™ biolistic particle delivery system with 1100 psi rupture disks, macrocarriers, and stopping screens (Bio-Rad).
2. Vortex mixer with multi-tube rack, microcentrifuge.
3. Vacuum pump, Büchner flask with fritted support.
4. Nylon filter membranes, pore size 0.45–0.8 μm, diameter 47 mm (e.g., Whatman® cat. no. 7408-004)
5. Absolute ethanol (molecular biology grade).
6. Gold particles: Resuspend 30 mg 1.0–1.6 μm diameter particles (e.g., Bio-Rad cat. no. 1652264) in 1 mL absolute ethanol, vortex for 3 min. Spin 1 min at 14000 × g, discard supernatant, wash twice with 1 mL water, and resuspend in 1 mL 50% sterile glycerol. Store at 4 °C.
7. 2.5 M $CaCl_2$: Dissolve 3.675 g $CaCl_2 \times 2H_2O$ (Sigma-Aldrich, C-2536) in 10 mL H_2O, freeze 0.5 mL aliquots at −20 °C. Store opened aliquot at 4 °C.
8. Protamine sulfate (1 mg/mL): Dissolve 1–2 mg of protamine sulfate (Sigma-Aldrich cat. no. P4505) in a corresponding volume of water and freeze 0.5 mL aliquots at −20 °C. Store opened aliquot at 4 °C.
9. Constructs expressing proteins of interest with fluorescent tags (0.25–1 μg/μL). Suitable promoters include LAT52 and UBQ10.
10. Constructs expressing genetically encoded lipid sensors (e.g., PI4P, PI(4,5)P_2, PA, and PS) for colocalization with proteins of interest [19].
11. Constructs expressing lipid-modifying enzymes (e.g., phospholipases D and C, PI4P-5-kinases, SAC phosphatases) for perturbation experiments [7, 16].
12. Solid PTNT medium plates (*see* Subheading 2.5).

2.7 Pollen Tube Imaging and Data Analysis

1. Microscopic imaging chambers with bottom coverslips with a thickness suitable for confocal imaging (e.g., LAB-TEK cat. no. 155360).
2. Micro spatula with flat end.
3. Confocal laser scanning or spinning disk microscope equipped with: 10× objective, 40× and 63× oil or water-immersion objective, laser lines appropriate for fluorophores used (e.g., 488 nm for GFP, 514 nm for YFP, 561 nm for mCherry), and filters enabling detection of fluorophores in appropriate detection windows (e.g., 506–538 nm for GFP, 528–556 nm for YFP, 588–626 nm for mCherry). Bright-field channel is optional but beneficial.
4. Fiji/ImageJ software for quantitative fluorescence analysis (including membrane–cytoplasm intensity ratio measurement, line-scan analysis, and colocalization) [20].

3 Methods

3.1 Expression and Purification of Recombinant Proteins in E. coli

1. Prepare plasmids encoding affinity-tagged proteins of interest (e.g., GST fusions, *see* **Note 3**). Transform recombinant vectors into a BL21-based expression strain suitable for eukaryotic proteins, such as ArcticExpress DE3 (Agilent).
2. Plate transformants on LB agar containing appropriate antibiotics and incubate overnight at 37 °C.
3. Pick a single bacterial colony and inoculate 10 mL of liquid LB medium containing the same antibiotics (*see* **Note 4**). Incubate overnight at 37 °C at 160 rpm.
4. Dilute the grown culture into 1 L of LB medium supplemented with the same antibiotics and grow at 37 °C until the culture attains an optical density at 600 nm (OD_{600}) of approximately 0.8.
5. Transfer the bacterial culture to room temperature and induce expression of the tagged protein (e.g., addition of lactose to a final concentration of 1 g/l for pGEX vectors, *see* **Note 5**). Incubate for a time and at a temperature that allows optimal expression of the target protein; typically, 20 h at 15 °C and 160 rpm.
6. Centrifuge the culture at 7000 × g and 4 °C for 10 min. Discard the supernatant and freeze the pellet at −20 °C.
7. Thaw the pellet on ice, add cold purification buffer (50 mL), and resuspend the cells.
8. Lyse bacterial cells using a probe sonicator while keeping the samples on ice to avoid warming. Clear the lysate by centrifugation at 20000 × g and 4 °C for 1 h.

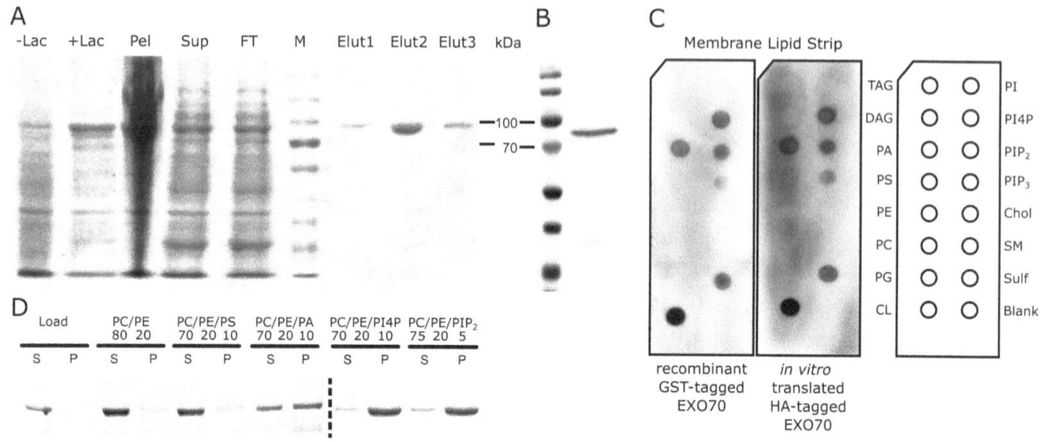

Fig. 2 (**a**) Purification of Arabidopsis EXO70 isoform induced in *E. coli*. –Lac non-induced control culture, +Lac lactose induced culture, Pel sonicated bacterial pellet, Sup supernatant used on Glutathione Sepharose 4B column, FT flow-through, M protein marker, Elut1, Elut2, Elut3, multiple elutions. (**b**) Western blot analysis of in vitro translated HA-tagged EXO70 isoform from Arabidopsis. (**c**) Comparison between protein-lipid overlay assays performed with recombinant GST-tagged Arabidopsis EXO70 isoform purified from *E. coli* lysates or in vitro generated EXO70 isoform fused to an HA epitope tag. Dark spots indicate binding of a tagged protein to a specific lipid. The scheme on the right depicts lipid layout. TAG triglyceride, DAG diacylglycerol, PA phosphatidic acid, PS phosphatidylserine, PE phosphatidylethanolamine, PC phosphatidylcholine, PG phosphatidylglycerol, CL cardiolipin, PI phosphatidylinositol, PI4P phosphatidylinositol 4-phosphate, PIP_2 phosphatidylinositol 4,5-bisphosphate, PIP_3 phosphatidylinositol 3,4,5-trisphosphate, Chol cholesterol, SM sphingomyelin, Sulf 3-sulfogalactosylceramide. (**d**) Results from co-sedimentation lipid-binding assays of GST-tagged Arabidopsis EXO70 isoform purified in *E. coli*. The phospholipid compositions of LUVs are provided as percentages. S supernatant, P pellet

9. Load the supernatant onto a column containing Glutathione Sepharose™ 4B, discard the flow-through, wash with 50 mL of cold purification buffer to remove unbound proteins, and elute bound proteins with 1 mL elution buffer containing reduced glutathione. Repeat the elution **steps 2–4** times. Keep elution fractions separately (*see* **Note 6**).
10. Evaluate the purity and quality of the purified protein by SDS–PAGE and Coomassie brilliant blue staining (Fig. 2a). Determine protein concentration, and store proteins on ice for immediate use in protein–lipid overlay and liposome co-sedimentation assays or at −80 °C for longer-term storage.

3.2 Preparation of In Vitro Translated Protein

1. Thaw a 30 μL aliquot of wheat-germ expression mix (e.g., TNT SP6 High-Yield Wheat Germ Protein Expression System; Promega cat. no. L3260) on ice.
2. Add 4–10 μg plasmid DNA and nuclease-free water to a final reaction volume of 50 μL (*see* **Notes 3** and 7). Mix gently by pipetting up and down.
3. Incubate the reaction at 25 °C for 2 h.

4. Verify protein integrity and in vitro translation efficiency by Western blotting using an antibody against the affinity tag (Fig. 2b).
5. Use in vitro translated proteins immediately. For liposome co-sedimentation, clear the translation mix by centrifugation at 72000 × g and 4 °C for 20 min to remove sedimenting contaminants.

3.3 Protein–Lipid Overlay Assay

1. Prepare 3 mL fresh blocking solution and transfer it into a small container (*see* **Note 8**).
2. Place the membrane lipid strip carefully into the container using gloves and tweezers.
3. Incubate for 30 min at room temperature with gentle agitation.
4. Add affinity-tagged protein of interest to a final concentration of 1–5 μg/mL and incubate with the lipid strip on a shaker for 1.5 h at room temperature (*see* **Notes 9** and **10**).
5. Wash the membrane three times for 10 min each in PBST.
6. Incubate the membrane with primary antibody against the affinity tag diluted in blocking solution for 1 h at room temperature (*see* **Notes 9** and **11**).
7. Wash the membrane three times for 10 min each in PBST.
8. Incubate the membrane with a secondary HRP-conjugated antibody diluted in blocking solution for 50 min at room temperature (*see* **Note 11**).
9. Wash the membrane three times for 10 min each in PBST.
10. Apply ECL solution gently onto the membrane, incubate for approximately 4 min, and image the chemiluminescent signal using an appropriate imaging system (Fig. 2c, *see* **Note 12**).

3.4 Liposome Co-Sedimentation Assay

1. Mix phospholipid stocks in the desired molar ratios in a 1.5 mL tube (total amount 400–800 nmol per reaction). For example, use DOPC:DOPE (80:20) for neutral liposomes and include 5–20% anionic lipids (e.g., DOPA or $DOPIP_2$) in place of DOPC for negatively charged liposomes.
2. Evaporate the organic solvent in a vacuum concentrator to obtain a dry lipid film.
3. Add 500 μL of extrusion buffer to the dried lipids and slowly hydrate the pellet at room temperature on a rotator for 90 min.
4. Disperse the hydrated lipid suspension using bath sonication for 5–10 min (*see* **Note 13**).
5. Assemble the lipid extruder (Fig. 3a, b) with a 200 nm polycarbonate membrane according to the manufacturer's instructions (*see* **Notes 14** and **15**).

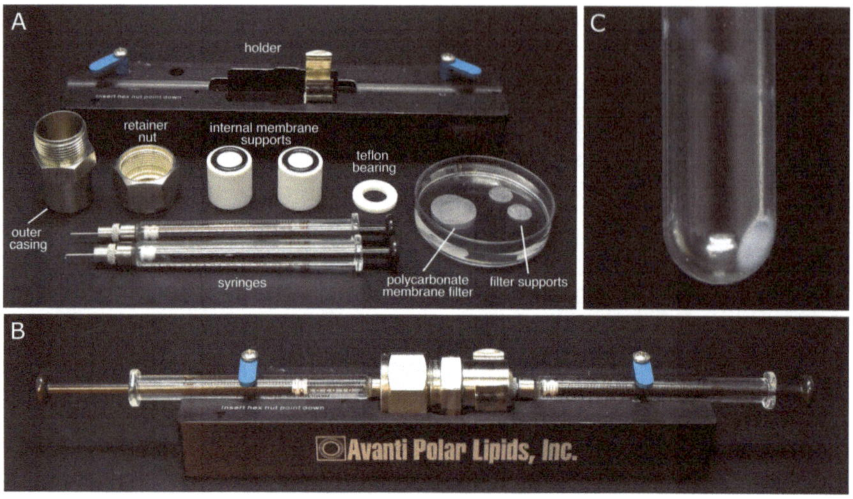

Fig. 3 (**a**) Disassembled Avanti® mini-extruder with individual parts. (**b**) The fully assembled Avanti® mini-extruder with inserted syringes. (**c**) A pellet of liposomes after centrifugation

6. Load the lipid suspension into the syringes and extrude the sample through the membrane 20 times to obtain large unilamellar vesicles (LUVs) of uniform diameter (Fig. 3b, *see* **Notes 16** and **17**).
7. Mix the LUV suspension with 500 μL of binding buffer.
8. Centrifuge at room temperature for 20 min at 74000 × g, discard the supernatant, and resuspend the pellet (Fig. 3c) in 500 μL binding buffer (*see* **Note 18**). Repeat this washing step three times to remove the extrusion buffer from the exterior of the LUVs (*see* **Note 19**).
9. Resuspend the final LUV pellet in 100 μL binding buffer.
10. Add 5 μg of recombinant protein or 25–50 μL of in vitro translated protein to the LUV suspension and adjust the volume to 200 μL with binding buffer. Incubate on a shaker at room temperature for 45 min (*see* **Note 20**).
11. Centrifuge at room temperature for 20 min at 74000 × *g* to separate LUV-bound protein (pellet) from unbound protein (supernatant). Keep supernatant (*see* **step 14** in this section).
12. Wash the pellet in 400 μL binding buffer to remove residual unbound protein, repeating the washing step two to three times.
13. Resuspend the pellet containing lipid-bound proteins in 50 μL 6× SDS loading buffer.
14. Precipitate unbound proteins from the supernatant by adding three volumes of cold acetone and incubating for 1 h at −20 °C. Centrifuge at 20000 × g and 4 °C for 30 min, discard

the supernatant, and resuspend the pellet in 50 μL 6× SDS loading buffer.

15. Denature all samples by heating at 70 °C for 10 min, then analyze lipid-bound and unbound proteins by SDS–PAGE and Coomassie brilliant blue staining (Fig. 2d). Quantify band intensities using image analysis software (*see* **Notes 21** and **22**).

3.5 Preparation of DNA Macrocarriers and Transient Biolistic Transformation

1. Thoroughly vortex gold particle stock for about 3 min to ensure an even suspension (*see* **Note 23**).
2. For co-expression experiments, premix the two DNA plasmids (*see* **Note 24**).
3. Coat gold particles with DNA under continuous vortexing in a 1.5 mL tube by adding: 25 μL gold suspension, 2.5–5 μL total plasmid DNA (0.25–1 μg/μL; LAT52 promoter preferred, UBQ10 optional), 25 μL 2.5 M $CaCl_2$, and 10 μL protamine (1 mg/mL). Vortex for 3 min (*see* **Note 23**).
4. Centrifuge at room temperature for 30 s at 14000 × g, carefully remove the supernatant.
5. Add 200 μL absolute ethanol, vortex for 3 min, centrifuge again, remove the supernatant, and resuspend the pellet in 18 μL absolute ethanol (*see* **Note 25**).
6. Pipette the suspension evenly onto a macrocarrier mounted in the macrocarrier holder. Allow ethanol to evaporate completely in a vibration-free environment to obtain a uniform particle coating (*see* **Note 26**).
7. Transfer pollen to solid 1 × PTNT plates: Prepare liquid 1 × PTNT medium by diluting 2 × PTNT in a ratio of 1:1 with sterile water (use 5 mL per sample).
8. If using freeze-stored pollen stock, weigh 1 mg per sample, allow it to recover for 10 min at room temperature before adding 5 mL of liquid 1 × PTNT medium. If fresh pollen is available, collect fully opened tobacco flowers with pink corollas, excise 5–10 anthers for one sample with sterile tweezers, and place them into 5 mL liquid 1 × PTNT.
9. Vortex the suspension thoroughly to obtain a homogeneous pollen slurry. If fresh anthers were used, remove them using a sterile filter sieve or tweezers.
10. Pre-wet a polyamide or nylon membrane (0.45–0.8 μm pore size) mounted on a vacuum filtration system. Filter 5 mL pollen suspension through the membrane so that pollen grains are retained on the surface.
11. Using tweezers, invert the membrane onto the solid PTNT plate, gently press for 1–2 s, then lift to leave a uniform monolayer of pollen grains on the medium surface (*see* **Note 27**). For

transient biolistic transformation, continue immediately to pollen bombardment. For experiments using stably transformed pollen, incubate plates at 21 °C in the dark until pollen tubes reach the desired developmental stage.

12. Assemble the gold-coated macrocarrier, stopping screen, and 1100 psi rupture disk in the PDS-1000/He biolistic device according to the manufacturer's instructions.
13. Place PTNT pollen plates into the second slot below the macrocarrier within the chamber.
14. Evacuate the chamber to 28 in Hg, fire the shot, release the vacuum immediately, seal plates with parafilm, and incubate them upside down at 21 °C in the dark for 5–9 h to allow expression of the introduced constructs.

3.6 Imaging, Data Analysis, and Quantification

1. Turn on the confocal microscope and allow the lasers to warm up.
2. Using a scalpel, cut a ~1 × 2 cm section of solid PTNT medium containing pollen tubes, scoop it with a flat-ended spatula, and carefully transfer upside down to a microscopy imaging chamber with a coverslip bottom, so the pollen tubes face the coverslip.
3. Locate suitable pollen tubes using a 10× objective and then acquire high-resolution images with a 40× water immersion or 63× oil or water immersion objective. Keep imaging parameters (laser power, acquisition time, detector gain, averaging, pinhole, zoom, etc.) constant for all samples and treatments.
4. For estimating the growth rate, acquire a second image of the same pollen tube 1 min after the first image while minimizing phototoxicity.
5. Export images as TIFF files with grayscale lookup tables preserved and open them in Fiji/ImageJ.
6. For growth-rate determination, merge the first and second images into an RGB composite using the "Merge Channels" function.
7. Measure the displacement of the pollen tube apex between the first and second image with the straight-line tool; the measured distance indicates the growth rate in μm/min (*see* **Note 28**).
8. For PM-to-cytoplasm fluorescence ratio analysis, draw three horizontal lines of interest (ROIs) at standardized distances from the apex and add them to the ROI Manager (Fig. 4a, *see* **Note 29**).
9. For each ROI, use the "Plot Profile" function to obtain intensity profiles across the pollen tube tip (Fig. 4b).

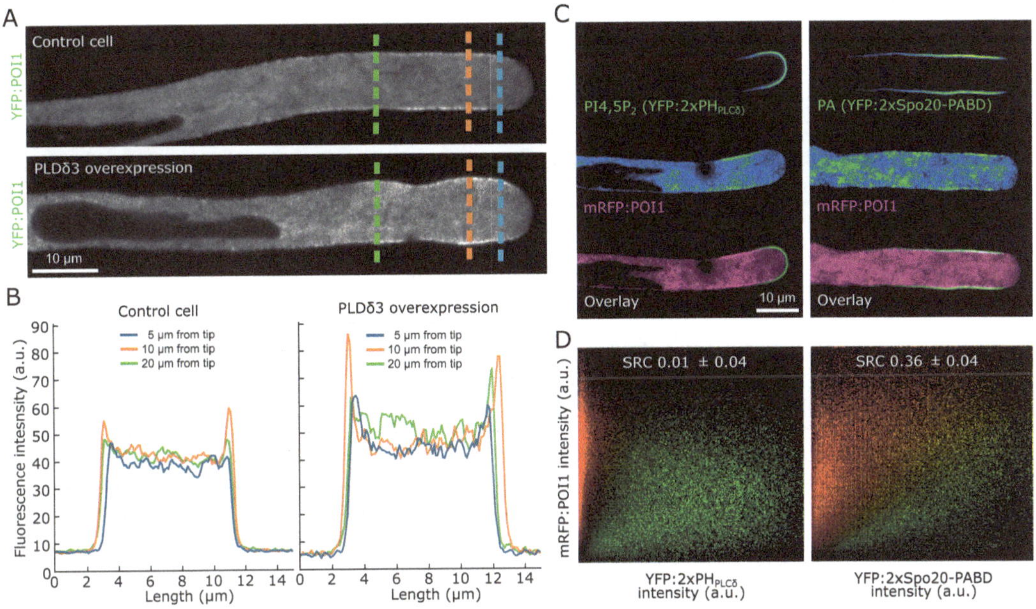

Fig. 4 (**a**) Elevation of PA levels by the overexpression of the tobacco phospholipase Dδ3 (PLDδ3) results in increased recruitment of the YFP-tagged protein of interest (YFP:POI1) to the plasma membrane in tobacco pollen tubes. Colored dashed lines mark the site of intensity profiles measured at 5, 10, and 20 µm distance from the pollen tube tip. (**b**) Quantification of relative levels of PM-associated and cytoplasmic fluorescence displayed by YFP:POI1 in tobacco pollen tubes overexpressing PLDδ3 compared to control tubes expressing YFP:POI1 alone. Data represent intensity profiles measured along the dashed lines shown in the upper panel, measured at 5, 10, and 20 µm distance from the pollen tube tip. (**c**) Colocalization analysis of mRFP:POI1 with the genetically encoded sensors for $PI4,5P_2$ (YFP:2 × $PH_{PLC\delta}$) and PA (YFP:2 × xSpo20-PABD) at the PM of tobacco pollen tubes. (**d**) Quantification of the colocalization measured as the Spearman correlation coefficient (SRC) between mRFP:POI1 and the respective lipid sensor at the PM of tobacco pollen tubes and visualized as a color scatter plot. Scale bar = 10 µm

10. Determine PM fluorescence as the mean of the two peak maxima at both plasma membrane edges.
11. Determine cytoplasmic fluorescence as the mean intensity within a 4 µm-wide central region between the two peaks.
12. Calculate the PM/Cyt fluorescence ratio for each ROI as PM fluorescence divided by cytoplasmic fluorescence.
13. Average the values from the three ROIs to obtain a PM/Cyt ratio for each pollen tube.
14. For colocalization analyses of proteins of interest with lipid biosensors (Fig. 4c), calculate Pearson or Spearman correlation coefficients using the Fiji Coloc2 plug-in (Fig. 4d).

4 Notes

1. It is important to prepare extrusion and binding buffers fresh before use.
2. Because PTNT–Phytagel medium is highly viscous and begins to solidify quickly, pre-warming both 2 × PTNT and Phytagel and pipetting slowly but consistently ensures homogeneous plates. An overly hot medium may cause excessive evaporation or bubble formation.
3. For expression and purification of recombinant proteins in *E. coli*, we use a glutathione S-transferase (GST) affinity tag. However, other tags, such as maltose-binding protein (MBP) or 6 × histidine tags, may also work well. For expression of recombinant proteins in in vitro coupled transcription/translation systems, we have found that a hemagglutinin (HA) tag is the most suitable. However, the FLAG tag can also be used.
4. Inoculation can be done by washing the colony (pipetting up and down) or by leaving the toothpick/pipette tip in the liquid LB medium.
5. 0.1–1 mM isopropyl β-D-1-thiogalactopyranoside (IPTG) can also be used for the induction of protein expression. We prefer to use lactose over IPTG as it is a cost-effective variant with lower innate toxicity to the bacterial culture [21]. The concentration of the inducer can be optimized to achieve the best protein expression.
6. The number of elution repetitions can be adjusted based on the protein of interest.
7. The coding sequence of the gene of interest must be cloned into a suitable vector carrying an SP6 promoter (e.g., pTNT vector, when TNT® SP6 High-Yield Wheat Germ Protein Expression System is used). The cloning is described in detail in ref. [22].
8. To reduce the volume of reagents, a sealed plastic bag can be used, provided that the lipid-spotted part of the membrane does not directly touch the plastic surface.
9. This step can also be done overnight at 4 °C.
10. If an in vitro translated protein is used, the amount of protein can differ according to its size and complexity. Typically, at least half (25 μL) of the reaction must be used.
11. The incubation time can be adjusted according to the specific antibody.
12. The strips are sensitive to touch, and they should be kept with the lipid side facing up. Therefore, an imager equipped with a

top camera must be used. The risk of non-specific signals is increased when strips are oriented with lipids facing down.

13. After sonication, the sample will be cloudy or opaque.
14. The polycarbonate membrane filter must be placed precisely at the center of the internal membrane support to avoid wrinkles and possible rupture of the filter.
15. The two hexagonal parts of the retainer nut and extruder outer casing should be aligned, and both parts should be tightly fitted.
16. Polycarbonate membranes are not intended for multiple samples and should be replaced after each extrusion. Also, needle-syringe connections should be re-tightened after each extrusion.
17. After the extrusion, the sample's opacity will decrease.
18. Small volume (1–2 mL), thick-walled transparent polycarbonate tubes should be used. The side of the tube where the pellet is expected should be marked for easier identification. Liposomes will sediment into a small, highly translucent or slightly opaque pellet (Fig. 3c).
19. It is necessary to remove the supernatant immediately after the centrifugation is finished, otherwise the pellet rapidly resuspends.
20. Check that the protein of interest does not sediment in the absence of LUVs before performing the LUVs co-sedimentation assay. See the "Load" sample in Fig. 2d for an example of this type of control.
21. The gel analyzer tool in the open-source software Fiji [20] can be used.
22. When an in vitro translated protein is used for the co-sedimentation assay, visualize the lipid-bound and unbound fractions by performing Western blotting with an antibody recognizing the affinity tag [16].
23. Gold particles sediment rapidly even in 50% glycerol; always vortex thoroughly ($\geq$3 min) immediately before taking aliquots to ensure consistent particle density in every preparation. Avoid pipetting from the surface or bottom of the stock, as particle concentration is uneven.
24. Determine the mixing ratio experimentally. Generally, use only a small amount (0.5–1 μg of plasmids expressing phospholipid markers to prevent the perturbation of phospholipid signaling due to overexpression of the lipid-binding domain.
25. Ensure that the pellet is fully resuspended in ethanol by vigorous vortexing and pipetting along the tube walls. Incomplete resuspension leads to particle aggregation, clumping on

macrocarriers, and drastically reduced transformation efficiency.

26. Allow macrocarriers to air-dry in a vibration-free environment. Vibrations or airflow cause uneven particle distribution, resulting in "hot spots" or empty regions on the macrocarrier and uneven transformation.
27. Maintain monolayer coverage during membrane imprinting. Overly dense pollen layers reduce germination and cause skewed transformation rates; very sparse layers reduce the effective sampling area for imaging-based quantification.
28. Use only rapidly growing pollen tubes ($\geq$3 μm/min) for quantification, as slower tubes may be stressed or affected by protein overexpression or phototoxicity. Low growth rates correlate strongly with aberrant PM/cytoplasm ratios.
29. Always use identical ROI geometry (three horizontal lines at defined distances from the apex, fixed line width, consistent vertical reference) when extracting PM and cytoplasmic fluorescence. Even small deviations introduce bias into PM/Cyt ratio measurements.

Acknowledgments

M.P. and P.P. acknowledge support from Czech Science Foundation grant GA22-35916S and GA23-07733, respectively. E.Š. was supported by the Charles University Grant Agency (project no. GAUK290123).

References

1. Sekereš J, Pleskot R, Pejchar P, Žárský V, Potocký M (2015) The song of lipids and proteins: dynamic lipid-protein interfaces in the regulation of plant cell polarity at different scales. J Exp Bot 66:1587–1598. https://doi.org/10.1093/jxb/erv052
2. Fritz C, Reimann TM, Adler J, Knab J, Schulmeister S et al (2025) Plasma membrane and cytoplasmic compartmentalization: a dynamic structural framework required for pollen tube tip growth. Plant Physiol 197:kiae558. https://doi.org/10.1093/plphys/kiae558
3. Potocký M, Pleskot R, Pejchar P, Vitale N, Kost B et al (2014) Live-cell imaging of phosphatidic acid dynamics in pollen tubes visualized by Spo20p-derived biosensor. New Phytol 203:483–494. https://doi.org/10.1111/nph.12814
4. Serrano N, Pejchar P, Soukupová H, Hubálek M, Potocký M (2022) Comprehensive analysis of glycerolipid dynamics during tobacco pollen germination and pollen tube growth. Front Plant Sci 13:1028311. https://doi.org/10.3389/fpls.2022.1028311
5. Yperman K, Papageorgiou AC, Merceron R, De Munck S, Bloch Y et al (2021) Distinct EH domains of the endocytic TPLATE complex confer lipid and protein binding. Nat Commun 12:3050. https://doi.org/10.1038/s41467-021-23314-6
6. Zhou Y, Yang Y, Niu Y, Fan T, Qian D et al (2020) The tip-localized phosphatidylserine established by Arabidopsis ALA3 is crucial for Rab GTPase-mediated vesicle trafficking and pollen tube growth. Plant Cell 32:3170–3187. https://doi.org/10.1105/tpc.19.00844
7. Pejchar P, Sekereš J, Novotný O, Žárský V, Potocký M (2020) Functional analysis of phospholipase Dδ family in tobacco pollen tubes.

Plant J 103:212–226. https://doi.org/10.1111/tpj.14720

8. Žárský V, Potocký M (2010) Recycling domains in plant cell morphogenesis: small GTPase effectors, plasma membrane signaling and the exocyst. Biochem Soc Trans 38:723–728. https://doi.org/10.1042/BST0380723
9. Sekereš J, Pejchar P, Šantrůček J, Vukasinovic N, Žárský V et al (2017) Analysis of exocyst subunit EXO70 family reveals distinct membrane polar domains in tobacco pollen tubes. Plant Physiol 173:1659–1675. https://doi.org/10.1104/pp.16.01709
10. Synek L, Pleskot R, Sekereš J, Serrano N, Vukasinovic N et al (2021) Plasma membrane phospholipid signature recruits the plant exocyst complex via the EXO70A1 subunit. Proc Natl Acad Sci USA 118:e2105287118. https://doi.org/10.1073/pnas.2105287118
11. Dowler S, Kular G, Alessi DR (2002) Protein lipid overlay assay. Sci STKE 129:6. https://doi.org/10.1126/stke.2002.129.pl6
12. Han X, Yang Y, Zhao F, Zhang T, Yu X (2020) An improved protein lipid overlay assay for studying lipid–protein interactions. Plant Methods 16:33. https://doi.org/10.1186/s13007-020-00578-5
13. Julkowska MM, Rankenberg JM, Testerink C (2013) Liposome-binding assays to assess specificity and affinity of phospholipid–protein interactions. In: Munnik T, Heilmann I (eds) Plant lipid signaling protocols. Humana Press, Totowa, pp 261–271. https://doi.org/10.1007/978-1-62703-401-2_24
14. Rosano GL, Ceccarelli EA (2014) Recombinant protein expression in Escherichia coli: advances and challenges. Front Microbiol 5:172. https://doi.org/10.3389/fmicb.2014.00172
15. Garenne D, Haines MC, Romantseva EF, Freemont P, Strychalski E et al (2021) Cell-free gene expression. Nat Rev Methods Primers 1:49. https://doi.org/10.1038/s43586-021-00046-x
16. Scholz P, Pejchar P, Fernkorn M et al (2022) DIACYLGLYCEROL KINASE 5 regulates polar tip growth of tobacco pollen tubes. New Phytol 233:2185–2202. https://doi.org/10.1111/nph.17930
17. Škrabálková E, Pejchar P, Potocký M (2024) Exploring lipid–protein interactions in plant membranes. J Exp Bot:erae199. https://doi.org/10.1093/jxb/erae199
18. Yperman K, Wang J, Eeckhout D, Winkler J, Vu L et al (2021) Molecular architecture of the endocytic TPLATE complex. Science Adv 7:eabe7999. https://doi.org/10.1126/sciadv.abe7999
19. Platre MP, Noack LC, Doumane M, Bayle V, Simon MLA et al (2018) A combinatorial lipid code shapes the electrostatic landscape of plant endomembranes. Dev Cell 45:465–480. https://doi.org/10.1016/j.devcel.2018.04.011
20. Schindelin J, Arganda-Carreras I, Frise E, Kaynig V, Longair M et al (2012) Fiji: an open-source platform for biological-image analysis. Nat Methods 9:676–682. https://doi.org/10.1038/nmeth.2019
21. Dvorak P, Chrast L, Nikel PI, Fedr R, Soucek K et al (2015) Exacerbation of substrate toxicity by IPTG in Escherichia coli BL21(DE3) carrying a synthetic metabolic pathway. Microb Cell Factories 14:201. https://doi.org/10.1186/s12934-015-0393-3
22. Kubátová Z, Pejchar P, Potocký M, Sekereš J, Žárský V et al (2019) Arabidopsis trichome contains two plasma membrane domains with different lipid compositions which attract distinct EXO70 subunits. Int J Mol Sci 20:3803. https://doi.org/10.3390/ijms20153803

Chapter 14

Evaluation of Pectin and Arabinogalactan Protein Distribution in Olive Pollen Tube Cell Walls Using Immunofluorescent Labeling

Aslıhan Çetinbaş-Genç

Abstract

The pollen tube is widely recognized as a suitable model for investigating the structure and spatial organization of cell wall components during polarized growth. This chapter describes the application of an established immunofluorescent labeling protocol for the localization of two major cell wall components, pectins and arabinogalactan proteins, using specific monoclonal antibodies from the JIM series. JIM5 and JIM7 were employed to detect de-esterified and esterified homogalacturonan regions of pectin, respectively, while JIM8 and JIM13 were used to label distinct epitopes of arabinogalactan proteins. The protocol includes pollen germination, paraformaldehyde fixation, enzymatic digestion with cellulysin (for arabinogalactan protein detection only), and sequential antibody incubation, followed by confocal microscopy imaging using FITC filter settings. This approach enables precise visualization of the distribution patterns of pectins and arabinogalactan proteins in the pollen tube wall and provides a reliable framework for further studies on cell wall architecture in plant reproductive tissues.

Key words Arabinogalactan proteins, Cell wall, Immunofluorescent labeling, JIM antibodies, Pectin

1 Introduction

The pollen tube stands out as an ideal model for studying cell wall biosynthesis due to its characteristic fast tip growth [1]. At the dome-shaped apex of the pollen tube, the cell wall must be sufficiently flexible to allow relaxation following the fusion of new vesicles, yet strong enough to counterbalance the turgor pressure that drives growth. In contrast, along the shank (the cylindrical part of the pollen tube), the cell wall must exhibit a higher degree of mechanical rigidity that prevents lateral expansion under turgor pressure and maintains the cylindrical shape of the tube [2]. This functionality is made possible by the distinct spatial distribution of pollen tube cell wall components along the tube axis. The pollen tube cell wall is composed not only of structural polysaccharides

Giampiero Cai et al. (eds.), *Pollen and Pollen Tube Analysis: Methods and Protocols*, Methods in Molecular Biology, vol. 3042,
https://doi.org/10.1007/978-1-0716-5308-1_14,

such as pectin, callose, and cellulose, but also of additional molecules that play critical roles in cell wall function, including arabinogalactan proteins (AGPs). Together, these components contribute to the growth, mechanical stability, and cellular integrity of the pollen tube [3]. Pectins regulate the mechanical properties of the pollen tube cell wall by providing flexibility at the apex and rigidity along the shank. Callose, which is absent from the tube apex, accumulates in the shank region, contributing to the mechanical stiffness of the pollen tube. Cellulose supports the structural integrity of the cell wall along the entire tube axis, including the apex. Although AGPs are not classified as mechanical components, they interact with pectin, callose, and cellulose, thereby playing indirect yet crucial roles in modulating the mechanical properties of the cell wall [4].

During pollen tube growth, the spatial distribution of cell wall components is continuously remodeled. This dynamic reorganization directly influences the tube's flexibility, growth direction, and elongation rate [5]. When evaluating the dynamics of pollen tube growth, it is not sufficient to simply identify the components present in the cell wall; it is also necessary to understand where these components are located, at what concentrations, and how they are organized within the cell wall. At this point, immunofluorescent labeling emerges as a powerful technique that provides spatial localization information on the cellular level [6]. Although callose and cellulose are fundamental structural components that ensure the mechanical integrity of the pollen tube wall, their analysis via immunofluorescent labeling is not as standardized or widely used as pectins and AGPs. While some specific antibodies are available for callose, antibody-based labeling methods for cellulose have only been used in a limited number of studies, with most investigations relying on dye-based techniques [7–9]. Pectins and AGPs, on the other hand, can be visualized with high specificity using well-characterized and widely used monoclonal antibodies, such as the JIM (John Innes Monoclonal) series, which was developed at the John Innes Centre in the United Kingdom. These antibodies enable detailed investigation of the localization, distribution, and dynamic changes of these components [10]. Monoclonal antibodies such as JIM5 and JIM7 are commonly used to label de-esterified and esterified pectin regions, respectively [11, 12]. AGPs are hydroxyproline-rich, membrane-bound glycoproteins that play a role in signal transduction between the plasma membrane and the cell wall [13]. By interacting with other cell wall components, these proteins influence the structural organization of the pollen tube wall, thereby contributing to tube elongation [14]. AGPs represent a highly complex group of molecules due to the diversity of their epitope structures, and this complexity necessitates the use of different monoclonal antibodies for their characterization [15]. Among the most commonly used antibodies, JIM8 recognizes carbohydrate epitopes containing glucuronic acid, while JIM13 binds to the β-D-GlcA-(1,3)-α-D-GalA-(1,2)-α-L-Rha epitope [16, 17].

This chapter presents a detailed, step-by-step protocol for the localization of pectins and arabinogalactan proteins (AGPs) in the pollen tube cell wall using highly specific JIM-series monoclonal antibodies. Although olive pollen tubes are used as a model system, the protocol is broadly applicable to a wide range of pollen species. The method, based on the selective binding of JIM antibodies to their respective epitopes, enables the visualization of cell wall components in a configuration that closely resembles their native state. It is designed as a reliable, reproducible, and adaptable tool for future studies investigating pollen tube wall architecture.

2 Materials

Unless otherwise specified, use dH_2O for the preparation of all solutions. Dispose of waste materials in accordance with waste management regulations.

2.1 Stock Solutions

1. 0.5 M PIPES (pH 6.9): Add 75 mL of dH_2O to a glass beaker. Weigh 15.11 g of PIPES and transfer it into the beaker. Adjust the pH to 6.9 using NaOH (*see* **Note 1**), and stir the mixture in a magnetic stirrer until a clear solution is obtained. Once the PIPES has completely dissolved, bring the final volume up to 100 mL with dH_2O. The stock PIPES solution can be stored at +4 °C for up to 1 month (*see* **Note 2**).
2. 100 mM EGTA (pH 7–8): Add 75 mL of dH_2O to a glass beaker. Weigh 3.80 g of EGTA and transfer it to the beaker. Adjust the pH to a value between 7 and 8 using NaOH (*see* **Note 3**), and stir the solution on a magnetic stirrer until a clear solution is obtained. Once the EGTA is fully dissolved, adjust the final volume to 100 mL with dH_2O. The stock EGTA solution can be stored at +4 °C for up to 3 months (*see* **Note 2**).
3. 100 mM $MgCl_2$: Add 75 mL of dH_2O to a glass beaker. Weigh 0.9521 g of $MgCl_2$ and transfer it into the beaker. Stir until the $MgCl_2$ is completely dissolved. Adjust the final volume to 100 mL with dH_2O. The stock $MgCl_2$ solution can be stored in aliquots at −20 °C for up to 1 year (*see* **Note 2**). A single aliquot designated for current use may be stored at +4 °C for up to 3 months.

2.2 Buffers and Reagents

1. PEM buffer: Using previously prepared stock solutions, mix 10 mL of stock PIPES, 1 mL of stock EGTA, and 2 mL of stock $MgCl_2$ in a glass cylinder. Add dH_2O to bring the final volume

to 100 mL. Prepare this solution freshly before use and store at room temperature (*see* **Note 4**).

2. PEM buffer containing 3% paraformaldehyde: Add 100 mL of PEM buffer to a glass conical flask. Weigh 3 g of powdered paraformaldehyde (PFA) and transfer it into the flask. Seal the mouth of the flask with parafilm. Place the flask in a 60 °C water bath (*see* **Note 5**) and gently swirl until a clear solution is obtained (*see* **Note 6**). Once the solution has cooled, transfer it to a storage bottle and keep it at +4 °C for up to 3 months (*see* **Note 7**).
3. PEM buffer containing 1.5% cellulysin: Add 10 mL of PEM buffer to a glass beaker. Weigh 0.15 g of cellulysin, transfer it into the beaker, and stir until fully dissolved. Aliquot the solution into 1.5 mL microcentrifuge tubes wrapped in aluminum foil. Keep one aliquot at room temperature for immediate use. Store the remaining aliquots at −20 °C for up to 6 months (*see* **Note 8**).
4. Solutions of primary antibodies: Wrap a 1.5 mL microcentrifuge tube with aluminum foil. Dilute the primary antibody in PEM buffer to the desired working concentration (*see* **Note 9**) [18, 19]. Prepare freshly just before use and keep at room temperature. While a 1:5 to 1:200 dilution is commonly used for JIM series antibodies, we advise testing a dilution series if signal intensity or background levels are suboptimal (*see* **Note 10**) (Table 1).
5. Solution of secondary antibodies: Wrap a 1.5 mL microcentrifuge tube with aluminum foil. Dilute the Alexa Fluor 488-conjugated goat anti-rat IgG (H+L) secondary antibody

Table 1
Details of antibodies

Primary antibody	Target molecule/ Epitope	Host/Type	Cell wall digestion	Dilution used	Secondary antibody/ Host/Type	Dilution used	λex/λem (nm)
JIM5	De-esterified homogalacturonan	Rat/ Monoclonal	Not required	1:5	Anti-rat/ Goat/ Polyclonal	1:50	488/519
JIM7	Esterified homogala cturonan						
JIM8	AGP epitope containing glucuronic acid		Required	1:50			
JIM13	AGP epitope: β-D-GlcA-(1,3)-α-D-GalA-(1,2)-α-L-Rha						

in PEM buffer to the desired working concentration. Prepare the solution freshly just before use and keep it at room temperature. We advise testing a dilution series if signal intensity or background levels are suboptimal (*see* **Note 10**) [20] (Table 1).

3 Methods

Unless otherwise specified, perform all procedures at room temperature. The immunofluorescent labeling protocol applied for the localization of pectins and AGPs in olive pollen tube cell walls is illustrated in Fig. 1.

3.1 Fixation

1. Germinate 2.5 mg of pollen in 5 mL of pollen germination medium in sterile 60 mm Petri dishes (*see* **Note 11**).
2. Transfer the germinated pollen suspension from the Petri dish into 10 mL centrifuge tubes (*see* **Note 12**).
3. Centrifuge the tubes at 2500 g for 10 min. After centrifugation, discard the supernatant and proceed with the pollen pellet remaining at the bottom of the tube (*see* **Note 13**).
4. Add 1 mL of room temperature PEM buffer containing 3% PFA to each tube and allow fixation for 30 min.
5. Centrifuge the tubes at 2500 g for 10 min. After centrifugation, discard the supernatant and continue working with the pollen pellet.
6. Add 1 mL of room temperature PEM buffer to each tube and gently invert to resuspend. Centrifuge the tubes again at 2500 g for 10 min. Discard the supernatant but keep the pellet. Repeat this washing step twice (*see* **Note 14**).

3.2 Cell Wall Digestion

This step is required only for JIM8 and JIM13 applications. Omit this step for JIM5 and JIM7 (*see* **Note 15**) [21].

1. Add 500 μL of room temperature PEM buffer containing 1.5% cellulysin to each tube and incubate in the dark for 7 min (*see* **Note 16**).
2. Add 1 mL of room temperature PEM buffer to each tube and gently invert to resuspend. Centrifuge the tubes at 2500 g for 10 min. After centrifugation, remove the supernatant and proceed with the pollen pellet. Repeat the washing step twice.

3.3 Primary Antibody Treatment

1. Add 500 μL of the primary antibody solution (JIM5, JIM7, JIM8, or JIM13) to each tube (*see* **Note 17**). For the control, add PEM buffer rather than the primary antibody solution to one of the tubes. Incubate overnight at +4 °C in the dark (*see* **Note 18**).

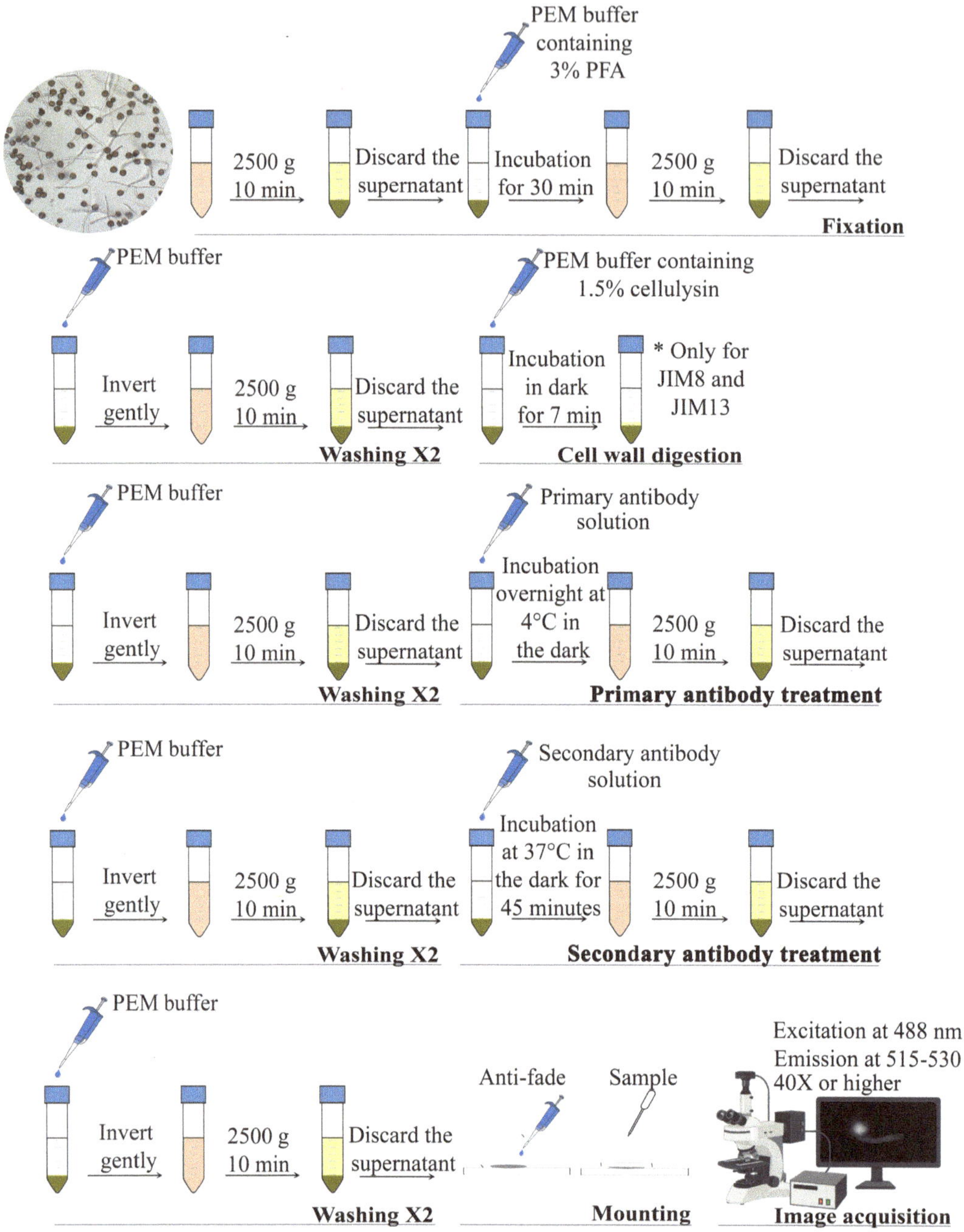

Fig. 1 Schematic representation of the Immunofluorescent labeling protocol for detecting cell wall components, specifically pectins and AGPs, in olive pollen tubes. The procedure includes fixation, cellulysin treatment (applied only for AGP-targeting antibodies JIM8 and JIM13), and immunofluorescent = labeling with JIM5 and JIM7 (targeting esterified and de-esterified homogalacturonan regions of pectin, respectively) and JIM8 and JIM13 (targeting specific AGP epitopes). Fluorescence detection is performed using a FITC filter

2. Centrifuge the tubes at 2500 g for 10 min. After centrifugation, discard the supernatant and proceed with the pollen pellet.
3. Add 1 mL of room temperature PEM buffer to each tube and gently invert to resuspend. Centrifuge the tubes again at 2500 g for 10 min. Discard the supernatant but keep the pellet. Repeat the washing step twice.

3.4 Secondary Antibody Treatment

1. Add 500 μL of Alexa Fluor 488 goat anti-rat IgG (H+L) secondary antibody solution to each tube (*see* **Note 17**). Incubate in the dark at 37 °C for 45 min (*see* **Note 18**).
2. Centrifuge the tubes at 2500 g for 10 min. After centrifugation, discard the supernatant and proceed with the pollen pellet.
3. Add 1 mL of room temperature PEM buffer to each tube and gently invert to resuspend. Centrifuge the tubes again at 2500 g for 10 min. Discard the supernatant but keep the pellet. Repeat the washing step twice (*see* **Note 19**).

3.5 Mounting

Unless otherwise specified, always keep samples and preparations protected from light.

1. Using a micropipette, draw 15 μL of glycerol-based anti-fade mounting medium (*see* **Note 20**). Place a drop onto a clean and dry microscope slide (*see* **Note 21**) and allow it to spread evenly.
2. Using a micropipette, draw 15 μL of the pollen suspension and carefully place it onto the drop of mounting medium, ensuring no air bubbles are introduced.
3. Gently place a clean coverslip over the sample, avoiding air bubbles, and allow 1–2 min for the preparation to stabilize (*see* **Note 22**).

3.6 Image Acquisition

1. Add a drop of immersion oil onto the surface of the slide (*see* **Note 23**) and adjust the objective lens for magnifications of 40× or higher (*see* **Note 24**).
2. Set up an optimal microscope configuration and exposure time, and maintain these settings consistently throughout the analysis.
3. Examine the preparations using a confocal microscope by exciting at a wavelength of 488 nm and detecting fluorescence in the 515–530 nm emission range with FITC-compatible settings, or alternatively visualize them using an epifluorescence or basic fluorescence microscope equipped with an appropriate FITC filter set (Fig. 2) (*see* **Note 25**).

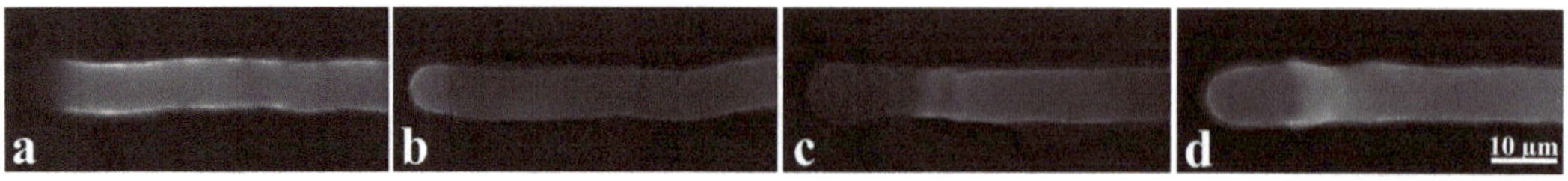

Fig. 2 Representative confocal images of olive pollen tubes immunofluorescent labeled with monoclonal antibodies. (**a**) JIM5 labeling was absent at the apex but prominent along the shank, indicating localization of de-esterified pectins in the subapical region. (**b**) In contrast, JIM7, which recognizes esterified pectins, showed strong labeling at the apex that gradually decreased toward the shank. JIM8 (**c**) and JIM13 (**d**), both markers for AGPs, displayed weak fluorescence at the tip but were strongly expressed along the shank. Notably, the distribution patterns of these AGPs, particularly those detected by JIM8 and JIM13, may vary depending on the species and even among different cultivars

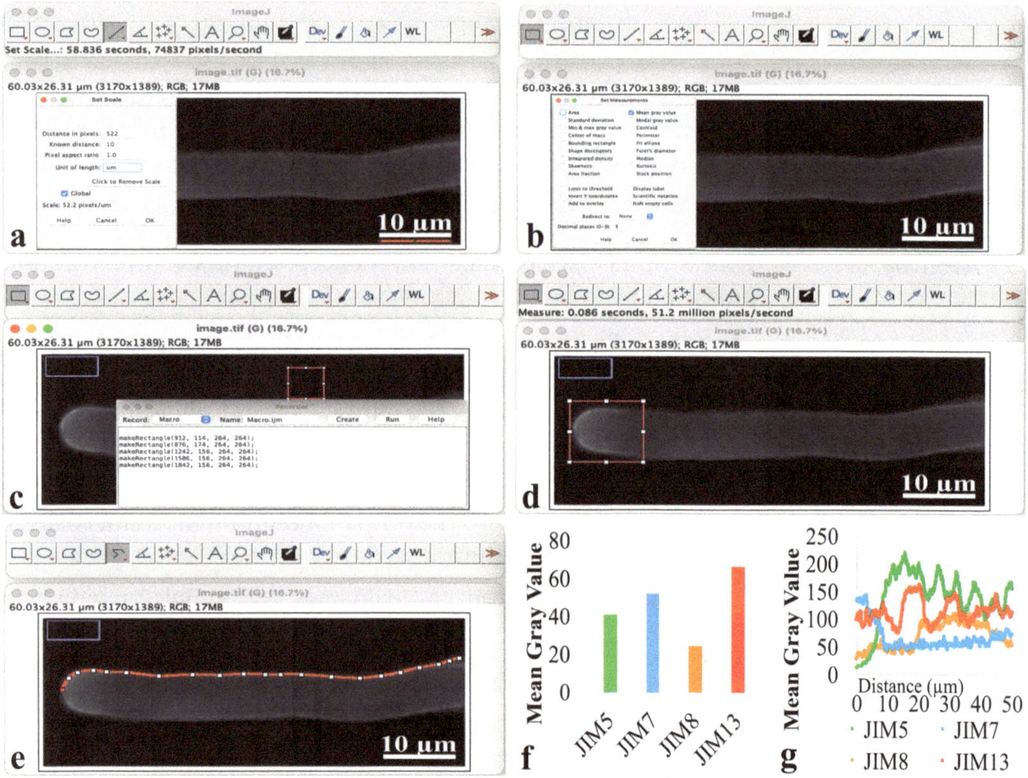

Fig. 3 FI analysis workflow in pollen tubes using ImageJ. (**a**) Calibration settings. (**b**) Macro configuration. (**c**) Creating a macro. (**d**) Region of interest at the tube tip for FI measurement. (**e**) Segment along the *tube* for FI measurement. (**f**) FI of antibodies at the tube tip. (**g**) FI of antibodies along the tube

3.7 Image Analysis: Calibration

1. Select an image that includes a scale bar (*see* **Note 26**).
2. Open the ImageJ software. ImageJ is a free, open-source program that can be downloaded and installed on any computer.
3. To open the image, go to File > Open in the menu, or simply drag and drop the image into the program window.

4. From the toolbar, select the "Straight Line" tool. Draw a line over the scale bar (*see* **Note 27**).
5. Go to Analyze > Set Scale in the menu. In the window that appears, enter the known length corresponding to the line you drew in the "Known distance" field. In the "Unit of length" field, type "um" (instead of "μm") (*see* **Note 28**). Do not modify the values in the "Distance in pixels" and "Pixel aspect ratio" fields (*see* **Note 29**). Check the "Global" box (*see* **Note 30**) and click "OK" to complete the calibration (*see* **Note 31**) (Fig. 3a).

3.8 Macro Creation

1. Open a randomly selected image in ImageJ. A macro should be created using the black background areas that do not contain any fluorescent signal (*see* **Note 32**).
2. Go to Analyze > Set Measurements in the menu. Make sure that only the "Mean Gray Value" option is selected in the dialog box that appears (*see* **Note 33**) (Fig. 3b).
3. Click Plugins > Macros > Record from the menu. Keep the "Recorder" window open.
4. Select the "Rectangle" tool from the toolbar and mark a background region in the image that does not show any fluorescent signal. Each selection will automatically appear in the "Recorder" window. Repeat this process for at least five different black background regions (*see* **Note 34**).
5. Once all selections are complete, click the "Create" button in the "Recorder" window. Save the newly opened "Macro.ijm" window by going to File > Save As in the menu. This will generate a macro that can be used to eliminate background noise (*see* **Notes 35** and **36**) (Fig. 3c).

3.9 Measurements

To measure the fluorescence intensity (FI) in the 100 μm^2 area at the tip of the pollen tube, follow the steps below.

1. Open the image you want to measure in ImageJ.
2. Select the "Rectangle" tool from the toolbar and draw a fixed square measuring 10 μm × 10 μm, starting from the tip of the pollen tube and extending backward. This square should be positioned symmetrically along the center of the tube, starting just behind the tip of the pollen tube (*see* **Note 37**) (Fig. 3d).
3. Click Analyze > Measure from the menu.
4. The mean value displayed in the "Results" window represents the mean FI within the selected region (*see* **Note 38**).
5. After completing the measurements for all groups, you can copy the data from the "Results" window and paste it into any desired document or analysis software.

To measure the FI along the edges of the pollen tube within a 50 μm segment starting from the central tip of the tube and extending backward along the shank, follow the steps below.

1. Open the image you want to measure in ImageJ.
2. Right-click on the "Line Tool" icon in the toolbar and select "Segmented Line" from the available options (*see* **Note 39**).
3. Double-click the "Segmented Line" tool to open the settings window. Set the "Line width" to 2 μm and check the "Spline fit" box (*see* **Notes 40** and **41**).
4. Draw a segmented line starting from the center of the pollen tube tip and extending 50 μm backward along the length of the tube (*see* **Note 42**) (Fig. 3e).
5. Go to Analyze > Plot Profile from the menu (*see* **Note 43**).
6. A plot window will appear showing the FI measured along the drawn line. The X-axis represents distance, while the Y-axis indicates FI values.
7. Click the "Data" tab below the graph, then select "Copy All Data" to copy the entire dataset. You can paste it into any document or analysis software to generate a graph.

4 Notes

1. The pH of the PIPES solution must exceed 6.5 for the compound to dissolve efficiently.
2. Before use, ensure that no visible precipitates or color changes are present in the solution.
3. EGTA dissolves readily when the solution pH is within the range of 7–8.
4. Although some researchers store PEM buffer at +4 °C for up to 1 week, it is strongly recommended to prepare it freshly before use and maintain it at room temperature. This helps avoid potential loss of stability due to EGTA chelating Mg^{2+} ions, as well as to prevent microbial contamination.
5. At temperatures above 60 °C, paraformaldehyde more readily volatilizes, which may lead to a decrease in its effective concentration in the solution. At 60 °C, complete dissolution typically occurs within 45–50 min.
6. As paraformaldehyde is highly toxic, all weighing, mixing, and dissolution steps must be conducted under a chemical fume hood.
7. As a more practical alternative, some researchers store PEM buffer containing 3% PFA in aliquots at −20 °C for up to one year. In such cases, filtration prior to use is recommended.

8. It typically takes 20–25 min for a 1.5 mL aliquot removed from −20 °C to reach room temperature.
9. Many researchers store primary antibodies in aliquots at −20 °C. It is recommended to centrifuge thawed primary antibodies at 10,000 g for 10 min and use the supernatant for dilution in PEM buffer.
10. In our system, a 1:50 dilution was optimal for JIM5 and JIM7, while a 1:5 dilution yielded the most specific and robust signal for JIM8 and JIM13. A 1:50 dilution also provided strong and specific fluorescence when using Alexa Fluor 488-conjugated goat anti-rat IgG (H+L) as the secondary antibody. Nonetheless, due to potential variability in antibody lot composition, supplier formulations, and epitope abundance within the target tissue, empirical titration is strongly recommended to determine the optimal working dilution under specific experimental conditions.
11. It is recommended to use a pollen germination medium optimized for the specific pollen type under investigation. Germination duration and temperature may vary depending on the pollen species and experimental objectives.
12. When transferring germinated pollen solution into tubes, gently swirl the Petri dish on the benchtop to collect the pollen toward the center. Then, pipette the solution from the center and transfer it into the tubes. Repeat this step a few times to minimize the risk of leaving pollen grains behind in the Petri dish.
13. Following this centrifugation step, in some cases, a small amount of pollen or pollen tubes may remain in the supernatant. If observed, it is not recommended to increase centrifugation speed or duration, as this may cause damage to the pollen tubes.
14. If the initial steps take longer than expected or if a pause in the protocol is necessary, samples may be stored overnight at +4 °C in PEM buffer after **step 6**. However, to obtain optimal results, it is recommended to proceed through the protocol without interruptions.
15. The epitopes targeted by JIM8 and JIM13 are located beneath the cell wall. Therefore, partial enzymatic digestion of the wall with cellulysin is required to allow antibody access. In contrast, the epitopes recognized by JIM5 and JIM7 are exposed on the outer surface of the cell wall, making wall digestion unnecessary for their detection.
16. As cellulysin treatment renders pollen tubes more fragile, handle samples with extra care in all subsequent steps.

17. The indicated volumes are optimized for a single sample (2.5 mg of germinated pollen). If the amount of biological material is altered, total volumes may be adjusted accordingly.
18. Incubation time may vary depending on antibody reactivity. It is strongly recommended to conduct a preliminary trial with a single sample prior to performing the full experiment.
19. After the final centrifugation step, do not completely remove the supernatant. Leave approximately 30 μL of buffer in the tube to avoid sample loss.
20. In addition to preserving fluorescence signals and preventing sample dehydration, the selected anti-fade mounting medium should also limit the movement of pollen grains and pollen tubes. Therefore, a glycerol-based anti-fade mounting medium is recommended. Due to its viscosity, glycerol embeds the sample, preventing movement and ensuring sharper, more stable fluorescence signals under the microscope. Furthermore, glycerol reduces refractive index differences, thereby minimizing light scattering and enhancing optical clarity. For this reason, glycerol-based anti-fade mounting media are especially advantageous for high-magnification imaging.
21. To ensure even spreading of the glycerol-based anti-fade mounting medium and proper distribution of samples, thus avoiding signal overlap, use long microscope slides and coverslips. This enables the sample to be spread over a wider area and provides clearer, more distinguishable signals under the microscope.
22. If delays occur during sample preparation, completed slides may be stored in a light-protected box at +4 °C overnight. However, for the best results, it is recommended to examine the samples immediately after preparation.
23. Use immersion oil with a refractive index of 1.5, low autofluorescence, and minimal background signal.
24. Since the aim is to visualize signals in the cell wall rather than the wall structure itself, a minimum magnification of 40× is recommended.
25. Scanning and imaging settings for confocal, epifluorescence, and basic fluorescence microscope systems should be optimized based on the specific configuration of the microscope and its associated hardware. Epifluorescence and basic fluorescence microscope systems offer lower spatial resolution compared to confocal microscopy and tend to produce higher background fluorescence due to the detection of out-of-focus light.
26. If your imaging system does not automatically embed a scale bar into the images, capture an image of a millimeter scale slide

using the same objective magnification used for the experimental images, and use this image during calibration.

27. When using a millimeter scale slide, note that the smallest divisions are usually spaced 10 μm apart.
28. Since the "μm" character cannot be entered technically, use "um" instead. Make sure the unit of measurement is correctly defined.
29. When you draw a length using the "Straight Line" tool, ImageJ automatically calculates its length in pixels and enters this value into the "Distance in pixels" field. If you modify this value, the physical length associated with the digital length (in pixels) will no longer match, resulting in an inaccurate calibration. Additionally, in most microscope and digital camera systems, pixels are square; therefore, the "Pixel aspect ratio" should remain at 1.0. If you change this value, the measured distances will yield inaccurate results.
30. Checking the "Global" box ensures that the calibration is applied to all images captured at the same magnification and opened during the same session.
31. The calibration is valid only for the current ImageJ session. Once the program is closed or a new session is started, the calibration is reset. Always check whether the calibration is still active at the beginning of each session. If necessary, repeat the calibration using the same procedure before starting your analysis.
32. Since all images were acquired using the same imaging settings as described in **Note 25**, the choice of image does not affect the measurement results, and a macro generated from a single image is sufficient. However, for more precise background correction, signal-free regions from multiple images may also be selected.
33. "Mean Gray Value" represents the average FI within the selected area. "Integrated Density" is the product of this intensity and the area, reflecting the total signal. For determining background levels, the average value is sufficient; therefore, only "Mean Gray Value" should be selected.
34. To accurately interpret the fluorescence signal, background noise must be eliminated. The average values obtained from non-fluorescent regions of the image are used as references for background correction. Since the background signal may not be uniformly distributed across the image, measurements should be taken from at least five different locations to obtain a more reliable and representative baseline level.
35. Different antibodies may exhibit varying fluorescence intensities and distribution patterns. Therefore, creating separate

macros for each primary antibody ensures more accurate and comparable measurements.

36. Macro settings are deleted when ImageJ is closed. If you wish to reuse the same macro, you can load the previously saved "macro.ijm" file by going to Plugins > Macros > Install after launching the program.
37. Using a fixed square of 10 μm × 10 μm for measurements ensures a standardized region of interest across all samples, thereby enhancing statistical comparability between groups. This size is small enough to capture localized fluorescence at the pollen tube tip, yet large enough to provide meaningful data for analysis. It preserves both signal intensity and spatial precision. However, this dimension should be adjusted based on the pollen tube morphology of the species under study. In species with narrower tubes, the area may need to be reduced, whereas in species with wider tubes, it should be optimized to accurately represent the signal.
38. When the region of interest size is kept constant across all samples, "Mean Gray Value," which represents the average fluorescence intensity, provides accurate and comparable results between groups. However, if the region of interest size varies from sample to sample, it is recommended to use "Integrated Density," which accounts for both intensity and area, to ensure reliable comparisons.
39. Pollen tubes typically do not extend in a perfectly straight line; they are often naturally curved or bent. The "Segmented Line" tool allows for accurate tracing of the actual path along the tube by accommodating these curvatures. This enables the FI profile to be measured in a more precise and biologically meaningful manner.
40. The selected line width should be set to encompass the cell wall. This value may vary depending on the species under study and the objective magnification used during image acquisition.
41. The "Spline fit" option is selected to ensure that the segmented line is drawn as a smooth, natural curve rather than as sharp, angular segments. This allows the line to more accurately follow the true morphology of curved structures such as pollen tubes, resulting in an intensity profile that better reflects the underlying biological structure.
42. The segmented line is drawn starting from the apical tip of the pollen tube and extending 50 μm backward along the shaft, as this region encompasses the most active cellular processes involved in tube growth, such as vesicle trafficking, cell wall synthesis, and calcium gradient formation. Measuring the distribution of FI along this line reveals the spatial profile of biological activity at the tube tip. A 50 μm length includes

both the apical and subapical regions, allowing for comparative analysis. However, this distance may need to be adjusted depending on the species studied and the objective magnification used during image acquisition.

43. The "Plot Profile" command presents the average FI values along the drawn line as a function of distance. These values correspond to the "Mean Gray Value" at each point along the line. "Integrated Density" is not calculated, as the goal is not to assess total signal but rather to analyze the regional distribution of intensity. This analysis reveals how the signal changes from the apex toward the base and is particularly useful for identifying biological patterns such as signal gradients or accumulation along the apical-to-basal axis.

Acknowledgments

Grateful acknowledgment is extended to Claudia Faleri and Giampiero Cai for their devoted support throughout the completion of the study. The study was carried out at the Department of Life Sciences, University of Siena, with fellowship support from TÜBİTAK (2219 International Postdoctoral Research Program).

References

1. Calabrese EJ, Agathokleous E (2021) Pollen biology and hormesis: pollen germination and pollen tube elongation. Sci Total Environ 762: 143072
2. Hepler PK, Rounds CM, Winship LJ (2013) Control of cell wall extensibility during pollen tube growth. Mol Plant 6(4):998–1017
3. Cai G, Parrotta L, Cresti M (2017) The cytoskeleton of pollen tubes and how it determines the physico-mechanical properties of cell wall. In: Pollen tip growth: from biophysical aspects to systems biology. Springer, pp 35–62
4. Mollet JC, Leroux C, Dardelle F, Lehner A (2013) Cell wall composition, biosynthesis and remodeling during pollen tube growth. Plants 2(1):107–147
5. Geitmann A (2010) How to shape a cylinder: pollen tube as a model system for the generation of complex cellular geometry. Sex Plant Reprod 23:63–71
6. Dardelle F, Lehner A, Ramdani Y, Bardor M, Lerouge P, Driouich A, Mollet JC (2010) Biochemical and immunocytological characterizations of Arabidopsis pollen tube cell wall. Plant Physiol 153(4):1563–1576
7. Meikle PJ, Bonig I, Hoogenraad NJ, Clarke AE, Stone BA (1991) The location of (1→ 3)-β-glucans in the walls of pollen tubes of Nicotiana alata using a (1→ 3)-β-glucan-specific monoclonal antibody. Planta 185:1–8
8. Liesche J, Ziomkiewicz I, Schulz A (2013) Super-resolution imaging with Pontamine Fast Scarlet 4BS enables direct visualization of cellulose orientation and cell connection architecture in onion epidermis cells. BMC Plant Biol 13:1–9
9. Knox JP (2008) Revealing the structural and functional diversity of plant cell walls. Curr Opin Plant Biol 11(3):308–313
10. Bush MS, Marry M, Huxham MI, Jarvis MC, McCann MC (2001) Developmental regulation of pectic epitopes during potato tuberisation. Planta 213:869–880
11. Willats WG, Limberg G, Buchholt HC, van Alebeek GJ, Benen J, Christensen TM et al (2000) Analysis of pectic epitopes recognised by hybridoma and phage display monoclonal antibodies using defined oligosaccharides, polysaccharides, and enzymatic degradation. Carbohydr Res 327(3):309–320
12. Clausen MH, Willats WG, Knox JP (2003) Synthetic methyl hexagalacturonate hapten

inhibitors of anti-homogalacturonan monoclonal antibodies LM7, JIM5 and JIM7. Carbohydr Res 338(17):1797–1800

13. Jiao HJ, Wang HW, Kun RAN, Dong XC, Ran DONG, Wei SW, Wang SM (2023) Identification and functional analysis of arabinogalactan protein expressed in pear pollen tubes. J Integr Agric 22(3):776–789
14. Geitmann A, Steer M (2006) The architecture and properties of the pollen tube cell wall. In: The pollen tube: a cellular and molecular perspective. Springer Berlin Heidelberg, Berlin Heidelberg, pp 177–200
15. Lora J, Hormaza JI (2018) Pollen wall development in mango (Mangifera indica L., Anacardiaceae). Plant Reproduction 31(4): 385–397
16. Yates EA, Valdor JF, Haslam SM, Morris HR, Dell A, Mackie W, Knox JP (1996) Characterization of carbohydrate structural features recognized by anti-arabinogalactan-protein monoclonal antibodies. Glycobiology 6(2): 131–139
17. Pattathil S, Avci U, Baldwin D, Swennes AG, McGill JA, Popper Z et al (2010) A comprehensive toolkit of plant cell wall glycan-directed monoclonal antibodies. Plant Physiol 153(2): 514–525
18. Knox JP, Linstead PJ, King J, Cooper C, Roberts K (1990) Pectin esterification is spatially regulated both within cell walls and between developing tissues of root apices. Planta 181:512–521
19. Li YQ, Faleri C, Geitmann A, Zhang HQ, Cresti M (1995) Immunogold localization of arabinogalactan proteins, unesterified and esterified pectins in pollen grains and pollen tubes of Nicotiana tabacum L. Protoplasma 189:26–36
20. Biagini G, Faleri C, Cresti M, Cai G (2014) Sucrose concentration in the growth medium affects the cell wall composition of tobacco pollen tubes. Plant Reproduction 27:129–144
21. Parrotta L, Faleri C, Del Duca S, Cai G (2018) Depletion of sucrose induces changes in the tip growth mechanism of tobacco pollen tubes. Ann Bot 122(1):23–43

Chapter 15

Evaluating Self-Incompatibility in Citrus by Histological and Molecular Assays

Stefania Bennici, Gaetano Distefano, and Alessandra Gentile

Abstract

Self-incompatibility (SI) is a system used by flowering plants to prevent self-fertilization and promote outbreeding, ensuring genetic diversity. The system is regulated molecularly by the single polymorphic *S*-locus that expresses the male and female determinants that interact, inhibiting pollen germination or pollen tube growth when pollen from the same plant or genetically similar individuals reaches the stigma. Knowing the self-incompatible phenotype among cultivars is fundamental to ensuring fruit set or obtain seedless fruits in parthenocarpic species. Here we describe three techniques to measure self-incompatibility: (i) the self-incompatibility index (ISI); (ii) fluorescence microscopy to observe pollen germination and pollen tube growth after pollination; (iii) molecular analysis to assess diversity at the *S*-locus and expression of *S*-alleles.

Key words Aniline blue staining, Pistils, Pollen tube growth, Semi-in vivo, PCR-analysis, *S*-locus, Consensus primers

1 Introduction

Self-incompatibility (SI) is a genetic mechanism that prevents self-fertilization in sexually reproducing plants, ensuring genetic diversity through outcrossing [1]. SI works by a prezygotic barrier that recognizes self-pollen and inhibits its growth or germination on the stigma, leading to the inability to produce seeds or zygotes. SI is classified into two main types: the Sporophytic SI (SSI), where the incompatibility phenotype is determined by the sporophyte (the diploid plant) that produces the pollen inhibiting the pollen germination on the surface of the stigma; and Gametophytic SI (GSI), where pollen's genetic makeup determines the reaction inhibiting the pollen tube growth within the style [2]. The SSI is typical of species belonging to families of Brassicaceae, Asteraceae, Convolvulaceae, Betulaceae, Caryophyllaceae, Sterculiaceae, and Polemoniaceae. While the GSI is the most widespread among plants and is found in several species belonging to the families of Solanaceae,

Giampiero Cai et al. (eds.), *Pollen and Pollen Tube Analysis: Methods and Protocols*, Methods in Molecular Biology, vol. 3042, https://doi.org/10.1007/978-1-0716-5308-1_15,

Rosaceae, Plantaginaceae, and Rutaceae. SI is generally controlled by a single highly polymorphic locus, the *S*-locus, which contains two tightly linked male and female genes (pollen and pistil *S*-determinants). Both have multiple alleles determining self- and inter-compatibility or incompatibility within a species [3].

Knowing the SI behavior among cultivars and genotypes is essential to select compatible pollinators and plan an orchard able to ensure fruit set (e.g., almond, pear, cherry, etc.). Differently, among parthenocarpic species (such as mandarins or mandarin-like varieties), the absence of fertilization due to SI guarantees the production of seedless fruits, highly appreciated by consumers.

Historically, the SI is measured by the SI Index (ISI) calculated as the ratio between the amount of fruits set resulting from self-pollination respect that resulting from cross-pollination. High ISI (closer to 1.0) suggests the plant is highly self-compatible (self-fertility), while low ISI (closer to 0.0) suggests greater SI [4, 5].

Histological analysis with aniline blue staining is a procedure used to monitor the pollen germination and the pollen tube growth behavior in the pistil after self-pollination and is one of the most widespread procedures used to measure SI. Aniline blue selectively stains callose, a polysaccharide component of the pollen tube wall, allowing for the visualization of pollen tubes within the stigma and style, providing insights into pollen viability, compatibility, and the success of germination and growth in different plant species [6–10]. Here, we present a semi *in vivo* method to evaluate the pollen tube growth rate in "Comune" clementine (*Citrus clementina* Hort. ex Tan.) used as self-incompatible reference genotypes [11].

Another method to measure SI is the molecular analysis of the diversity of genes at the *S*-locus, which controls the incompatibility mechanism. This involves identifying the specific alleles present in both pollen and pistil tissues, by using a process that often uses polymerase chain reaction (PCR) and DNA and/or RNA sequencing approaches to detect polymorphisms and gene structure and expression [12–15]. The expression analysis of the female determinant, *S-RNase*, gives indications on the self-incompatible and self-compatible phenotypes [14, 16–18]. Generally, the expression of both *S*-alleles results in the self-incompatible phenotype, while the lack of expression of even just one *S*-allele results in the self-compatible phenotype [16]. Here we describe the *S*-genotyping procedure in "Comune" clementine performed by consensus PCR analysis followed by Sanger sequencing of purified amplicons to define the complete *S*-allele composition. Also, here we report expression analysis of *S*-alleles in clementine associated with manifestation of the self-incompatible phenotype [14].

In this chapter, we describe three different techniques that can be used to measure SI in citrus: (i) the SI index (ISI), which measures the fruit set from cross-pollination versus self-pollination; (ii) fluorescence microscopy with aniline blue staining to observe pollen germination and pollen tube growth after self-pollination; (iii) molecular techniques to analyze diversity and expression of the *S*-alleles.

2 Materials

All solutions should be prepared using deionized water of 18 MΩ.

2.1 Plant Materials

1. Plant material can be provided by the nursery.
2. Trees, older than 10 years, of "Comune" clementine (*Citrus clementina* Hort. *ex* Tan.) used as self-incompatible reference genotypes.
3. Commercial growing substrate consisting of acidic peat, uncomposted vegetable soil improver, and silica sand, and characterized by the following physical properties: pH in H_2O 6.5, electrical conductivity 0.30 dS/m, dry bulk density 500 kg m^{-3}, total porosity 85% v/v.
4. 45 cm pots.
5. Fertilizer 20-10-10 NPK (20% nitrogen–10% phosphorus–10% potassium).
6. Growth chamber at 25 °C under 16 h light/8 h dark cycle and 50% relative humidity (RH) conditions.

2.2 Self-Incompatibility Index (ISI) Measurement

1. Forceps.
2. Small brush.
3. Transparent nylon bag with laces.
4. 5 cm Petri dish.

2.3 Aniline Blue Staining Assay

1. Forceps.
2. Small brush.
3. Transparent nylon bag with laces.
4. 5 cm Petri dish.
5. 15 mL Falcon tubes.
6. Scalpels.
7. Beakers (10 mL).
8. Pasteur pipette.
9. Glass microscope slides, 25 × 75 mm.
10. Micro cover glasses, 24 × 60 mm.
11. Fixative FFA solution: formalin/glacial acetic acid/70% ethanol, 1:1:18, v/v/v [19].
12. Distilled water.

13. 5% sodium sulfite.
14. Aniline blue solution: 0.1% aniline blue in 0.1 M K_3PO_4 (*see* **Note 1**).
15. Fluorescence microscope with I3 filter excitation 450 to 490 nm (DM 2500; Leica Microsystems, Wetzlar, Germany) with Leica DFC 320 digital camera.

2.4 Molecular Analysis of the S-RNase

1. Annotated genomic sequence of *S-RNases*. The genomic sequence information for the *S-RNases* from *C. maxima* [S_1-S_9-*RNases* (MN652897.1, MN652898.1, MN652899.1, MN652900.1, MN652901.1, MN652902.1, MN652903.1, MN652904.1, MN652905.1) and S_{16}-*RNase* (LC575207.1), *C. reticulata* [S_{10}-S_{11}-*RNases* (MN652906.1 and MN652907.1), *Atalantia buxifolia* [S_{12}-*RNase* (MN652908.1)], *C. cavaleriei* [S_{13}-*RNase* (MN652909.1)], *C. medica* [S_{14}-*RNase* (MN652910.1)], *C. tamurana* [S_{15}-*RNase* (LC575202.1)], *C. hassaku* [S_{17}-*RNase* (LC575209.1)], and *C. sinensis* [S_m-*RNase* (MN652911.1)] used in this study can be found at the National Center for Biotechnology (NCBI).
2. Sequence alignment software, for example, Clustal Omega.
3. Primers design software, for example, Primer3 software.
4. Young leaves of "Comune" and "Monreal" genotypes.
5. Liquid nitrogen.
6. Mortar and pestle.
7. TissueLyser II (Qiagen).
8. Stainless steel (5 mm).
9. Eppendorf tubes (1.5 and 2.0 mL).
10. Plant DNA extraction kit, for example, the ISOLATE II Plant DNA Kits (Bioline).
11. Polymerase PCR kit, for example, the BIOTAQ™ DNA Polymerase (Bioline).
12. DNA oligonucleotides primers (Table 1).
13. DNA quantification equipment, for example, Nanodrop 2000 spectrophotometer (Thermo Scientific).
14. Thermocycler for PCR analysis, for example, thermal cyclers Mastercycler Nexus Gradient (Eppendorf).
15. Agarose gel electrophoresis equipment and supplies.
16. Gel purification kit, for example, the ISOLATE II PCR and Gel Kit (Bioline).
17. Scalpels.
18. Pistils of "Comune" and "Monreal" genotypes.

Table 1
Primers used

Primer	Target	Sequence (5′-3′)	Product size (bp)	PCR
4FWC2	C2 of *S-RNase*	TWCAYGGYCTCTGGCCVGT	~ 280	Consensus
5RVC3	C3 of *S-RNase*	TRCCRTGYWBWABCCACT		
S7RTFW	S_7-*RNase*	TTTCGTCCTACATGGCCTCTGG	150	qRT-PCR
S7RTRV		TTCGAAATGAGACTCGGCCAGT		
S11RTFW	S_{11}-*RNase*	TCTGGCCCGTTGACTCTAATGG	146	
S11RTRV		GCCAGCTTTACTCTGAACAGCG		

19. Plant RNA extraction kit, for example, the Spectrum™ Plant Total RNA Kit (Sigma-Aldrich).
20. DNase I kit, for example, The On-Column DNase I Digestion Set (Sigma-Aldrich).
21. cDNA synthesis kit, for example, the High-Capacity cDNA Reverse Transcription Kit (Applied Biosystems).
22. Thermocycler for qRT-PCR assays, for example, the Rotor-Gene Q for (Qiagen) with the Rotor-Gene Q 2.3.5 software (Qiagen) for data analysis.

3 Methods

3.1 Plant Materials

1. Grow "Comune" clementine plants under the following conditions prior to conducting the ISI measurement, the aniline blue staining assay, and the molecular analysis of the *S-RNase*.
2. The nursery plants are transferred to 45 cm pots containing growing substrate and fertilizer 20–10–10 and placed in a growth chamber at 25 °C under 16 h light/8 h dark cycle and 50% relative humidity (RH) conditions. Remove apical buds to increase the number of flowers produced.

3.2 Self-Incompatibility Index (ISI) Measurement

Day 1

1. Use terminal flowers for pollination (*see* **Note 2**).
2. Emasculate 50 flowers 1 day before anthesis by using forceps, and gently close with a transparent nylon bag with laces (*see* **Note 3**).
3. Collect anthers from about 50 flowers 1 day before anthesis. Left anthers to dehisce on paper in a growth chamber at 25 °C under 16 h light/8 h dark cycle and 50% relative humidity (RH) conditions for 1 day. Collect pollen in a 5 cm Petri dish and use it for pollination on day 2 (*see* **Note 4**).

Day 2

4. Carefully remove the nylon bag from the emasculated flowers. Hand-pollinate pistils from "Comune" clementine with pollen from "Comune" (self-pollination) and pollen from the self-compatible donor (cross-pollination) using a small brush.

2 months after pollination

5. Count the number of fruits resulting from self- and cross-pollinations.
6. Apply the following formula:

$$\mathrm{ISI} = \frac{\text{n.fruits from self} - \text{pollination}}{\text{n.fruits from cross} - \text{pollination}}$$

An ISI value closer to 1 suggests the plant is highly self-compatible (self-fertility), while an ISI value closer to 0 suggests the plant is highly self-incompatible.

3.3 Aniline Blue Stain Assay

Day 1

1. Use terminal flowers for pollination.
2. Emasculate 50 flowers 1 day before anthesis by using forceps, and gently close with a transparent nylon bag with laces.
3. Collect anthers from about 50 flowers 1 day before anthesis. Left anthers to dehisce on paper in a growth chamber at 25 °C under 16 h light/8 h dark cycle and 70% relative humidity (RH) conditions for 1 day. Collect pollen in a 5 cm Petri dish and use it for pollination on day 2.

Day 2

4. Carefully remove the nylon bag from the emasculated flowers. Hand-pollinate pistils of "Comune" clementine using a small brush.

Day 10

5. Collect 10 self-pollinated pistils of "Comune" clementine. Fix pistils in FAA solution in a 15 mL Falcon tube for at least 24 h and store at 4 °C until microscope observation (*see* **Note 5**).
6. Monitor pistils on the squash preparations [11]. Wash pistils fixed three times in distilled water (1 h for each wash) in a 10 mL beaker, then soften with 5% sodium sulfite overnight.
7. Soften pistils in 5% sodium sulfite solution in a microwave for 45 s (*see* **Note 6**). Cut pistils longitudinally and split into two parts on a microscope slide (*see* **Note 7**). Stain with 0.1% aniline

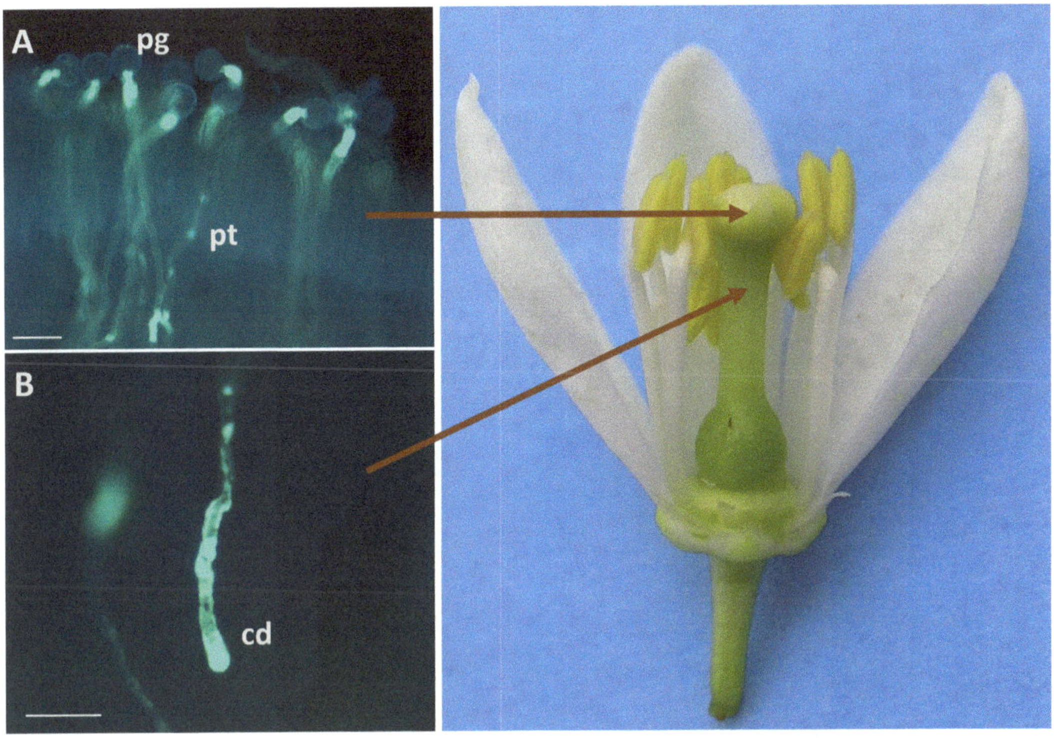

Fig. 1 Aniline blue staining of self-pollinated flowers of 'Comune' clementine (self-incompatible). (**a**) Pollen grain (pg) germination; pollen tube (pt) development in an incompatible combination; (**b**) pollen tube growth arrest and heavy callose deposition (cd) at the tip level. Scale bar 50 μm

blue in 0.1 M K_3PO_4 using a Pasteur pipette and squash using a micro cover glass.

8. Observe preparation under a fluorescence microscope equipped with UV excitation filters 450 to 490 nm. SI reaction occurs in the upper part of the style, highlighted by a heavy accumulation of callose at the tip level of the pollen tube (Fig. 1).

3.4 Molecular Analysis of the S-RNase

Molecular analysis of the *S-RNase* includes PCR and sequencing to analyze genomic *S*-alleles diversity and qRT–PCR to analyze *S*-alleles expression.

3.5 S-Genotyping

S-genotyping is performed by PCR-amplification of the two *S*-alleles of the genotype using consensus primers designed on conserved domains of *S-RNase* alleles reported in literature, then followed by Sanger sequencing of purified amplicons. The strategy allows defining the *S*-genotype of each cultivar or variety.

1. Align genomic sequences of *S-RNases* deposited in the NCBI database using Clustal Omega (Fig. 2).

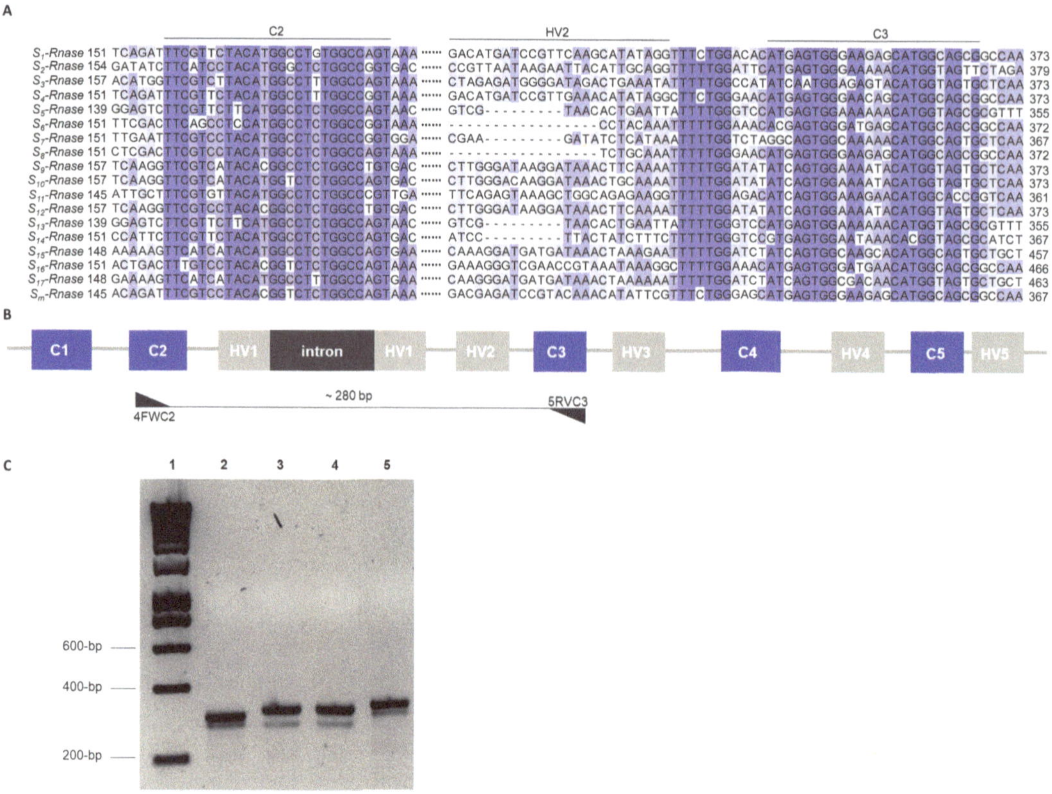

Fig. 2 (**a**) Nucleotide sequence alignment of *S-RNases* from *C. maxima* [S_1-S_9-*RNases* (MN652897.1, MN652898.1, MN652899.1, MN652900.1, MN652901.1, MN652902.1, MN652903.1, MN652904.1, MN652905.1) and S_{16}-*RNase* (LC575207.1), *C. reticulata* [S_{10}-S_{11}-*RNases* (MN652906.1 and MN652907.1), *Atalantia buxifolia* [S_{12}-*RNase* (MN652908.1)], *C. cavaleriei* [S_{13}-*RNase* (MN652909.1)], *C. medica* [S_{14}-*RNase* (MN652910.1)], *C. tamurana* [S_{15}-*RNase* (LC575202.1)], *C. hassaku* [S_{17}-*RNase* (LC575209.1)], and *C. sinensis* [S_m-*RNase* (MN652911.1)]. The alignment was generated with Clustal Omega. Nucleotides are color coded according to their conservancy: dark violet are fully conserved, while those marked in light violet present less similar. Five conserved domains (C1-C5) and five hypervariable regions (HV1-HV5) are indicated. (**b**) Schematic representation of the citrus *S-RNase* shows 5 conserved (C1-C5, in dark violet) and hypervariable domains (HV1-HV5, in light grey) and an intron (in grey). Arrows at the bottom indicate PCR *consensus* primers used to amplify C2 and C3 domains, and the respective size of PCR amplicons. (**c**) Visualization on agarose gel of amplified PCR product using *consensus* primers; 1, HyperLadder™ 1 kb (Bioline), 2, mandarin, 3, clementine, 4, sweet orange, 5, mandarin

2. Design degenerate primers (Fig. 2, Table 1) on conserved domains using Primer3 software.
3. Collect young leaves of "Comune" clementine and store in −80 °C until the extraction. Homogenize 100 mg of nitrogen-frozen leaf tissue in 2 mL Eppendorf tube containing stainless steel using TissueLyser II (Qiagen) at 18.5 Hz for 2 min.
4. Extract genomic DNA using the DNA extraction kit, and quantify using DNA quantification equipment, store at −20 °C until PCR analysis.

Table 2
Consensus primer PCR reaction mixture

Component	Volume (μL)
NH4 Reaction Buffer (10×)	5
$MgCl_2$ (50 mM)	2
dNTPs mix (2 mM)	5
Forward primer: 4FWC2 (5 μM)	2.5
Reverse primers: 5RVC3 (5 μM)	2.5
Template DNA (100 ng/μL)	2.5
BIOTAQ	0.5
ddH_2O	30
Total	50

Table 3
Consensus primer PCR cycling conditions

Step	Temperature	Time	Cycles
Initial denaturation	94 °C	2 min	1
Denaturation Annealing Extension	94 °C 52 °C 72 °C	1 min 1 min 2 min	15
Denaturation Annealing Extension	94 °C 53 °C 72 °C	1 min 1 min 2 min	20
Final extension	72 °C	30 min	1

5. Using DNA extracted from the leaf as a template, PCR-amplify the two *S*-alleles (Tables 2 and 3).
6. Separate amplicons (two per genotype) by electrophoresis on 2% agarose gel (*see* **Note 8**).
7. Visualize amplicons under UV light, then excise amplicons from the agarose gel using a scalpel (*see* **Note 9**).
8. Purify the amplicons with a commercial gel purification kit.
9. Subject the purified amplicons to Sanger sequencing using consensus primers (Table 1).
10. Evaluate sequence diversity with respect to sequences deposited in the NCBI database.

3.6 S-Alleles Expression

The analysis of expression of the *S-RNases* is performed by qRT–PCR using *S*-alleles-specific primers on female tissues. The analysis allows establishing an self-incompatible phenotype if both *S*-alleles are expressed, while the self-compatible phenotype is established if even just one *S*-allele does not express itself. Here we report the analysis of expression of *S*-alleles in the self-incompatible "Comune" and the self-compatible "Monreal," both showing the S_7–S_{11} genotype, by using pistils [14].

1. Collect virgin pistils of "Comune" and "Monreal" genotypes 24 h after anthesis and store in −80 °C until the extraction.
2. Homogenize 100 mg of nitrogen-frozen pistils using a mortar and pestle with liquid nitrogen.
3. Extract and purify RNA using the Spectrum™ Plant Total RNA Kit and the On-Column DNase I Digestion Set (Sigma–Aldrich), then evaluate quantity and integrity and quantify by spectrophotometer and agarose gel electrophoresis. Store samples at −80 °C until analysis.
4. Synthetize cDNA from 1 μg of extracted RNA using the High-Capacity cDNA Reverse Transcription Kit (Applied Biosystems).
5. Using cDNA synthesized, perform qRT–PCR assays (Tables 4 and 5) using primers reported in Table 1 in the Rotor-Gene Q and analyze expression with the Rotor-Gene Q 2.3.5 software (Qiagen).
6. The transcript level was calculated by standard curve and normalized against citrus *Elongation factor 1-alpha* (*EF-1α*) AY498567 gene as an internal control [20].

Table 4
qRT-PCR reaction mixture

Component	Volume (μL)
NH4 Reaction Buffer (10×)	2
$MgCl_2$ (50 mM)	0.8
dNTPs mix (2 μM)	1.6
Forward primer(5 μM)	1.2
Reverse primers (5 μM)	1.2
Template cDNA	2
SYTO9	0.6
BIOTAQ	0.2
ddH_2O	10.4
Total	20

Table 5
qRT-PCR cycling conditions

Step	Temperature	Time	Cycles
Initial denaturation	95 °C	5 min	1
Denaturation Annealing Extension	95 °C 59 °C 72 °C	5 s 20 s 2 min	35
Final extension	72 °C	30 min	1

4 Notes

1. The solution should be stored in the dark at 4 °C.
2. To ensure the fruit set is fundamental to pollinate terminal flowers, so before performing pollination, remove all lateral flowers; also, to avoid undesired pollinations, remove open flowers.
3. Covering the flowers to be pollinated is essential to avoid undesired pollination, alternatively, cotton tissue can be used.
4. Pollen can be stored at 4 °C for a few weeks, or at −20 °C for long periods. Prolonged times reduce germination.
5. Pistils fixed in FAA can be stored at 4 °C for several months.
6. The time needed to soften depends on the size of the pistil and the power of the microwave; in general, it is essential not to heat too much to prevent tissue damage preventing correct analysis.
7. As an alternative to the squash method, it is possible to use the slices method described by Montalt et al. [7], cutting the pistils transversely into sections, after washing in water and before aniline blue staining. The squash method allows following the elongation of the pollen tube along the pistil, while the second method facilitates counting the pollen tubes.
8. To allow the separation of the two bands, it is useful to prepare a large agarose gel (25 cm long), more concentrated (2%), and thick (about 1 cm) to load 50 uL of amplicon and conduct the electrophoresis slowly at 100 V for about 2.5 h.
9. Cut with a clean scalpel, minimizing the amount of gel surrounding the band and avoiding cutting the amplicon, clean the scalpel with ethanol for each band cut.

References

1. De Nettancourt D (2001) Incompatibility and incongruity in wild and cultivated plants, vol 3. Springer, Berlin
2. Newbigin E, Anderson MA, Clarke AE (1993) Gametophytic self-incompatibility systems. Plant Cell 5:1315–1324
3. McCubbin AG, Kao TH (2000) Molecular recognition and response in pollen and pistil interactions. Annu Rev Cell Dev Biol 16:333–364. https://doi.org/10.1146/annurev.cellbio.16.1.333
4. Montemurro C, Dambruoso G, Bottalico G, Sabetta W (2019) Self-incompatibility assessment of some italian olive genotypes (*Olea europaea* L.) and cross-derived seedling selection by ssr markers on seed endosperms. Front Plant Sci 10:1–13. https://doi.org/10.3389/fpls.2019.00451
5. Androulakis II, Loupassaki MH (1990) Studies on the self-fertility of some olive cultivars in the area of Crete. Acta Hortic:159–162
6. Livia de JV, de Santana JRF, Alfredo ACA et al (2015) Use of aniline blue stain to observing pollen tubes development in different *Manihot* Mill. species. African J Agric Res 10:1805–1809. https://doi.org/10.5897/ajar2014.8976
7. Montalt R, Cuenca J, Vives MC et al (2019) Influence of temperature on the progamic phase in *Citrus*. Environ Exp Bot 166:103806. https://doi.org/10.1016/j.envexpbot.2019.103806
8. Aloisi I, Distefano G, Antognoni F et al (2020) Temperature-dependent compatible and incompatible pollen-style interactions in *Citrus clementina* Hort. ex Tan. show different transglutaminase features and polyamine pattern. Front Plant Sci 11:1–13. https://doi.org/10.3389/fpls.2020.01018
9. Karabıyık Ş (2022) Effects of temperature on pollen viability and in vivo pollen tube growth in *Citrus sinensis*. J Appl Bot Food Qual 95:100–104. https://doi.org/10.5073/JABFQ.2022.095.013
10. Distefano G, Hedhly A, Las Casas G et al (2012) Male-female interaction and temperature variation affect pollen performance in *Citrus*. Sci Hortic (Amsterdam) 140:1–7. https://doi.org/10.1016/j.scienta.2012.03.011
11. Distefano G, Las Casas G, La Malfa S et al (2009) Pollen tube behavior in different mandarin hybrids. J Am Soc Hortic Sci 134:583–588
12. Sanzol J, Sutherland BG, Robbins TP (2006) Identification and characterization of genomic DNA sequences of the S-ribonuclease gene associated with self-incompatibility alleles S1 to S5 in European pear. Plant Breed 125:513–518. https://doi.org/10.1111/j.1439-0523.2006.01269.x
13. Gouta H, Ksia E, Laaribi I et al (2020) Efficiency of *S*-genotyping for diversity screening and self-incompatible group identification of almond cultivars within the Mediterranean basin. J Hortic Sci Biotechnol 96:338–343. https://doi.org/10.1080/14620316.2020.1850214
14. Bennici S, Poles L, Di Guardo M et al (2024) The origin and the genetic regulation of the self-compatibility mechanism in clementine (*Citrus clementina* Hort. ex Tan.). Front Plant Sci 15:1–11. https://doi.org/10.3389/fpls.2024.1360087
15. Bennici S, Di Guardo M, Distefano G et al (2020) Deciphering *S-RNase* allele patterns in cultivated and wild accessions of Italian Pear Germplasm. Forests 11:1–13. https://doi.org/10.3390/f11111228
16. Honsho C, Ushijima K, Anraku M et al (2021) Association of T2/S-RNase with self-incompatibility of Japanese citrus accessions examined by transcriptomic, phylogenetic, and genetic approaches. Front Plant Sci 12:1–17. https://doi.org/10.3389/fpls.2021.638321
17. Hu J, Xu Q, Liu C et al (2021) Downregulated expression of *S2-RNase* attenuates self-incompatibility in "Guiyou No. 1" pummelo. Hortic Res 8. https://doi.org/10.1038/s41438-021-00634-8
18. Liang M, Cao Z, Zhu A et al (2020) Evolution of self-compatibility by a mutant *Sm-RNase* in citrus. Nat Plants 6:131–142. https://doi.org/10.1038/s41477-020-0597-3
19. Johansen DA (1940) Plant microtechnique, New York
20. Distefano G, Caruso M, La Malfa S et al (2009) Histological and molecular analysis of pollen–pistil interaction in clementine. Plant Cell Rep 28:1439–1451. https://doi.org/10.1007/s00299-009-0744-9

Chapter 16

Comparing In Vivo and In Vitro Pollen Tube Attraction Assays for Arabidopsis and Maize

Yifang Sui, Wantong Qu, Daniel Bouyer, Junyi Chen, and Thomas Dresselhaus

Abstract

Pollen tube attraction is a crucial step in the reproduction of angiosperms, ensuring that the male gametophytes, which carry two sperm cells, accurately target the compatible female gametophytes, or embryo sacs, in order to achieve double fertilization. Various experimental systems have been developed to study this process in vivo and in vitro, particularly in the model plant *Arabidopsis thaliana* (Arabidopsis). However, fewer methods have been described for other plant species, including grasses, which generate much thicker reproductive tissues and organs. Here, we present and compare optimized protocols for tracing individual pollen tubes in Arabidopsis, which are required, for example, for studies in which wild-type and mutant pollen tubes exhibit different growth speeds, as well as for studying maize pollen tube growth and for characterizing potential attraction molecules. Together, these protocols contribute to the necessary tools to investigate the conserved and divergent mechanisms of pollen tube attraction in both eudicot and monocot species, offering valuable insights into reproductive signaling pathways that are relevant to both basic biology and applied breeding.

Key words Pollen tube, Pollination, Guidance, Peptide signaling, Arabidopsis, Maize

1 Introduction

In flowering plants (angiosperms), successful fertilization depends on the precise growth and guidance of pollen tubes toward and into the ovules, where two sperm cells are released to achieve double fertilization [1]. This process involves a continuous exchange of molecular signals between the growing pollen tube and the surrounding female tissues. It culminates in the pollen tube being attracted and recognized by the synergid cells located adjacent to the two female gametes (the egg and the central cell). Disruption to these signaling events results in failed fertilization and seed set, making pollen tube growth and guidance a central focus of plant reproductive biology [2].

Giampiero Cai et al. (eds.), *Pollen and Pollen Tube Analysis: Methods and Protocols*, Methods in Molecular Biology, vol. 3042,
https://doi.org/10.1007/978-1-0716-5308-1_16,

Over the past two decades, studies, particularly those focusing on the dicot model plant *Arabidopsis thaliana* (Arabidopsis) and the grass crop *Zea mays* (maize), have revealed that species-specific chemoattractant peptides, such as the LURE family in dicots and the ZmEA1 peptide in maize, are secreted from synergid cells and act as key guidance cues [3]. It has been shown, among other things, that pollen tubes express dedicated receptor-like kinases that perceive these signals and translate them into localized calcium, ROS (reactive oxygen species), and cytoskeletal responses that direct tube growth [1, 4]. Despite this progress, the dynamic nature of pollen tube growth and attraction deep inside the female tissues of the pistil or ovary, the observation that many more guidance molecules are probably yet to be discovered given the large number of potential signaling peptides generated by cells of the female gametophyte [5, 6], and the fact that the underlying molecular processes are additionally affected by environmental stress [7, 8], has made experimental access to these processes challenging.

To overcome these limitations, a combination of (high-throughput) in vivo and in vitro attraction assays has been developed for the eudicot model Arabidopsis. In vivo assays can be performed within the intact pistil/ovary and preserve the full physiological context, allowing the assessment of mutant phenotypes or the contribution of specific tissues to guidance. In contrast, semi-in vivo and in vitro assays reconstitute the terminal stages of pollen tube guidance using isolated ovules or beads containing attraction molecules placed near growing pollen tubes under defined conditions. This approach also enables high-resolution imaging, quantitative analysis of turning behaviour and functional testing of purified attractant peptides or receptor mutants [9–11]. Although technical advances have also been made to observe pollen tube growth and guidance in cereals such as wheat [12], proper protocols have not yet been fully described. For Arabidopsis, additional protocols are also required to study pollen tube growth and attraction in mutants exhibiting slower growth behavior, for example. Here, we present methods for studying the growth of single Arabidopsis pollen tubes in vivo, followed by a semi-in vivo pollen tube targeting assay using the dragging method. For maize, we present a collection of in vivo, semi-in vivo, and in vitro methods to study pollen tube growth and guidance in a grass species. Together, these complementary methods provide a powerful framework for dissecting the molecular mechanisms of pollen tube guidance across angiosperms, linking genetic, biochemical, and cellular perspectives on plant fertilization.

2 Material

2.1 Plants

1. *Arabidopsis thaliana* (Arabidopsis) Col-0 (wild type).
2. Arabidopsis *ptd/+* mutant (*pollen tube defective* in Col-0 background).
3. Arabidopsis *pAtLURE1.2::ZmPOI-GFP* transgenic plants (Col-0 background).
4. *Zea mays* (Maize) B73 inbred line (wild type).
5. Maize *pZmRALF1::GUS* transgenic plants (B104 background) [5].

2.2 Sterilization of Arabidopsis Seeds (Vapor-Phase)

1. Sodium hypochlorite (NaClO).
2. Concentrated (~37%) hydrochloric acid (HCl).
3. 1.5 mL microcentrifuge tubes.
4. Two metal tube racks.
5. Transparent plastic storage box with lid (~22 cm × 33 cm × 22 cm).
6. Glass beaker (100 mL).
7. Graduated glass pipette (10 mL).
8. Pipette controller (for 10 mL glass pipettes).

2.3 Plant Growth (Arabidopsis)

1. Germination medium: Murashige and Skoog (MS) medium [13], adjusted to a pH of 5.8, supplemented with 1% (w/v) sucrose, 0.5 g/L MES, 0.8% (w/v) agarose, and the appropriate antibiotics for the respective seed line. The medium was then autoclaved and poured into square petri dishes (~90 mm).
2. 70% (v/v) ethanol.
3. Parafilm paper.
4. Nitrile gloves.
5. Forceps.
6. Plant labels.
7. Soil, pots (~5 cm × 5 cm), trays (~50 cm × 30 cm), and humidity lids.
8. Clean bench.
9. Growth chamber (21/18 °C day/night, 16 h light/8 h dark photoperiod).

2.4 Emasculation and Restriction Pollination of Arabidopsis

1. Fine forceps.
2. Nitrile gloves.
3. Colored labeling tape.
4. Stereomicroscope.

5. Eyelash tool: a 1.0 mL syringe with a needle tip modified by inserting an eyelash, optionally secured with tape.
6. A custom double-layered slide that is designed to enable observation at two distinct focal planes.

2.5 Aniline Blue Staining of Arabidopsis Pistils

1. Fixative solution: mix acetic acid and ethanol in a ratio of 1:3 (v/v).
2. Ethanol series for rehydration: 70%, 50%, and 30% ethanol solutions.
3. Alkaline treatment solution: 8 M NaOH.
4. Aniline blue solution: 0.1% (w/v) aniline blue dissolved in 108 mM K_3PO_4 (pH ~ 11). After dissolving, store at 4 °C overnight. Then filter the solution through one spoon of active carbon powder placed on filter paper in a funnel for decolorization. After filtration, add glycerol to achieve a final concentration of 2% (v/v). Store at 4 °C.
5. Arabidopsis pollen donor and recipient plants.
6. Distilled water.
7. Fine forceps.
8. 1.5 mL microcentrifuge tubes.
9. Pipette and gel-loading pipette tips (1–200 μL).
10. Microscope slide.
11. Micro cover glass, 18 mm × 18 mm.
12. Use nail polish to preserve the slide samples.
13. Zeiss ApoTome.2 described in Subheading 2.8.

2.6 Pollen Germination Medium (PGM) for Semi-In Vitro (SIV) Pollen Tube Targeting Assays [11, 14]

1. PGM: 5 mM $CaCl_2$, 1 mM $MgSO_4$, 5 mM KCl, 0.01% (w/v) H_3BO_3, and 10% (w/v) sucrose. Adjust the pH to 7.5 with 0.1 M KOH (*see* **Note 1**). Add 1.3% low melting point agarose to the stock when in use.
2. 35 mm glass-bottom culture dish (14 mm well).
3. Parafilm paper.
4. Growth chamber at 21 °C under a 16 h light/8 h dark cycle.

2.7 Preparation of Arabidopsis Stigmas and Ovules

1. Glass slide.
2. Double-sided tape.
3. Fine forceps.
4. Dissecting needle.
5. 1.0 mL syringe needle.
6. Stereo microscope.
7. Humidifier.

2.8 Imaging

1. Zeiss ApoTome.2 fluorescence microscope system (Axio Imager M2 stand, HXP 200 C illuminator, Carl Zeiss).
2. Zeiss LSM 980 confocal laser scanning microscope with Airyscan 2 detector (Carl Zeiss).
3. Glass needles (~40 μm tip diameter) were used for dragging ovules during imaging (*see* **Note 2**).

2.9 Plant Growth (Maize)

Greenhouses for maize plants: 26 °C, 24,000 lux illumination, 16 h light/8 h dark cycle, and a relative humidity of about 60%.

2.10 Aniline Blue Staining of Maize Silks

1. Fixative solution: mix acetic acid and ethanol (100%) in a ratio of 1:3 (v/v).
2. Ethanol series for rehydration: 70%, 50%, and 30% ethanol.
3. Clearing solution: 10% chloral hydrate and 5 M NaOH.
4. Aniline blue solution: dissolve 0.1% (w/v) aniline blue in 0.1 M K_3PO_4 (pH 8.0) and then store at 4 °C overnight.
5. Fluorescence microscope (e.g., Leica DM6 B, Leica).

2.11 GUS Staining

GUS solution: 50 mM $Na_2HPO_4{\cdot}12H_2O$ (pH 7.2), 50 mM $NaH_2PO_4{\cdot}2H_2O$ (pH 7.2), 2 mM $K_3[Fe(CN)_6]$, 2 mM $K_4[Fe(CN)_6]$, 0.2% Triton X-100, 2 mM X-Gluc.

2.12 Maize In Vitro Pollen Tube Growth

1. Pollen germination medium (PGM) / pollen tube growth medium for *Zea mays* pollen [15]: 13% (w/v) sucrose, 0.001% (w/v) H_3BO_3, 10 mM $CaCl_2$, 0.05 mM KH_2PO_4, 6% PEG 4000.
2. Solid PGM for *Zea mays* pollen: add 0.7% (w/v) low-gelling-temperature agarose into the PGM.
3. Glass-bottom culture dish.

2.13 Maize Pollen Tube Attraction Assay

1. Silicone oil.
2. Gelatin.
3. Vortex Mixer.
4. Fine steel needle.
5. Inverted fluorescence microscope (EVOS FL Auto, Life Technologies).

3 Methods

3.1 Growing Arabidopsis Plant Material

1. Sterilization of *Arabidopsis* seeds with chlorine gas [16] (*see* **Note 3**): Place the opened 1.5 mL microcentrifuge tubes containing dry seeds in a sealed sterilization container. Generate the chlorine gas by adding 2 mL concentrated HCl to 50 mL of sodium hypochlorite in a 100 mL beaker (*see* **Note 4**) placed

inside the container. Seal the container immediately with parafilm and incubate at room temperature for 1 h. Vent the container in a fume hood for 3 h, then close the seed tubes and transfer them to a laminar flow hood for subsequent use.

2. Plating of sterilized seeds: In a laminar flow hood, evenly spread sterilized seeds on MS medium plates. Cover and seal the plates with parafilm to prevent contamination (*see* **Note 5**).
3. Growth of Arabidopsis on MS plates: Stratify plates for 3 days (*see* **Note 6**) at 4 °C, then transfer them to a growth chamber (*see* Subheading 2.3, **item 9**). Grow seedlings on the MS plates for 8 days.
4. Transfer to soil: Transplant seedlings into pots filled with soil. Cover the pots with a transparent plastic lid (*see* **Note 7**) to maintain high humidity for several days and remove the lid after 1 week.

3.2 Single Pollen Pollination

1. Preparation of female flowers (*see* **Note 8**): One day prior to artificial pollination, emasculate the female parent by carefully removing all stamens with fine forceps, while retaining sepals and petals (*see* **Notes 9** and **10**). Flowers at developmental Stage 12 [17], characterized by the appearance of petal tips at the bud apex, are most suitable for emasculation (*see* Fig. 1a–c).
2. Preparation of male flowers: Under a stereo microscope, dissect the male flowers at Stage 13 [17], when anthers are fully dehiscent, and excise the anthers with fine forceps.
3. Preparation of pollen: Place the excised anthers on the circled area of an upper glass slide and gently tap to release and evenly distribute pollen grains (*see* Fig. 1d, e).
4. Pollination: Position the emasculated stigma near the edge of the upper slide. Using a 4× objective, focus until both stigma and pollen are clearly visible. With an eyelash tool, pick up a single pollen grain or a small pollen ball and carefully transfer it to the stigma surface (*see* Fig. 1e, f and **Note 11**).
5. Post-pollination labeling: Mark the pollinated flower immediately to distinguish it from untreated buds [16].

3.3 Tracking Individual Arabidopsis Pollen Tubes in the Transmitting Tract via Single-Pollen Pollination

3.3.1 Aniline Blue Staining

1. Day 1, emasculation, as described in Subheading 3.2, **step 1**.
2. Day 2, pollination, as described in Subheading 3.2, **step 4**.
3. Day 3, fixation: Collect pistils 24 h after pollination (time point adjustable according to experimental requirements) and place them in 1.5 mL microcentrifuge tubes containing fixative solution (~1 mL). Air in the capped tubes was evacuated with a 50 mL syringe fitted with an 18-gauge needle until bubbling ceased. The specimens were then incubated in fixative solution at room temperature (RT) for at least 2 h (*see* **Note 12**).

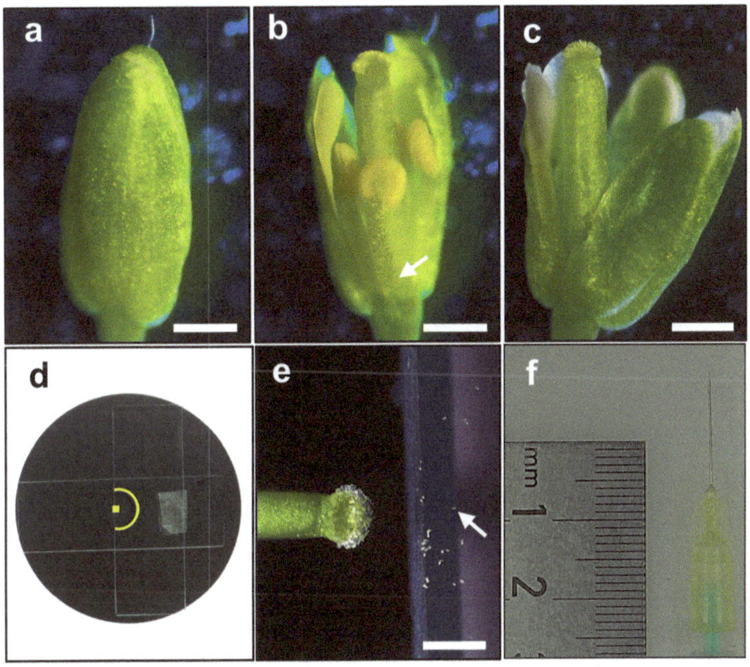

Fig. 1 Pollination using single pollen grains in Arabidopsis. (**a**) Stage 12 flower bud in which the petal tips become visible at the apex. (**b**) At stage 12, anthers have turned yellow, but dehiscence has not yet occurred. Stamens were excised with forceps from the position indicated by the white arrow. The image shows the anthers exposed after the removal of one sepal and one petal. (**c**) Emasculated flower bud with stamens removed. (**d**) Glass slide with two focal planes designed for simultaneous focusing of stigma (left) and pollen (right). (**e**) Experimental setup with stigma placed on the left side and pollen grains evenly spread on the upper slide on the right side. A limited amount of pollen was applied to facilitate the isolation of single pollen grains. Despite differences in size and plane, both stigma and pollen could be brought into focus under a 4× objective. (**f**) Eyelash tool is used for single pollen manipulation. Scale bars, 500 μm (a–c, e)

4. Day 4, rehydration: Replace the fixative solution with 70% ethanol and incubate the samples for 10 min at RT, followed by sequential treatment with 50% ethanol, 30% ethanol, and distilled water.
5. Day 5, alkaline treatment: Carefully transfer the specimens into a small Petri dish containing 8 M NaOH and incubate overnight at RT with the lid closed, sealing the dish with Parafilm paper.
6. Day 6, washing: Carefully replace the NaOH solution with distilled water, as the pistils become very soft at this stage. Incubate for 10 min at RT.

7. Day 7, aniline blue staining: Carefully replace the distilled water with aniline blue solution and incubate the specimens for at least 3 h (*see* **Note 13**) at RT in the dark by covering the dish with aluminum foil. Washing is not required after this treatment.

3.3.2 Microscopy

1. Transfer the pistils onto a slide containing a 20 μL drop of aniline blue solution.
2. Place a cover glass gently on the specimen from the side of the flower stem (*see* **Note 14**).
3. Specimens can be rapidly examined, for example, with a Zeiss ApoTome.2 fluorescence microscope to assess overall condition and to confirm the presence of more than one pollen tube. For high-resolution imaging, a confocal microscope (e.g., Zeiss LSM series) enables complete visualization of the growth path of individual pollen tubes (Fig. 2).

3.4 SIV Assay for Quantifying Competitive Pollen Tube Targeting Efficiency in Arabidopsis

3.4.1 PGM Preparation and Pollen Tube Growth [1, 14]

1. Prepare the stock solutions: 1 M $CaCl_2$, 1 M $MgSO_4$, 1 M KCl and store them at −20 °C.
2. Prepare 100 mL fresh liquid PGM solution each time by adding 500 μL $CaCl_2$, 500 μL KCl, 100 μL $MgSO_4$, 10 mg H_3BO_3, and 10 g sucrose to a 200 mL beaker, then adjusting the pH to 7.5 with 0.1 M KOH. This yields the following final concentrations: 5 mM $CaCl_2$, 1 mM $MgSO_4$, 5 mM KCl, 0.01% (w/v) H_3BO_3, and 10% (w/v) sucrose.
3. Add 2.5 mL of the liquid medium to a 10 mL beaker and add 32 mg of ultra-low melting point agarose (~1.3% final concentration).
4. Melt agarose in a microwave (400 W for 40–50 s).
5. Pipette 130 μL of the medium into the 14 mm well of a 35 mm glass-bottom dish and rotate the dish until the medium covers the well. Use the pipette tip to help spread the medium evenly on the dish.
6. Aspirate a total of 40 μL with a pipette from the edge of the glass-bottom well, leaving a final volume of 90 μL (Fig. 3b).
7. Place the prepared culture dish at room temperature (*see* **Note 15**).

3.4.2 Ovule Targeting Assay in Arabidopsis

1. Perform the procedures described in Subheading 3.4.1 in the early morning before starting the experiment.
2. Emasculate flowers at stage 12 [17] by removing the stamens 1 day in advance.
3. Perform hand pollination.

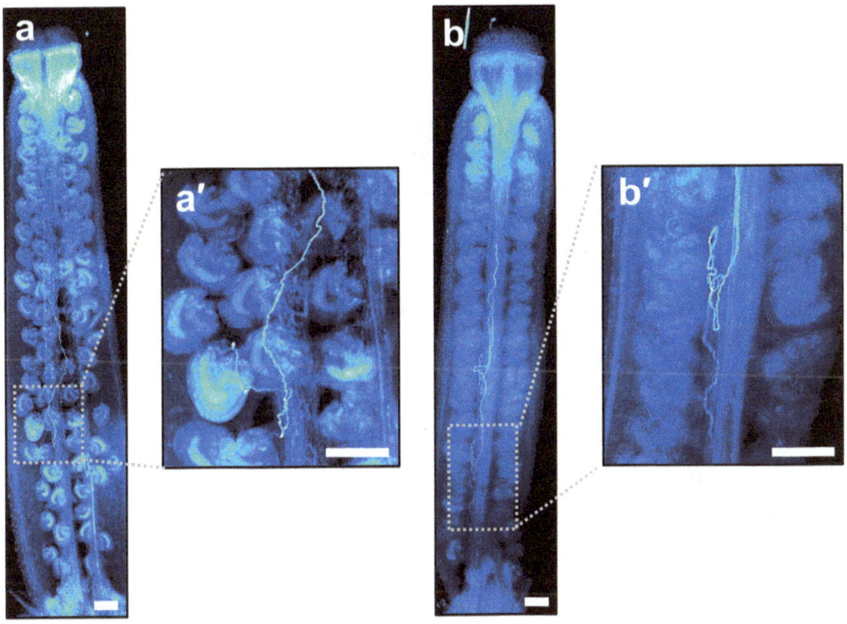

Fig. 2 In vivo tracking of single pollen tube behavior in Arabidopsis. Pistils stained with aniline blue (~24 h after pollination) and imaged using a Zeiss ApoTome.2 microscope with a 10× objective. (**a, a'**) In Col-0 wild-type pistils, pollen tubes directly reached and targeted the ovules (enlarged ovules indicate successful fertilization). (**b, b'**) In guidance-defective mutants, pollen tubes coiled within the transmitting tract and failed to target ovules ~24 h after pollination. Scale bars, 100 μm

4. Dissect the pistils and excise mature ovules (with funiculus attached) from the septum under high-humidity conditions to prevent desiccation and maintain ovule viability. Arrange the excised ovules in an arc on the medium with the micropyles oriented upward (*see* Fig. 3c and **Notes 16**, **17**, and **18**).
5. Excise stigmas 30–60 min after pollination and adjust the stigma-style on the medium so that the cut surface faces the ovule micropyles. Maintain a distance of approximately 2 ovule diameters between stigma and micropyles (*see* Fig. 3b, c).
6. Ensure that all dishes are properly closed. Place them inside a transparent container with a lid and surround the glass-bottom dishes with moistened tissue paper to maintain humidity (*see* Fig. 3a).
7. Incubate the dishes in a growth chamber (as described in Subheading 2.3, **item 9**) during the day for 5–6 h (*see* **Note 19**).
8. Observe pollen tube targeting under a Zeiss confocal microscope.

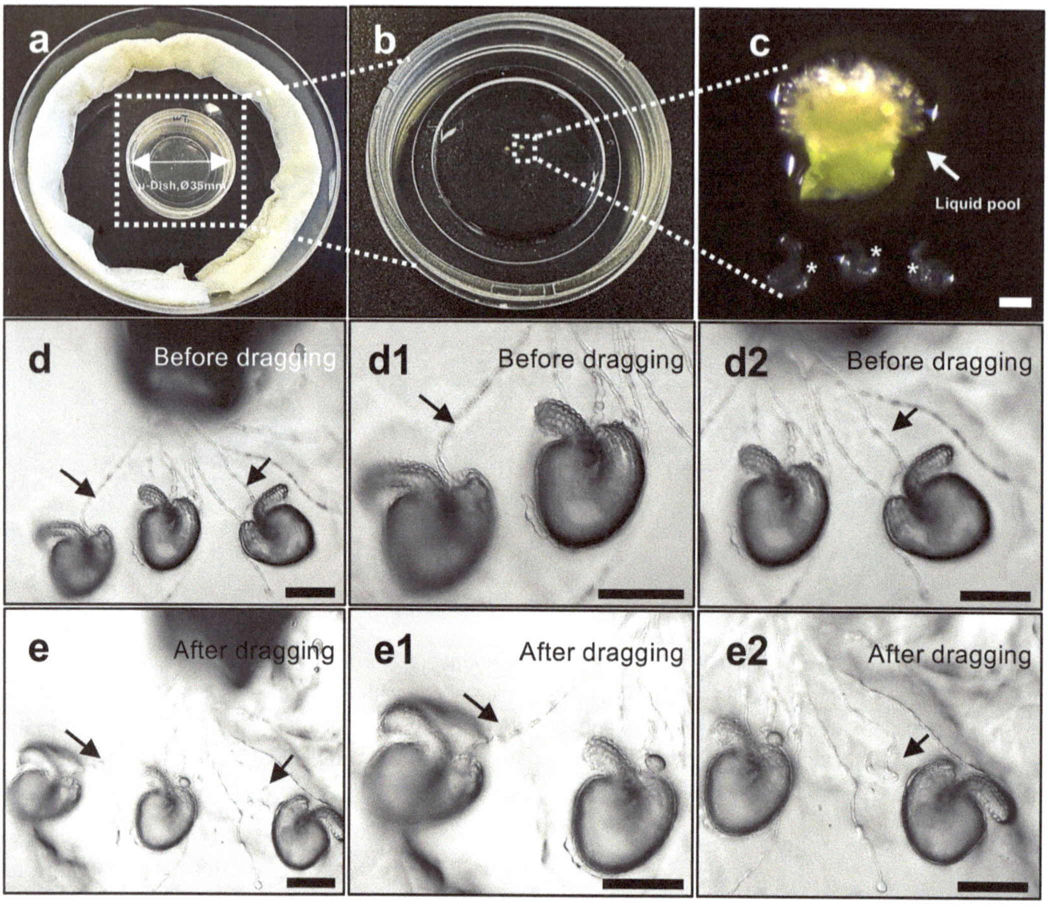

Fig. 3 Semi-in vivo pollen tube targeting assay using the dragging method. (**a**) A glass-bottom culture dish was placed in a transparent container with a lid, surrounded by moistened paper to maintain humidity. (**b**) 1–2 sets of samples were placed on the medium, each consisting of 3–4 ovules and one cut stigma. (**c**) Micropyles of the ovules were oriented toward the cut end of the stigma, with a spacing of 2–3 ovules (~250 μm). When properly placed, a small liquid pool formed around the samples. (**d**) Overview before dragging showing pollen tubes growing towards ovules (black arrows indicate pollen tubes potentially targeting an ovule). (**d1, d2**) Enlarged views of pollen tubes before dragging (left and right ovules, respectively). (**e**) Overview after dragging showing successful targeting by the left pollen tube and failure by the right pollen tube. (**e1**) Enlarged view after dragging (left ovule; black arrow indicates attachment of pollen tube to ovule and stretching into a straight line demonstrating successful targeting). (**e2**) Enlarged view after dragging (right ovule; black arrow indicates separation of pollen tube from ovule demonstrating unsuccessful targeting). Scale bars, 100 μm (c–e2), respectively

9. When pollen tubes appear to target the ovules, use the dragging method [18] to confirm targeting. Gently drag the ovule with a fine glass needle; if it remains attached to the pollen tube after displacement, targeting is considered successful (*see* Fig. 3d-e2 and **Note 20**).

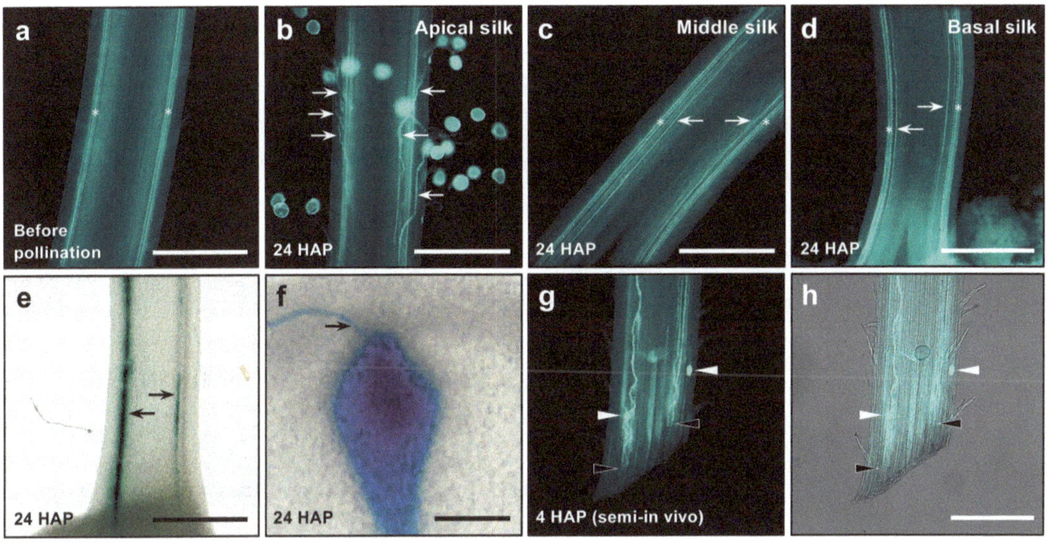

Fig. 4 Aniline blue and GUS staining to observe in vivo and semi-in vivo pollen tube growth of maize. (**a–d**) Fluorescence image of aniline blue-stained maize silks before (**a**) and 24 h after pollination (**b–d**). Xylem vessels of vascular tissue (asterisks) were strongly stained because they accumulate callose [20]. Invaded pollen tubes at apical silk (**b**) and pollen tubes that grew in the transmitting tract at middle (**c**), and basal (**d**) silks are indicated by white arrows. (**e**, **f**) GUS staining to visualize pollen tube growth (arrows) in the silk (**e**) and at the micropylar region of the ovule (**f**). (**g**, **h**) Aniline blue staining of pollinated cut silk showing that maize pollen tubes swell (white arrowheads) and/or stop growing (black arrowheads) when approaching the cutting edge, and do not exit at semi-in vivo conditions. HAP hours after pollination. Scale bars, 500 μm (a–e, g, h) and 50 μm (f), respectively

3.5 Monitoring Maize Pollen Tube Growth

Aniline blue staining allows for clear visualization of pollen tube growth within the silk after pollination (Fig. 4a–d).

3.5.1 Aniline Blue Staining

1. Cover maize ears with a paper bag before silk emergence. When silks emerge to a length of about 3 cm, pollinate them with freshly harvested pollen.
2. Cut pollinated silks with a sharp razor blade and collect them into 1 mL fixative solution in a 1.5-mL microcentrifuge tube (*see* **Note 21**).
3. Vacuum the samples for 20 min and incubate them at 4 °C overnight.
4. Add 70%, 50%, and 30% ethanol and 0.1 M K_3PO_4 (pH 8.0) in sequence, 5 min for each step.
5. Incubate the samples in 10% chloral hydrate at 65 °C for 5 min and then in 5 M NaOH at 65 °C for 10 min. After each step, samples were washed with 0.1 M K_3PO_4 (pH 8.0; *see* **Note 22**).
6. Stain the samples with aniline blue solution for more than 2 h in the dark. Stained samples can be stored at 4 °C for up to 1 week.
7. Observe and image under both brightfield and UV fluorescence settings on the fluorescence microscope.

3.5.2 GUS Staining

GUS staining enables efficient observation of pollen tubes in the silk tissue as well as at the micropyle (Fig. 4e, f). This assay requires pollination with pollen from a transgenic marker line that expresses GUS in the pollen tube.

1. Conduct pollination with freshly harvested *pZmRALF1::GUS* pollen.
2. 24 h after pollination, manually dissect silks or ovules. Incubate sliced silks or ovular sections containing embryo sacs at 37 °C in GUS solution in the dark, for a few hours or up to 1 day until the GUS signals appear.
3. Replace the GUS solution with 70% ethanol to decolorize the background until the GUS signals become clearly distinguishable.
4. Observe and image the GUS signals with a microscope under brightfield settings.

3.5.3 Semi-In Vivo Pollen Tube Growth of Maize

1. Weigh 7 mg of low-gelling-temperature agarose into 1 mL maize pollen tube growth medium in a 1.5-mL microcentrifuge tube. Heat the mixture for 20 min (on a heating block or in hot water at 90 °C) to ensure a complete solution.
2. Pipette approximately 200 μL of hot medium to the middle of a glass-bottom culture dish and allow the medium to solidify at room temperature (*see* **Note 23**).
3. Moisturize the surface of the solid PGM with liquid PGM and cover the culture dish with a lid.
4. After pollination with freshly harvested pollen, cut the pollinated silks with a sharp razor blade to a length of about 3 cm from the top silk region (*see* **Note 24**).
5. Quickly transfer the excised silk tissue onto the solid PGM, which has been moisturized with the liquid PGM.
6. Seal the culture dish with Parafilm and incubate at 28 °C in the dark for about 4–5 h.
7. Perform aniline blue staining as described in Subheading 3.5.1 to observe maize pollen tube growth behavior at semi-in vivo conditions (Fig. 4g, h).

3.6 In Vitro Pollen Tube Attraction Assay

Unlike in *Arabidopsis*, maize pollen tubes barely emerge from the silk under semi-in vivo conditions (Fig. 4g, h). Besides, they do not require a growth phase in the transmitting tract to become competent in responding to female-derived attracting signals [19]. We thus perform the maize pollen tube attraction assay under complete in vitro conditions using peptide-embedded beads.

3.6.1 Preparation of Peptide-Embedded Beads

1. Weigh 12–14 mg gelatin into 100 μL liquid PGM in a 1.5-mL microcentrifuge tube. Heat the mixture at 56 °C for 30 min and mix thoroughly by pipetting every 5 min during heating.
2. Sonicate or vigorously vortex the melting gelatin for 1 min (*see* **Note 25**).
3. Add 100 μL silicone oil to a 1.5-mL microcentrifuge tube.
4. In a 0.2-mL PCR tube, quickly mix 2 μL peptide of interest (POI) with 2 μL melting gelatin from **step 2**, then immediately transfer the mixture into the silicone oil from **step 3** using a 10 μL pipette (Fig. 5a, *see* **Note 26**).
5. Vortex vigorously for 30 s to disperse the mixture droplet into tiny beads (Fig. 5b) and immediately place the 1.5-mL microcentrifuge tube on ice.

3.6.2 Maize In Vitro Pollen Tube Attraction Assay

1. Prepare moisturized solid PGM as described in Subheading 3.5.3, **steps 1–3**.
2. Spread freshly harvested maize pollen grains on the surface of the solid PGM (*see* **Note 27**).

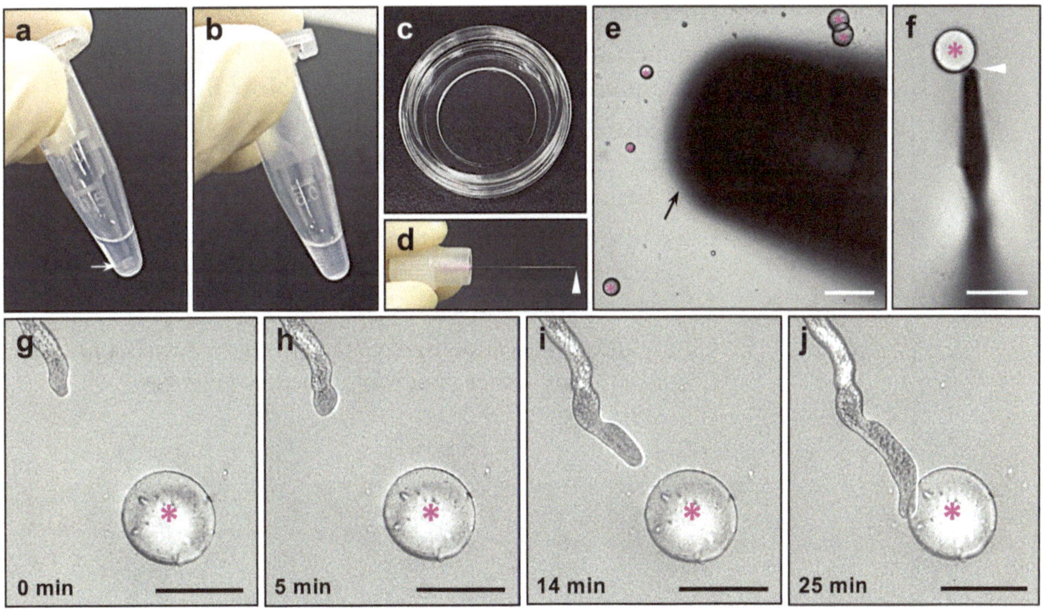

Fig. 5 In vitro assay for maize pollen tube attraction studies using peptide-embedded beads. (**a**) Addition of a liquid droplet on top of the peptide and gelatin mixture (indicated by the white arrow) into the silicone oil. (**b**) After vigorous vortexing, the liquid droplet has been dispersed into small suspending beads. (**c**) Glass-bottom culture dish for pollen tube growth studies and imaging. (**d**) A steel needle for bead positioning. White arrowhead points to the sharp needle tip. (**e**) Beads are added to the pollen tube growth medium with a pipette. The black arrow points towards the pipette tip. Purple asterisks indicate peptide-embedded beads. (**f**) Beads are positioned with the tip of a steel needle. (**g–j**) In vitro-grown maize pollen tube attracted by the bead. The peptide of interest (POI) is embedded in the beads. Scale bars, 100 μm (e) and 50 μm (f–j), respectively

3. Incubate at 28 °C for 30 min.
4. Add 0.3–0.5 μL peptide-embedded beads (from **step 5** in Subheading 3.6.1) to the pollen tube growth medium with a pipette (*see* **Note 28**, Fig. 5c–e).
5. Using a fine steel needle, place a bead to one side and in close proximity to the pollen tube tip (*see* **Note 29**, Fig. 5f).
6. Capture an image immediately to record the relative position between the pollen tube tip and the bead (Fig. 5g).
7. Repeat **steps 5** and **6** for additional pollen tubes.
8. Cover the culture dish with a lid to maintain humidity. Acquire images at 5-min intervals and monitor continuously for at least 30 min (*see* **Note 30**, Fig. 5h–j).

3.7 Pollen Tube Competition Assay (PTCA)

1. Germinate and grow maize pollen tubes on solid PGM as described in Subheading 3.6.2, **steps 1–3**.
2. Isolate mature ovules of emasculated Arabidopsis *pAtLURE1.2::ZmPOI-GFP* plants with a 1.0 mL syringe needle and collect them in the liquid PGM.
3. Transfer the excised ovules to the solid PGM using a pipette (*see* **Note 31**).
4. Symmetrically position two ovules on each side of the pollen tube tip using a fine steel needle (*see* **Note 32**).
5. Immediately capture images using the GFP and brightfield merged channels with an inverted fluorescence microscope (Fig. 6a).
6. Cover the culture dish with a lid to maintain humidity. Acquire images at 5-min intervals and monitor continuously for at least 30 min (*see* **Note 33**, Fig. 6b–d).
7. Gently drag the competitively attractive ovule with a fine steel needle to see whether penetration occurs (Fig. 6e).

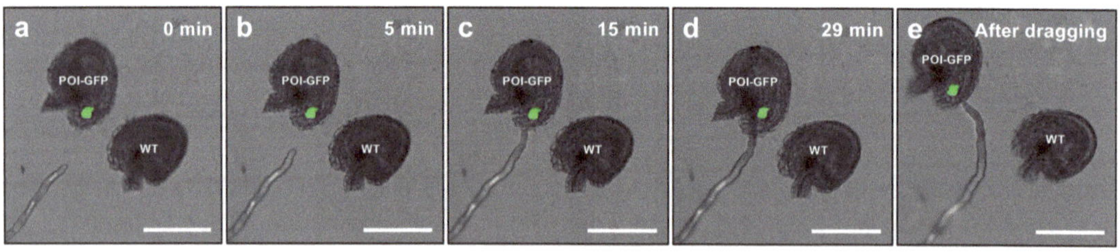

Fig. 6 Maize pollen tube competition assay using peptide-expressing Arabidopsis ovules. (**a–d**) A maize pollen tube is guided to the micropylar region of an Arabidopsis ovule expressing ZmPOI–GFP (green fluorescence-labeled). The fusion protein was expressed in Arabidopsis ovules under the control of the synergid-cell-specific *AtLURE1.2* promoter. (**e**) The "dragging" procedure shows guided pollen tube growth without penetration. Scale bars, 100 μm

4 Notes

1. Pollen germination is highly sensitive to the composition and freshness of the medium. Poor germination is often caused by deterioration of the medium [14]; therefore, freshly prepared medium should always be used to ensure stable pollen germination.
2. Alternatively, 1 μm dissecting needles (Tungsten needles, e.g., Item No. 10130-05, Fine Science Tools) were applied.
3. Chlorine gas sterilization must be performed in a chemical fume hood with appropriate personal protective equipment (lab coat, gloves, safety glasses).
4. Use a beaker of at least twice the total liquid volume (bleach + HCl) to avoid splashing during gas generation.
5. Ensure plates are sealed properly to prevent contamination during seedling growth.
6. Stratification at 4 °C is essential to synchronize seed germination.
7. High humidity is critical during the first days after transfer to soil; a transparent lid or plastic wrap can be used.
8. The first five flowers of Arabidopsis generally fail to develop normally; therefore, flowers emerging beyond the fifth silique are recommended for experimental use.
9. Keep the sepals and petals intact after emasculation, as they protect the stigma and help reduce the risk of unwanted pollen contamination.
10. Always wear gloves when handling the plant to avoid transferring heat from the hand, which may negatively affect the floral tissues.
11. To avoid pollen contamination between different genotypes, always change gloves, eyelash tool, and glass slide when switching to a new genotype.
12. Fixed samples can typically be stored at 4 °C overnight (up to 2 days).
13. Samples can be kept in the aniline blue solution overnight, but observation is recommended within 3 days.
14. If the pedicel is too long, it is recommended to trim it with a sharp razor blade to flatten the sample, which facilitates obtaining more homogeneous images.
15. For short-term storage at 4 °C, return the culture dish to room temperature or the growing chamber for at least 20 min before use to equilibrate the medium.

16. The number of ovules can be adjusted according to the experimental purpose.
17. It is recommended to place 1–2 stigmas per dish, as frequent lid opening leads to rapid desiccation of the medium, ovules, and stigmas.
18. To minimize drying, place a humidifier nearby during handling and perform the procedure within a simple humid chamber.
19. The culture dishes can also be placed in a climate chamber maintained at 21 °C, and pollen tube growth can be monitored using a confocal microscope after approximately 5 h of incubation.
20. Dragging should be carried out under high-humidity conditions. Use a humidifier near the stereomicroscope and briefly expose the medium to the humidified airflow before dragging to prevent rapid drying.
21. Rinse the silks 2–3 times with fixative solution to remove non-germinated pollen grains.
22. 10% chloral hydrate solution must be stored protected from light. Due to its instability after preparation, avoid using solutions for more than 1 week. As silks become extremely fragile following NaOH treatment, employ gentle manipulation techniques in subsequent steps to prevent mechanical damage.
23. Glass-bottom culture dishes may be briefly heated on the metal bath to prevent premature solidification before the medium fully covers the bottom.
24. As maize silks are highly susceptible to physical damage, use minimal force during cutting to avoid compression.
25. Avoid prolonged manipulation to prevent the gelatin from solidifying as it cools.
26. Mix the protein with the beads quickly to prevent the droplets from becoming difficult to disperse due to prolonged handling.
27. The morning-collected pollen is highly viable (germination rate > 90%) and thus ideal for subsequent experimental steps.
28. Apply the peptide-embedded beads to areas of low pollen tube density on the medium to facilitate clear observation and imaging.
29. For optimal imaging, use the needle gently when relocating the beads to preserve the integrity of the culture medium surface.
30. A pollen tube deflection angle greater than 20° is defined as evidence of attractive capability.
31. The transfer volume should not exceed 2 μL.

32. Select uniformly sized ovules and place them equidistant from the pollen tube tip.
33. A pollen tube is considered competitively attracted if it exhibits a deflection greater than 20° and successfully reaches the ovule's micropyle.

Acknowledgments

We thank Armin Hildebrand for plant care. The Deutsche Forschungsgemeinschaft (DFG) and the Agence National de la Recherche (ANR) are acknowledged for financial support via grants DR 334/19-1 to TD and ANR-21-CE12-0036-01 to DB. The research on maize pollen tube attraction assays is supported by grants from the National Natural Science Foundation of China (32570415 and 31970342) and by the Major Project of Hubei Hongshan Laboratory (2022hszd017) to JC.

References

1. Zhong S, Lan Z, Qu L-J (2025) Ingenious male–female communication ensures successful double fertilization in angiosperms. Annu Rev Plant Biol 76:401–431. https://doi.org/10.1146/annurev-arplant-083123-071512
2. Dresselhaus T, Franklin-Tong N (2013) Male-female crosstalk during pollen germination, tube growth and guidance, and double fertilization. Mol Plant 6(4):1018–1036. https://doi.org/10.1093/mp/sst061
3. Higashiyama T, Takeuchi H (2015) The mechanism and key molecules involved in pollen tube guidance. Annu Rev Plant Biol 66:393–413. https://doi.org/10.1146/annurev-arplant-043014-115635
4. Zhou LZ, Dresselhaus T (2019) Friend or foe: signaling mechanisms during double fertilization in flowering seed plants. Curr Top Dev Biol 131:453–496. https://doi.org/10.1016/bs.ctdb.2018.11.013
5. Chen J et al (2025) Fertilization-induced synergid cell death by RALF12-triggered ROS production and ethylene signaling. Nat Commun 16(1):3059. https://doi.org/10.1038/s41467-025-58246-y
6. Perkins CJ, Pryze K, Palanivelu R (2025) Peptide signalling in flowering plants: insights into reproductive thermotolerance. J Exp Bot. https://doi.org/10.1093/jxb/eraf192
7. Chaturvedi P et al (2021) Heat stress response mechanisms in pollen development. New Phytol 231(2):571–585. https://doi.org/10.1111/nph.17380
8. Gong W, Oubounyt M, Baumbach J, Dresselhaus T (2024) Heat-stress-induced ROS in maize silks cause late pollen tube growth arrest and sterility. iScience 27(7):110081. https://doi.org/10.1016/j.isci.2024.110081
9. Takahashi T, Mori T, Igawa T (2020) Restricted pollination for tracing individual pollen tubes in a pistil. Methods Mol Biol 2160:73–81. https://doi.org/10.1007/978-1-0716-0672-8_5
10. Mizuta Y et al (2024) Deep imaging reveals dynamics and signaling in one-to-one pollen tube guidance. EMBO Rep 25(6):2529–2549. https://doi.org/10.1038/s44319-024-00151-4
11. Zhong S, Wang Z, Qu LJ (2020) Semi-in vivo assay for pollen tube attraction. Methods Mol Biol 2160:83–92. https://doi.org/10.1007/978-1-0716-0672-8_6
12. Mishina K, Morita M, Matsumoto S, Sakuma S (2024) Optimizing visualization of pollen tubes in wheat pistils. Plants (Basel) 13(24). https://doi.org/10.3390/plants13243600
13. Murashige T, Skoog F (1962) A revised medium for rapid growth and bio assays with tobacco tissue cultures. Physiol Plant 15(3):473–497. https://doi.org/10.1111/j.1399-3054.1962.tb08052.x
14. Desnoyer NJ, Grossniklaus U (2023) Live imaging of Arabidopsis pollen tube reception and double fertilization using the semi-in vitro cum septum method. J Vis Exp 192:e65156. https://doi.org/10.3791/65156

15. Schreiber DN, Bantin J, Dresselhaus T (2004) The MADS box transcription factor ZmMADS2 is required for anther and pollen maturation in maize and accumulates in apoptotic bodies during anther dehiscence. Plant Physiol 134(3):1069–1079. https://doi.org/10.1104/pp.103.030577
16. Lindsey BE 3rd et al (2017) Standardized method for high-throughput sterilization of Arabidopsis seeds. J Vis Exp 128:56587. https://doi.org/10.3791/56587
17. Smyth DR, Bowman JL, Meyerowitz EM (1990) Early flower development in Arabidopsis. Plant Cell 2(8):755–767. https://doi.org/10.1105/tpc.2.8.755
18. Liu M et al (2021) AtLURE1/PRK6-mediated signaling promotes conspecific micropylar pollen tube guidance. Plant Physiol 186(2):865–873. https://doi.org/10.1093/plphys/kiab105
19. Márton ML, Fastner A, Uebler S, Dresselhaus T (2012) Overcoming hybridization barriers by the secretion of the maize pollen tube attractant ZmEA1 from Arabidopsis ovules. Curr Biol 22(13):1194–1198. https://doi.org/10.1016/j.cub.2012.04.061
20. Mason KN, Ekanayake G, Heese A (2020) Staining and automated image quantification of callose in Arabidopsis cotyledons and leaves. In: Anderson CT, Haswell ES, Dixit R (eds) Methods in cell biology, vol 160. Academic Press, pp 181–199

Chapter 17

Employing Biolistic Transformation to Investigate Subcellular Protein Localization and Interactions in Growing Tobacco Pollen Tubes

Patricia L. Seitz, Liang-Zi Zhou, and Thomas Dresselhaus

Abstract

The transformation of certain plant tissues, such as pollen grains, poses significant challenges when using conventional methods such as *Agrobacterium*-mediated or viral vector-based approaches. These tissues are often resistant to direct transformation, which complicates functional studies. This issue is further exacerbated in crop species, where stable transformation remains difficult and hinders many research projects. Consequently, transient biolistic transformation of *Nicotiana tabacum* (tobacco) pollen has become a valuable (heterologous) system for preliminary protein characterization, offering a rapid alternative to the lengthy process of stable transformation in crop plants. In this chapter, we present an optimized, robust protocol for the transient transformation of tobacco pollen, demonstrating its applicability to a variety of protein targets.

Key words Biolistic transformation, He-Gun, Pollen, Subcellular protein localization, *Nicotiana tabacum*

1 Introduction

Genetic transformation is an essential method in plant biology, as it enables the functional characterization of genes and promoter activity studies, as well as the production of genetically modified organisms. This addresses not only fundamental research questions, but also global agricultural challenges [1, 2]. Various transformation methods exist; however, not all methods are applicable to all organisms or desired tissues. A major issue is the difficulty of achieving *Agrobacterium*- or virus-mediated transformation in certain cell types, such as pollen, due to the robust cell wall and complex pollen coat, which form a strong barrier [2–5].

Moreover, pollen transformation provides a valuable opportunity to study gene function during male gametophyte development and cell signaling, as well as fertilization, and to deliver constructs

Giampiero Cai et al. (eds.), *Pollen and Pollen Tube Analysis: Methods and Protocols*, Methods in Molecular Biology, vol. 3042,
https://doi.org/10.1007/978-1-0716-5308-1_17,

for breeding purposes [3, 5]. Consequently, alternative approaches, such as biolistic delivery, have been developed to overcome these limitations and enable genetic manipulation of male gametophytes in plants [1, 6]. In particular, transient transformation systems provide a rapid and versatile platform to test subcellular localization or protein complex formation, for example, without the need for the time-consuming process of generating stable transgenic lines [4, 7] (*see* **Note 1**).

Nicotiana tabacum (tobacco) pollen is widely used in transient expression studies due to its ease of collection, high transformation efficiency, and well-characterized pollen tube growth. This makes tobacco pollen a valuable model for testing genetic constructs prior to their application in more challenging crop species, such as *Zea mays* (maize), *Triticum aestivum* (wheat), or *Glycine max* (soybean) [7]. Furthermore, pollen tubes serve as an effective single-cell model for visualizing protein localization, protein–protein interactions, and dynamic cellular processes in real time. The polarized localization of proteins, in particular, is a significant advantage of using polarized growing cells, such as pollen tubes.

Here, we present an optimized, reproducible protocol for biolistic transformation of tobacco pollen using the Bio-Rad Helium Gene Gun (PDS-1000/He). This method enables the efficient transient expression of fluorescently tagged proteins in growing pollen tubes, providing a practical tool for the functional study of gene constructs intended for use in crop transformation, for example.

2 Materials

2.1 Plant Material

1. *Nicotiana tabacum* (cv. Petite Havana SR1)

2.2 Chemicals

All the media and solutions were filter-sterilized, then stored at 4 °C.

1. 40% polyethylene glycol 3350 (PEG 3350).
2. 50% sucrose.
3. 1 M 2-(N-morpholino)ethanesulfonic acid potassium salt (MES·KOH, pH 6.0).
4. 1% H_3BO_3.
5. 2% $MgSO_4 \cdot 7H_2O$.
6. 1% KNO_3.
7. 1 M $Ca(NO_3)_2$.
8. plasmid DNA (concentration of 1 μg/μL).
9. 2.5 M $CaCl_2$ (filter sterilized and stored at −20 °C).

10. 1 mg/mL protamine (filter-sterilized and stored at −20 °C).
11. Ethanol p.a. (EtOH).
12. ddH_2O.

2.3 Equipment

1. Biolistic® Rupture Disks (1100 psi) (Fig. 1a).
2. Biolistic® Stopping screens (Fig. 1b).
3. Biolistic® Macrocarrier sheets (Fig. 1c).

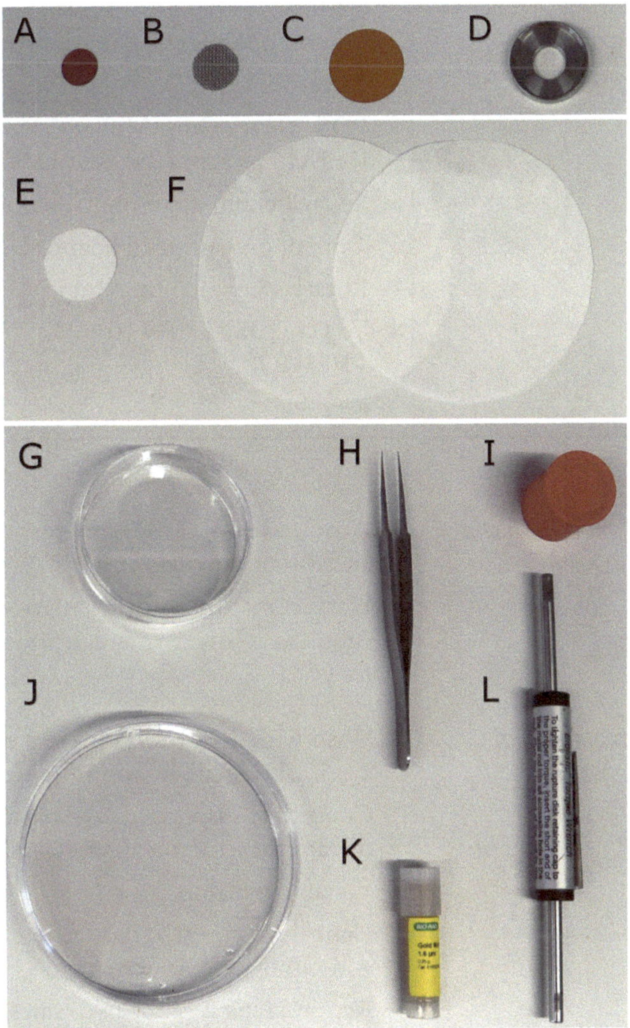

Fig. 1 Equipment needed for biolistic transformation of tobacco pollen using the PDS Gene Gun Model PDS-1000/He Biolistic® Particle Delivery System. (**a**) rupture disc (1100 psi), (**b**) stopping screen, (**c**) macrocarrier sheet, (**d**) macrocarrier holder, (**e**) Whatman™ paper (Ø 25 mm), (**f**) two pieces of filter paper (Ø 90 mm), (**g**) Petri dish (Ø 60 mm), (**h**) tweezers, (**i**) seating tool for macrocarriers, (**j**) Petri dish (Ø 90 mm), (**k**) gold microcarriers (Ø 1.6μm) and (**l**) torque wrench for rupture disc retaining cap

4. Biolistic® Macrocarrier holders (Fig. 1d).
5. Seating tool for macrocarriers (Fig. 1i).
6. Tweezers (Fig. 1h).
7. Gold microcarriers (Ø 1.6μm) (Fig. 1k).
8. Filter paper (cutted to Ø 90 mm; Fig. 1f).
9. Whatman™ paper (cutted to Ø 25 mm; Fig. 1e).
10. Petri dishes Ø 90 mm (Fig. 1j) and Ø 60 mm (Fig. 1g).
11. Torque-wrench for rupture disc retaining cap (Fig. 1l).
12. Microscope slides.
13. Cover slips.
14. PDS Gene Gun Model PDS-1000/He Biolistic® Particle Delivery System (Bio-Rad Laboratories GmbH).
15. 3-way helium metering (solenoid) valve.
16. Tank of pressurized Helium with Helium pressure regulators.
17. Rotary vane vacuum pump.
18. VSTG 35/230/110-CH, Autotransformer 230 VAC 110 VAC.

3 Methods

3.1 Plant Growth and Pollen Harvest

1. Grow *Nicotiana tabacum* plants under 16 h of light at 25 °C and 8 h of darkness at 21 °C until flowering stage.
2. Harvest mature anthers with tweezers, collect them in a 1.5 mL tube (*see* **Note 2**), and store them at −80 °C until further usage.

3.2 Preparation of Microcarriers

1. Wash with 70% EtOH p.a. 60 mg of gold particles, vortex for 1 min and incubate for 15 min at room temperature (RT).
2. Centrifuge suspended gold particles at 16.000 r.p.m. for 1 min, discard the supernatant and wash the pellet with 1 mL ddH_2O.
3. Vortex for 1 min.
4. Centrifuge the gold particles again for 16.000 r.p.m. for 1 min and remove the supernatant.
5. Repeat **steps 2–4** for two times.
6. Add 1 mL 50% glycerol and vortex for 1 min.
7. Aliquot 25μL of the gold suspension in 1.5 mL tubes.
8. Freeze the aliquots in −20 °C until further usage.

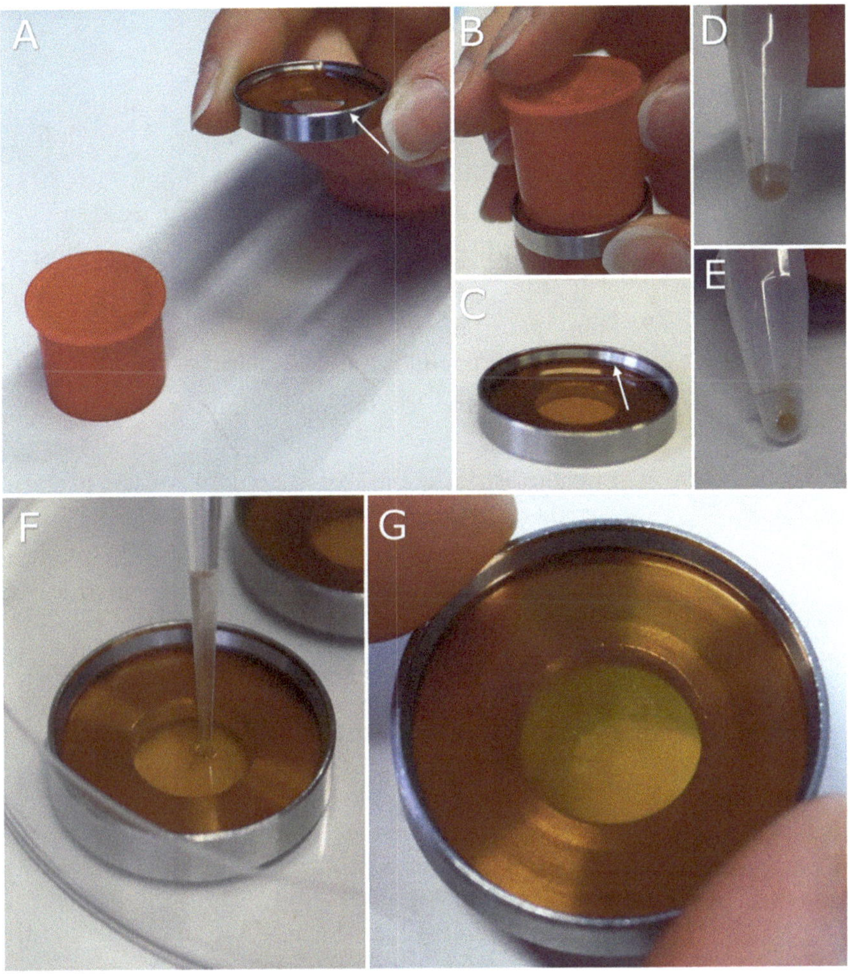

Fig. 2 Set-up to generate plasmid-coated gold microcarriers. (**a** and **b**) Insert the macrocarrier into the macrocarrier holder using the seating tool for macrocarriers. The macrocarrier bend should face upwards during insertion. (**c**) Check that the macrocarrier has been inserted correctly into the notch of the macrocarrier holder. (**d**) The microcarriers are evenly distributed after the vortexing steps. (**e**) The microcarriers are pelleted after the centrifugation steps. (**f**) The microcarriers are distributed onto the macrocarrier. (**g**) The microcarriers are distributed as uniformly as possible onto the macrocarrier, ready for biolistic transformation

3.3 Coating of Microcarriers with Plasmid

1. Prepare two macrocarriers per construct by inserting the macrocarrier into the macrocarrier holder with the help of the seating tool (Fig. 2a-c; *see* **Note 3**).
2. Pre-vortex aliquoted glycerol-suspended gold microcarriers (from Subheading 3.2) for 3 min at maximum speed.
3. Add in this order (*see* **Note 4**) to a new 1.5 mL tube:
   (a) 25μL of pre-vortexed gold microcarriers.
   (b) 2.5–5μL plasmid DNA (1μg/μL).

(c) 25μL 2.5 M $CaCl_2$.

(d) 10μL 1 mg/mL protamine.

4. Vortex the gold microcarrier suspension for 3 min at maximum speed (Fig. 2d).
5. Spin the gold microcarrier suspension down for 30 s at 12.000 r.p.m. and remove the supernatant (Fig. 2e).
6. Add 200μL pure EtOH p.a. and vortex at least for 3 min (Fig. 2d).
7. Spin gold microcarrier suspension down for 30 s at 12.000 r.p.m. and remove the supernatant (Fig. 2e).
8. Resuspend the coated gold microcarriers in 16μL pure EtOH p.a.
9. Equally distribute the coated gold microcarriers on two macrocarriers (Fig. 2f; *see* **Note 5**).
10. Let the EtOH evaporate at RT. Coated gold microcarriers will be visible as grey-dust looking residue on the macrocarrier (Fig. 1g).
11. Keep the two loaded macrocarriers inside a petri dish until further usage to prevent contamination. If the experiment is not continued immediately, loaded macrocarriers should be kept on ice.

3.4 Preparation of Pollen for Biolistic Transformation

1. Prepare fresh pollen germination medium (PGM) without PEG (400 mL/L 50% sucrose; 20 mL/L 1 M MES·KOH pH 6.0; 10 mL/L 1% H_3BO_3; 2% $MgSO_4 \cdot 7H_2O$; 10 mL/L 1% KNO_3; 3 mL/L 1 M $Ca(NO_3)_2$; 547 mL/L ddH_2O).
2. Prepare fresh PGM with PEG (600 mL/L 40% PEG 3350; 50 mL/L 50% sucrose; 20 mL/L 1 M MES·KOH pH 6.0; 10 mL/L 1% H_3BO_3; 2% $MgSO_4 \cdot 7H_2O$; 10 mL/L 1% KNO_3; 3 mL/L 1 M $Ca(NO_3)_2$; 297 mL/L ddH_2O). This is needed for later pollen germination and growth.
3. Prepare a petri dish (Ø 90 mm) as a tissue sample holder by putting in two filter papers (Ø 90 mm).
4. Add 1.6 mL of PGM without PEG onto the filter paper and let the medium distribute evenly (Fig. 3a; *see* **Note 6**).
5. Place a Whatman™ paper (Ø 25 mm) on top of the filter papers in the middle of the petri dish and let it completely soak with the PGM without PEG (Fig. 3b).
6. Close the petri dish to keep the humidity until further usage.
7. Take the previously collected tobacco anthers out of −80 °C.
8. Add 700μL PGM without PEG into the 1.5 mL tube. Pipette the medium gently up and down to release the pollen grain from the anthers (Fig. 3c).

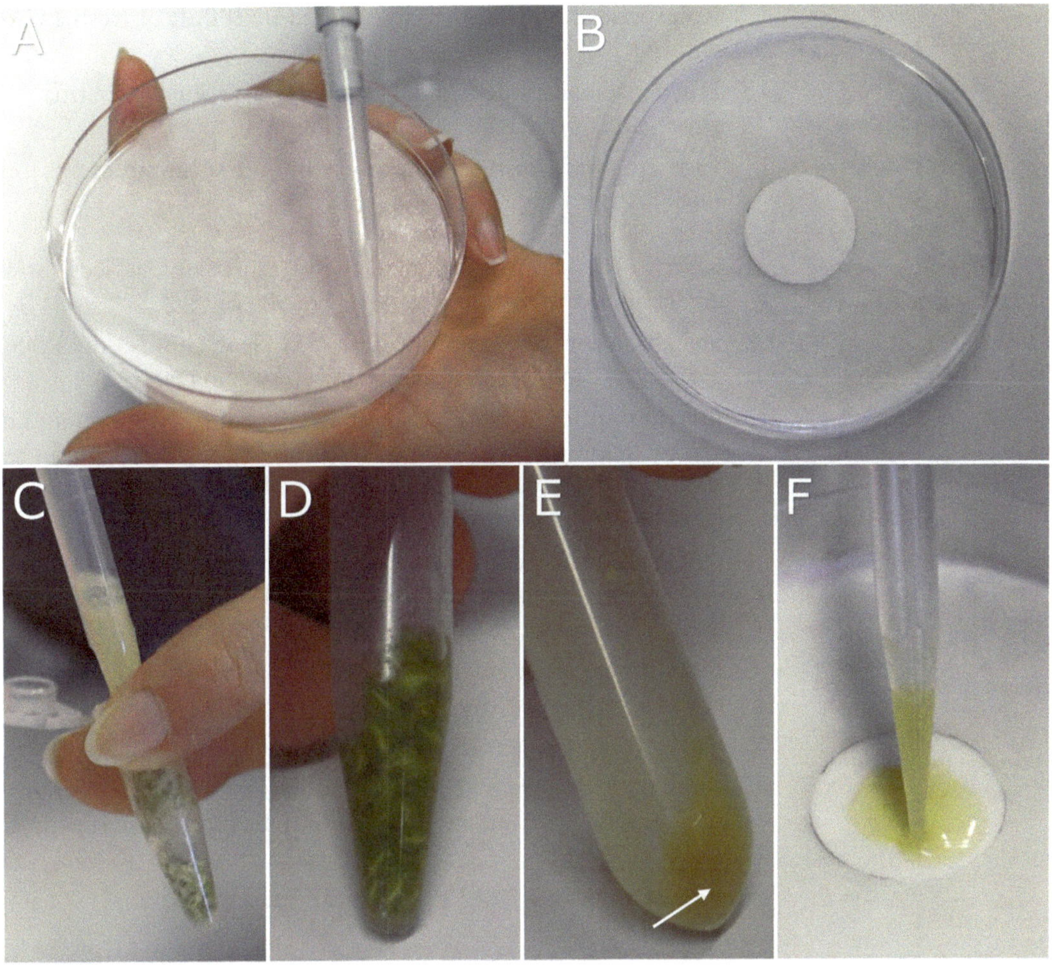

Fig. 3 Preparing tobacco pollen for biolistic transformation. (**a**) Petri dish containing two pieces of filter paper soaked in 1.6 mL of PGM without PEG. Avoid air inclusions between the filter papers and the Petri dish. (**b**) Place a Whatman™ paper (Ø 25 mm) on top of the filter papers. You may need to press down with tweezers to ensure it is completely saturated with PGM without PEG. (**c**) Collect tobacco pollen from the anthers using PGM without PEG. (**d**) The harvested anthers should contain little to no remaining pollen. (**e**) Centrifugate the pollen. The white arrow indicates the desired fresh, viable pollen. (**f**) Uniformly distribute the pollen onto the Whatman™ paper in the Petri dish

9. Transfer the released pollen grains into a new 2 mL tube. Repeat this step until all of the pollen is washed out of the anthers and transferred into the new tube (Fig. 3d; *see* **Note** 7). If necessary, a second 2 mL tube can be used.
10. Centrifuge the collected pollen shortly in the small table centrifuge to separate non-viable pollen from viable pollen (Fig. 3e).
11. Remove the supernatant containing non-viable pollen by careful pipetting.

12. Resuspend the pellet of viable pollen in 200μL of fresh PGM without PEG.
13. Distribute the pollen grain evenly on the small Whatman™ paper circle inside the petri dish (Fig. 3f).
14. Close the petri dish to keep everything in a humid condition until further usage.

3.5 Biolistic Transformation Procedure

1. Switch on Helium; check if all the measurement devices show correct build-up of He-pressure (Fig. 4a–c).
2. Switch on the vacuum pump and PDS Gene Gun Model PDS-1000/He Biolistic® Particle Delivery System (Fig. 4a).
3. Put the rupture disk (1100 psi) into the rupture disc retaining cap by helping yourself with the torque wrench. Be careful to avoid micro-damage of the rupture disk, which would lead to early burst under lower pressure than 1100 psi (Fig. 4d–f).
4. Insert the macrocarrier holder containing the loaded macro-carrier and the stopping screen into the microcarrier launch assembly (Fig. 4g–j; *see* **Note 8**).
5. Place the pollen grain-containing Petri dish on the target plate shelf (Fig. 4e; *see* **Note 9**).
6. Close the instrument door precisely.
7. Build up a vacuum to 25 Hg (*see* **Note 10**). Hold the vacuum, press the fire button, and hold until the rupture disc ruptures. Vent the vacuum.
8. Repeat **steps 3–7** until all macrocarriers are used up.

3.6 Incubation

1. Directly after the last shot, wash pollen grains into a Petri dish (Ø 35 mm) using 2.4 mL of PGM with PEG and put them immediately in the dark and on a horizontal shaker with 140 r.p.m. (Fig. 4k; *see* **Note 11**).
2. Let them incubate for 2–10 h, with the most optimal time point at 4–5 h.

3.7 Microscopy

1. Take 60μL of germinated pollen grains from the petri dish with a cut 200μL tip and distribute it on a microscope slide.
2. Add the cover slip.
3. Use a confocal microscope for imaging. Microscope settings depend on the fluorophores being used (Fig. 5; *see* **Notes 12** and **13**).
4. Up to three constructs can be transiently transformed into tobacco pollen tubes using this protocol. This can help with characterization of interactions or complex formations.

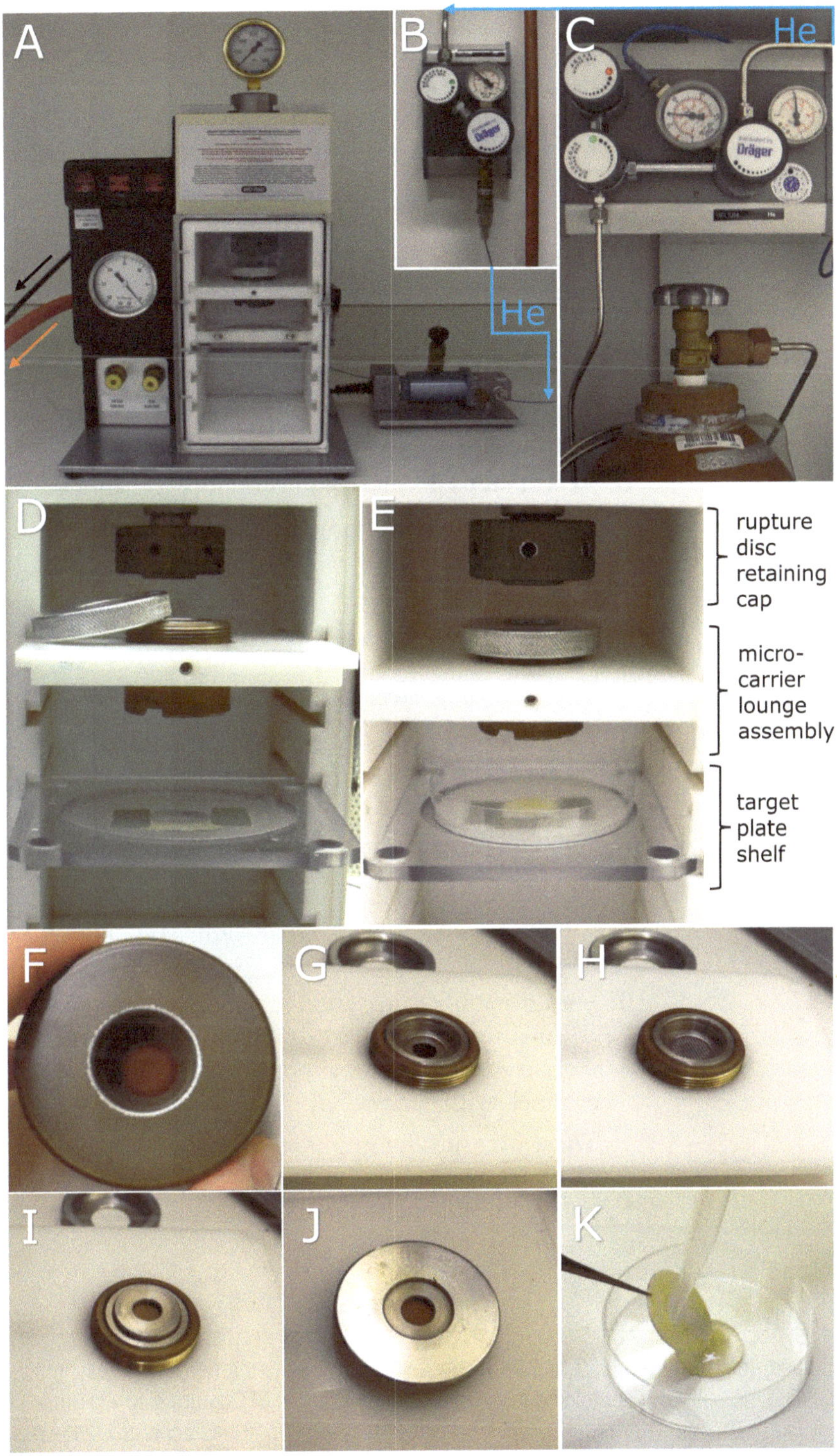

Fig. 4 Set-up of the PDS Gene Gun Model PDS-1000/He Biolistic® particle delivery system. (**a**) The PDS Gene Gun Model PDS-1000/He Biolistic® particle delivery system with a 3-way helium metering (solenoid) valve attached. The black arrow indicates the power supply connection with the inserted autotransformer, and the orange arrow indicates the rotary vane vacuum pump connection. (**b** and **c**) Helium regulators and pressurized helium tank. (**d**) Disassembled rupture disc retaining cap, microcarrier lounge assembly and target plate shelf.

3.8 Image Processing and Analysis

1. Open image file in ImageJ/Fiji.
2. Two different approaches of image processes and analysis are performed here. The first one being a fluorescence level analysis in form of a plot profile along a certain line of interest, and the second is a Pearson's correlation coefficient analysis to further confirm colocalization of proteins of interest.
3. For fluorescence level analysis, split fluorophore channels (*Image* > *Color* > *Split Channels*) and merge channels in their grey value afterward (*Image* > *Color* > *Merge Channels*).
4. Select the merged grey value image and perform plot profile analysis (*Analyse* > *Plot Profile*).
5. The plot profile window will pop up. Select "list," which gives you the plot values. Copy values in an Excel sheet.
6. Repeat **steps 4–5** for all fluorophores/channels.
7. Perform min–max–normalization to the plot values in excel sheet using following formula:

$$I_{\text{norm}} = \frac{I - I_{\min}}{I_{\max} - I_{\min}}$$

8. The according excel formula is

$$= (XQ - \text{MIN}(XY : XZ))/(\text{MAX}(XY : XZ) - (\text{MIN}(XY : XZ))$$

with X as column letter, Q as the current line of values, Y as the minimum line of values, and Z as the maximum line of values.

9. Plot the normalized intensities into one plot profile to compare the fluorescence signal along the region of interest (Fig. 5e).
10. For Pearson's correlation coefficient analysis, a two color channel image is needed. Consequently, the analysis for the triple transformation was performed by always comparing two channels with each other.
11. Therefore, split fluorophore channels (*Image* > *Color* > *Split Channels*) and select an ROI with ImageJ selection tools in one of the channels.

Fig. 4 (continued) (**e**) The assembled rupture disc retaining cap, microcarrier lounge assembly, and target plate shelf holding the Petri dish containing the tobacco pollen sample. (**f**) Rupture disc assembled into rupture disc retaining cap, fastened with torque wrench into system (e). (**g**) Open microcarrier lounge assembly unit. (**h**) Stopping screen inserted into microcarrier lounge assembly unit. (**i**) Macrocarrier inserted on top of stopping screen. (**j**) Closed microcarrier lounge assembly unit containing microcarrier-coated macrocarrier and stopping screen. (**k**) Transfer of transiently transformed tobacco pollen sample into small Petri dish for incubation

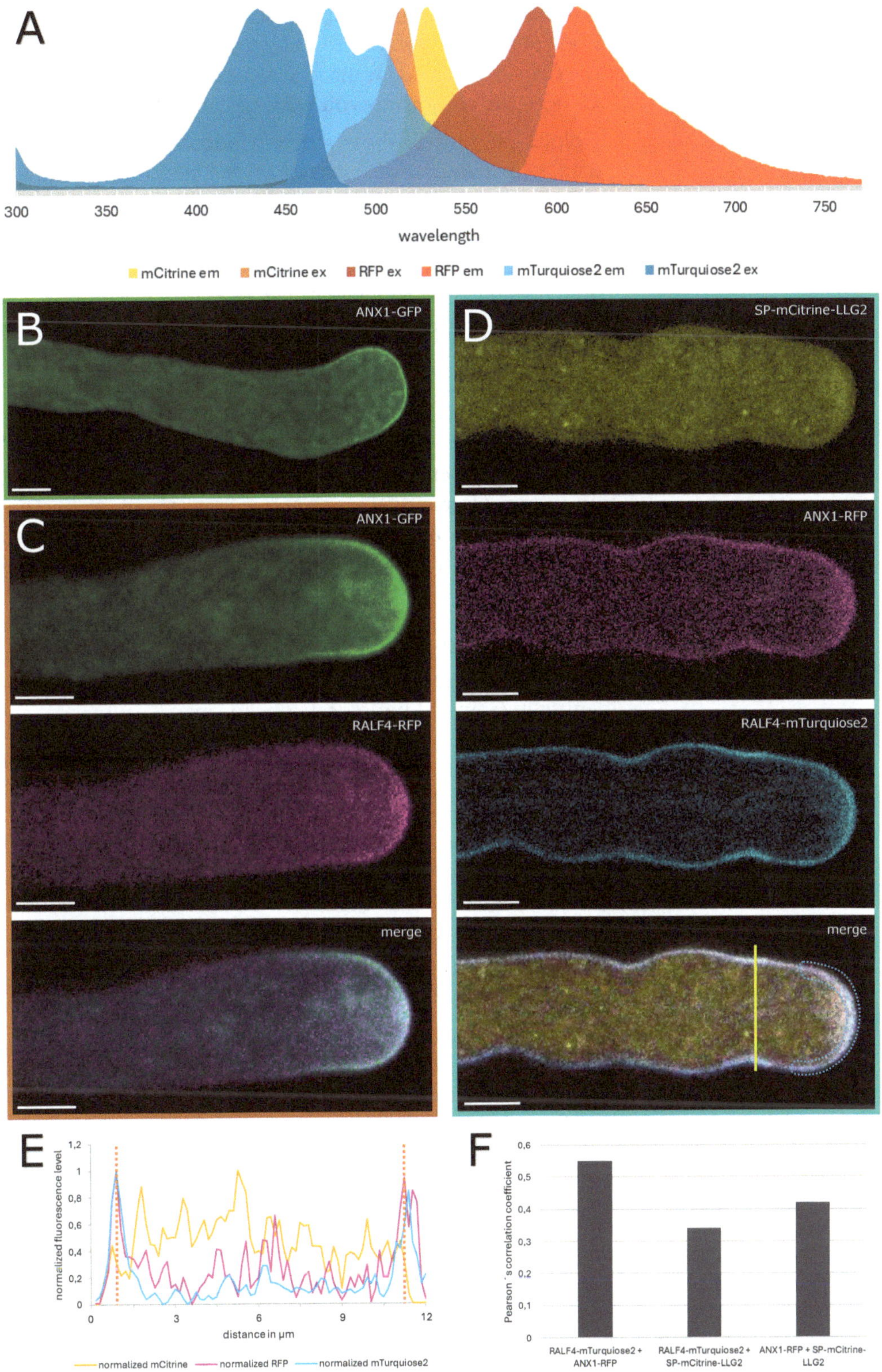

Fig. 5 Transiently transformed tobacco pollen tubes. (**a**) Excitation and emission spectra of mCitrine, RFP, and mTurquoise2. (**b**/green panel) Single transient transformation of ANX1-GFP. (**c**/orange panel) Double transient transformation of ANX1-GFP with RALF4-RFP. (**d**) Blue panel: triple transient transformation of SP-mCitrine-

12. Use the Coloc2-plugin for calculating the Pearson's correlation coefficient (*Analyze > Colocalization > Coloc2*). Choose the two channels you want to analyze alongside with the ROI. Choose the algorithms you want to calculate.
13. Coloc2-plugin will show a new window with the calculated data, which you can also export as *.pdf*-file for further analysis (Fig. 5f).
14. Image processing and analysis is exemplarily done for the pollen tube shown in Fig. 5d. For statistical analysis, several pollen tubes have to be analyzed to make conclusions about the protein localization [8].

4 Notes

1. Although the heterologous tobacco pollen tube transformation system offers certain advantages in terms of experimentation, it also presents notable limitations. For example, proteins from the organism under study may interact with native tobacco proteins, and overexpression artifacts cannot be ruled out. These factors contribute to variability in expression patterns and signal intensities across transient assays. Consequently, multiple bombardment replicates are essential to identify meaningful biological conclusions.
2. To directly get the right amount of tobacco anthers for one construct, fill the 1.5 mL tube right above the 0.5 mL mark.
3. When inserting the macrocarrier into the macrocarrier holder, ensure that the bend in the macrocarrier faces upwards (as indicated by the white arrow in Fig. 2a) and that it is inserted correctly into the notch in the macrocarrier holder (as indicated by the white arrow in the small image).
4. The order of adding is very important in this step. It is recommended not to change it.
5. The coated gold microcarriers will quickly aggregate due to gravity inside the pipette tip. Switching between the two macrocarriers during distribution ensures that the microcarriers are evenly distributed.

Fig. 5 (continued) LLG2, ANX1-RFP, and RALF4-mTurquoise2. All constructs are driven by the pollen-specific *LAT52* promoter. Merged images are shown at the bottom of panels c and d. Scale bars represent 5µm. (**e**) Fluorescence level analysis of the triple transient transformation along the yellow line in panel d; the orange dashed line indicates the position of the plasma membrane. (**f**) Pearson's correlation coefficient analysis of selected regions are marked with a blue dashed line in panel d

6. Avoid air inclusions between the filter papers, as these can interfere with the bombardment procedure under high pressure by displacing the Whatman™ paper carrying the pollen.
7. To avoid unnecessary loss of pollen, it is recommended that you do not change the tip during this procedure. It is also important to use 1000μL pipette tips to prevent pollen damage. If it is difficult to release the pollen from the anthers by pipetting up and down, you can also vortex carefully (never set the speed higher than 3–4 to avoid damaging the pollen).
8. The microcarrier lounge assembly must be inserted in a specific direction. Bear this in mind, as inserting it the wrong way round can bend the plexiglass front when vacuum is applied, leading to sealing loss in future experiments.
9. Ensure that the Whatman™ paper circle holding the pollen is positioned correctly in the middle and aligned with the firing channel. Remove the lid just before the first round of biolistic transformation begins, to prevent humidity loss.
10. The reported setup (1100 psi, 25 Hg, and 1.6μm gold microcarriers) is specifically designed for tobacco pollen. These parameters must be adjusted for biolistic transformation with other tissues.
11. After biolistic transformation, it is recommended that you only wash the pollen from areas where gold particles are clearly visible. This will increase the number of tubes that are successfully transformed when you examine them under a microscope later.
12. To select the appropriate fluorophores, especially for double and triple transformations, it is essential to analyze the excitation and emission spectra in order to prevent overlap and cross-excitation during imaging. Web-based programmes such as fpbase.org can be used for this purpose (*see* Fig. 5). These programmes can also provide additional information about the fluorophores used, which can be helpful during the imaging process.
13. To further characterize proteins, different cell staining can be performed (e.g., with propidium iodide (PI) or FM4–64). When designing your constructs, also keep the excitation and emission wavelengths of cell stains in mind to avoid interference while imaging.

Acknowledgments

This study was funded via SFB 924 of the Deutsche Forschungsgemeinschaft (DFG) to T. D.

References

1. Levengood H, Zhou Y, Zhang C (2024) Advancements in plant transformation: from traditional methods to cutting-edge techniques and emerging model species. Plant Cell Rep 43:273. https://doi.org/10.1007/s00299-024-03359-9
2. Anjanappa RB, Gruissem W (2021) Current progress and challenges in crop genetic transformation. J Plant Physiol 261:153411. https://doi.org/10.1016/j.jplph.2021.153411
3. Quiroz LF, Khan M, Gondalia N, Lai L, McKeown PC, Brychkova G, Spillane C (2025) Tissue culture-independent approaches to revolutionizing plant transformation and gene editing. Hortic Res 12:uhae292. https://doi.org/10.1093/hr/uhae292
4. Wang H, Jiang L (2011) Transient expression and analysis of fluorescent reporter proteins in plant pollen tubes. Nat Protoc 6:419–426. https://doi.org/10.1038/nprot.2011.309
5. Zhou LZ, Juranic M, Dresselhaus T (2017) Germline development and fertilization mechanisms in maize. Mol Plant 10:389–401. https://doi.org/10.1016/j.molp.2017.01.012
6. Yang A, Su Q, An L, Liu J, Wu W, Qiu Z (2009) Detection of vector- and selectable marker-free transgenic maize with a linear GFP cassette transformation via the pollen-tube pathway. J Biotechnol 139:1–5. https://doi.org/10.1016/j.jbiotec.2008.08.012
7. Müller AO, Blersch KF, Gippert AL, Ischebeck T (2017) Tobacco pollen tubes - a fast and easy tool for studying lipid droplet association of plant proteins. Plant J 89:1055–1064. https://doi.org/10.1111/tpj.13441
8. Seitz PL, Qu LJ, Dresselhaus T, Zhou LZ (2025) Spatial organization and trafficking dynamics of ANX/BUPS-RALF-LLG signalling complexes during pollen tube growth. Plant Reprod 38:18. https://doi.org/10.1007/s00497-025-00528-y

Chapter 18

Isolation of Proteins and Extracellular Vesicles Secreted by Pollen of *Actinidia chinensis* L. During In Vitro Germination

Chiara Suanno, Elisabetta A. M. Verderio, and Stefano Del Duca

Abstract

Extracellular vesicles (EVs) are membrane-bound particles secreted into the extracellular space by cells of any living kingdom. Plant-derived EVs (PDEVs) in particular are regarded as promising nutraceutical and drug carriers, but their purification is hindered by the contamination from intracellular and intramural vesicles, which is inherently associated with the existing PDEVs isolation protocols. In the last decade, pollen has been proven to release EVs, called pollensomes, during in vitro germination. When secreted in a liquid culture medium, pollensomes can be easily isolated with minimal cellular contamination, thus representing a promising source of purified PDEVs either for fundamental research studies or for biomedical applications. Here, we describe a straightforward protocol to isolate EVs smaller than 220 nm (called small EVs) from *Actinidia chinensis* L. germinated pollen, using sequential filtration and ultracentrifugation at 100,000× *g*. We also detail procedures to extract proteins secreted by pollen during germination, either via EVs or by EV-independent pathways, for further analyses of the pollen secretome.

Key words EVs, Exosomes, Pollensomes, Pollen tube, Secretome, Ultracentrifugation, Ultrafiltration

1 Introduction

Extracellular vesicles (EVs) are nano- or micro-sized particles delimited by a lipid bilayer membrane, released by cells in the extracellular space [1]. In the last decade, EVs have been isolated from different plant cells, including pollen [2–4]. Due to their low immunogenicity and the possibility to sustainably produce them on a large scale, plant-derived EVs (PDEVs) are regarded as promising nanocarriers for nutraceutical and biomedical applications [5]. However, most of the currently available methodologies to isolate PDEVs from sporophytic tissues cannot exclude contamination from intracellular or intramural vesicles [6]. On the contrary, in vitro cultures of pollen grains in liquid medium offer a promising model to isolate pollen EVs, called pollensomes, without contamination from cellular vesicles [7]. Here we present a simple protocol to isolate EVs smaller than 220 nm (called small EVs) from

Giampiero Cai et al. (eds.), *Pollen and Pollen Tube Analysis: Methods and Protocols*, Methods in Molecular Biology, vol. 3042, https://doi.org/10.1007/978-1-0716-5308-1_18,

Actinidia chinensis L. germinated pollen, by filtration and ultracentrifugation at 100,000× *g*. The particles isolated using this protocol can be classified as EVs according to the following features: (I) diameter ranging between 15 and 266 nm; (II) round shape compatible with EVs when observed by atomic force microscopy (AFM); (III) absence of cell compartment markers such as H^+-ATPase, UDP-glucose pyrophosphorylase (UGPase), and plant cytochrome oxidase subunit II (COXII); (IV) presence of EV markers, namely clathrin and BRO1-domain proteins [8]. Moreover, we illustrate how to extract proteins from pollen EVs and how to concentrate proteins secreted in the medium by EVs-independent secretion pathways, for subsequent proteomic studies.

2 Materials

Prepare all solutions at room temperature, using ultrapure water.

2.1 EVs Isolation

1. Germination medium (GM): 10% sucrose, 324 μM H_3BO_3, 1.27 mM $Ca(NO_3)_2$. Prepare 10 mL of 32.4 mM H_3BO_3 stock solution by dissolving 20 mg of H_3BO_3 in a final volume of 10 mL of pure water. Prepare 10 mL of 12.7 mM $Ca(NO_3)_2$ stock solution by dissolving 21 mg of $Ca(NO_3)_2$ in a final volume of 10 mL of pure water. To prepare 50 mL of GM, dissolve 5 g of sucrose, 500 μL of H_3BO_3 stock solution, and 2.5 mL of $Ca(NO_3)_2$ stock solution in a final volume of 50 mL of ultrapure water (*see* **Note 1**).
2. Pollen extraction buffer (PEB): 20 mM Tris–HCl pH 8.5, 2 mM DTT, 3 mM EDTA, 0.005% (v/v) Tween80, 1:100 protease inhibitors cocktail DMSO solution (Sigma-Aldrich). Prepare 50 mL of 100 mM Tris–HCl pH 8.5 stock solution by dissolving 0.606 g of Tris in 40 mL of ultrapure water. Adjust the pH of the solution to 8.5 with 1 M HCl, then make it up to volume. Prepare 2 mL of 2 M DTT stock solution by dissolving 617 mg of DTT in 2 mL of ultrapure water (*see* **Note 2**). Make 50 mL of PEB premix solution by dissolving 0.044 g of EDTA in 10 mL of 100 mM Tris–HCl pH 8.5, 50 μL of 2 M DTT, 2.5 μL of Tween80, and ultrapure water (*see* **Note 3**). Before use, prepare the PEB working solution by adding 10 μL of protease inhibitors cocktail DMSO solution to 1 mL of PEB premix solution (*see* **Note 4**).
3. Filtered Dulbecco's Phosphate Buffered Saline (particle-free PBS), 50 μL per sample.
4. Incubator with shaker.
5. Ultracentrifuge and rotor with a maximum g-force of at least 100,000 × *g*.

2.2 Protein Precipitation from EVF Fraction

1. Pure acetone at −20 °C, 1 mL per sample.
2. 50% (w/v) trichloroacetic acid (TCA), 4 mL per sample.

3 Methods

3.1 Instruments Setup

1. Pre-heat the incubator at 30 °C.
2. Pre-cool the centrifuge at 4 °C.
3. Pre-cool the ultracentrifuge rotor at 4 °C.
4. Create a humid chamber by lining a 10 cm Petri dish with blotting paper soaked in water. Fill a 6 cm Petri dish with water and place it in the larger Petri dish (Fig. 1).

3.2 EVs Isolation

1. Weigh 20 mg of frozen, dry *Actinidia chinensis* pollen per sample on an aluminum weighing boat and place the boat in the humid chamber. Close the Petri dish with the lid and put it in the incubator at 30 °C for 30 min, to rehydrate the pollen (Fig. 1).

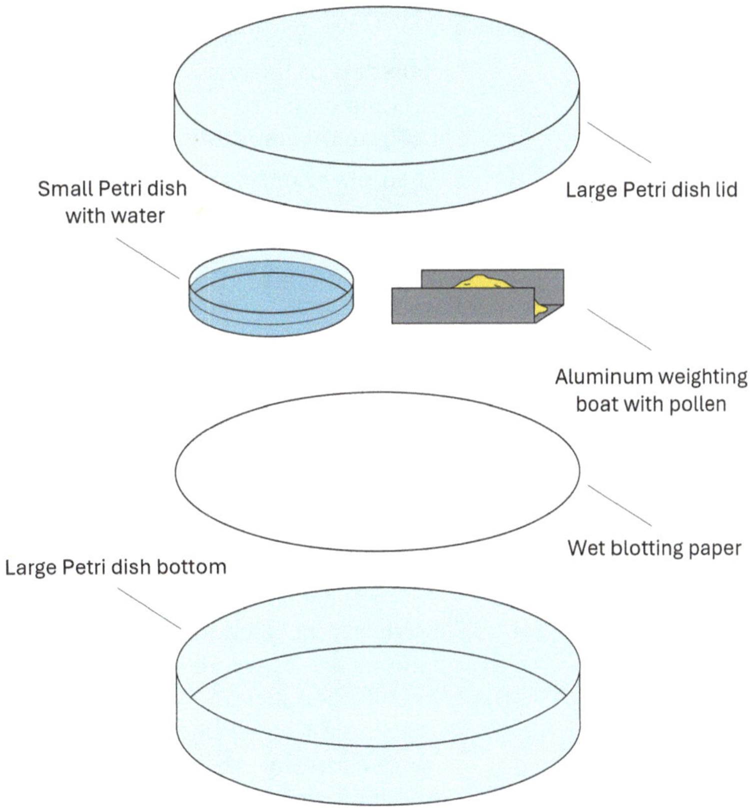

Fig. 1 Exploded-view drawing of the humid chamber for pollen hydration

2. Transfer 20 mL of GM to a glass culture tube for each sample, and then transfer the rehydrated pollen to the tubes using a spatula (*see* **Notes 5** and **6**). Cap the tube and shake it thoroughly to disperse the pollen until no pollen clump is visible, then pour it into a glass Petri dish and incubate at 30 °C, under gentle shaking, for 2 h (*see* **Note** 7).
3. After germination, slowly rotate the Petri dish on the bench to homogeneously distribute the pollen. Aspirate 15 μL of the sample with a micropipette from the center of the plate, at mid-depth, place it onto a glass slide, and close it with an 18 × 18 mm coverslip (*see* **Note 8**). Repeat this step twice, by aspirating 15 μL of the sample in two other areas of the Petri dish, distant from the center. Observe the slides in light microscopy at 200× magnification and calculate the germination rate over at least 100 pollen grains per slide, counting grains as germinated when their pollen tube length is equal to or greater than their diameter (*see* **Note 9**).
4. Carefully transfer the sample into centrifuge tubes using either a 5000 μL pipette or an electronic pipette controller with a 25 mL serological pipette. Aspirate the sample slowly to avoid pollen rupture, then tilt a clean 50 mL tube and gently pour the sample along the lower inner wall to minimize the impact velocity. Avoid bubble formation while pipetting, as their bursting could break the pollen tubes. For this purpose, it is recommended to use a 1,000 μL pipette with a wide-bore tip to aspirate the last 2 mL of the sample.
5. Spin the samples at 5,000× *g* for 15 min at 4 °C, to pellet the pollen grains. Using an electronic pipette controller with a 25 mL serological pipette, slowly aspirate the supernatant (S) for the first 15 mL, then transfer it to another tube and keep it on ice. Collect the remaining S with a 1,000 μL pipette for higher precision, and transfer it to the same tube. Take note of the volume retrieved, which will be around 19 mL due to sample losses in **steps 4** and **5** (*see* **Note 10**).
6. Gently resuspend the pelleted pollen in 100 μL of PBS using a 100 μL pipette with a wide-bore tip. Place 15 μL of the sample onto a microscope slide, close with an 18 × 18 mm coverslip, and check the integrity of the pelleted pollen in light microscopy, at 400× magnification (*see* **Note 11**).
7. Remove the plunger from a 20 mL needleless syringe and mount a 0.22 μm nitrocellulose syringe filter on its tip. Place the filter on a clean 50 mL tube and transfer S in the syringe. Insert the plunger into the syringe and depress it fully to filter S for the first time (F1) (Fig. 2). Remove the plunger and replace the filter with a clean one. Place the filter on a clean 50 mL tube, transfer F1 into the syringe, and filter the sample for the

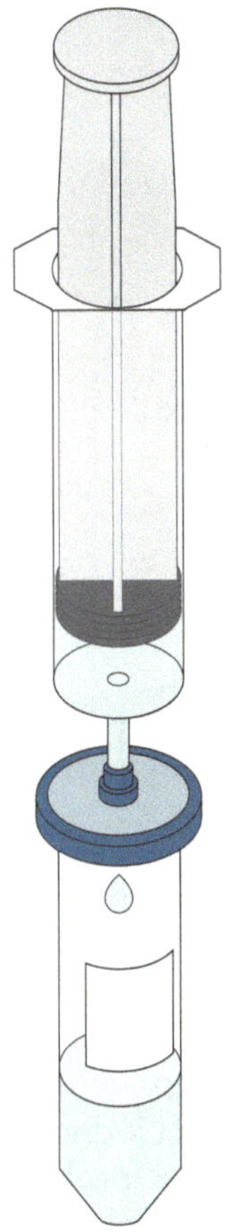

Fig. 2 Schematic representation of the filtration steps

second time (F2). Double filtration removes pollen debris and particles larger than small EVs (*see* **Note 12**).

8. Transfer F2 samples in ultracentrifuge tubes and balance them to the nearest 0.001 g with an analytical balance, adding particle-free PBS where necessary. Spin the samples in a pre-cooled ultracentrifuge at 100,000× *g*, 4 °C, for 60 min, to pellet the EVs (*see* **Note 13**).
9. Clean the lab bench and cover it with a sheet of blotting paper. Carefully remove the ultracentrifuge tubes from the rotor and place them in a rack on the bench. Any abrupt movement could cause the resuspension of the pellet.
10. Slowly recline the ultracentrifuge tube, keeping the pellet on the upper side (*see* **Note 14**). Aspirate the supernatant (EV-free fraction, EVF) with a sterile plastic Pasteur pipette pointed towards the lower inner wall of the tube, opposed to the pellet (Fig. 3). Do not shake or reverse the ultracentrifuge tube during this operation, to avoid resuspending the pellet. Transfer the EVF sample to a clean 50 mL tube and keep it on ice.

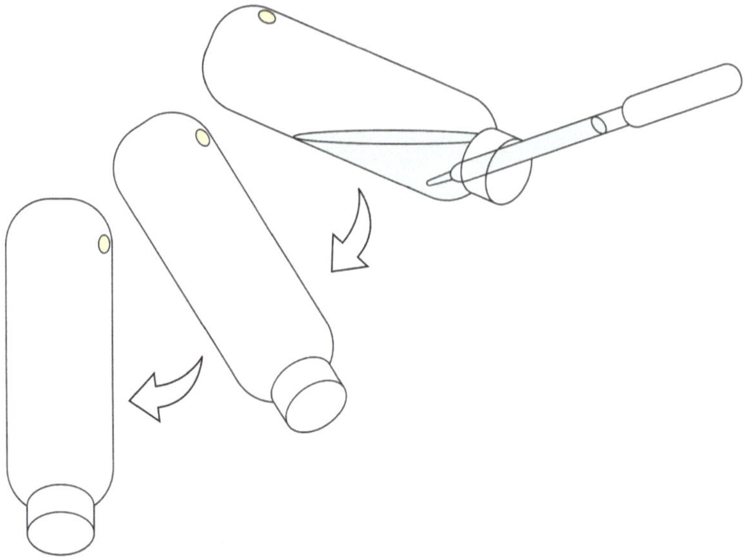

Fig. 3 Aspiration of the vesicle-free fraction (EVF) with a plastic Pasteur pipette after ultracentrifugation, and rotation of the test tube thereafter. The pellet is represented as a yellow dot

11. Continue the rotation of the ultracentrifuge tube until it is upside-down (Fig. 3), then place it on the blotting paper and let it drain upside-down for 5 min. Repeat **steps 10** and **11** for each tube. At this stage, avoid any vibration of the bench or movement of the tubes, as these can cause the pellet to detach.
12. Lift the ultracentrifuge tube, keeping it upside-down. Using a pipette, aspirate and discard any drops present near the tube opening in order to reduce contamination from EVF in the EV fraction. Then, turn the tube upright and place it in the rack. Repeat for each tube.
13. Pelleted EVs can be resuspended in different buffers depending on the purpose of the experiment. Use 100 μL of particle-free PBS for vesicle visualization or 50 μL of PEB for protein analysis (*see* **Note 15**). Keep the tube slightly tilted with the pellet on the lower side. Using a 100 μL micropipette, deposit the buffer where the pellet is expected. Do not fully depress the plunger but reach the first stop instead: this will reduce the formation of bubbles, thereby enhancing the recovery efficiency. While keeping the plunger on the first stop, move the pipette tip to the bottom of the vial and aspirate the buffer again. Repeat this operation for about 2 min. Rub the side of the tip against the pellet while pouring the buffer to facilitate pellet removal. Transfer the EVs sample to a 500 μL vial and take note of the volume (*see* **Note 16**).

3.3 Protein Precipitation from EVF Fraction

1. Under fume hood, add 50% TCA solution to the EVF sample to achieve a final concentration of 10% (w/v) TCA (*see* **Note 17**). Shake the mixture vigorously and incubate at 4 °C for 1 h. TCA selectively precipitates the proteins, allowing them to concentrate and to remove carbohydrates and lipids from the sample.
2. Spin the EVF sample at 15,000 × *g*, 4 °C, for 5 min. A white, fluffy pellet will form. Under the fume hood, carefully discard the supernatant using a pipette (*see* **Note 18**).
3. Resuspend the pellet in 1 mL of cold (−20 °C) pure acetone and vortex. Upon adding acetone, the solution will briefly turn white. Spin the EVF sample at 15,000× *g*, 4 °C, for 5 min, then carefully discard the supernatant with a micropipette. Repeat twice.
4. Leave the tubes open under the fume hood until the pellet has dried (about 10 min), then add 50 μL of PEB to each EVF sample to enhance comparability with EV samples. Keep it on ice for immediate use or store it at −20 °C.

4 Notes

1. H_3BO_3 and $Ca(NO_3)_2$ stock solutions can be stored at 4 °C for at least 1 month, while GM can be stored at −20 °C for several years. When high volumes of GM are needed on a regular basis, it is possible to prepare stock GM solutions that can be stored at −20 °C for years. However, due to the high sucrose content, concentrations higher than 2× can only be made as premix solutions without sucrose. This option is recommended for complex GMs containing numerous components in low concentrations, or for media that can be adapted to different pollen types by adjusting the sucrose percentage of the working solution.
2. 100 mM Tris–HCl pH 8.5 can be stored at 4 °C for at least 2 months, while 2 M DTT can be stored at −20°c for several years. Be aware that Tris pH varies according to the temperature of the solution, so it is important to adjust the pH at the intended working temperature. At room temperature (21 °C), around 1.6 mL of 1 M HCl is needed to prepare 50 mL of 100 mM Tris–HCl pH 8.5. If pH is lower than 8.5, EDTA will not dissolve completely in the PEB premix solution.
3. Tween80 is thick and difficult to pipette. For higher precision, it is recommended to: (I) use wide-bore or low-retention tips; (II) aspirate and dispense slowly; (III) wait for 2–3 s after aspirating and releasing; (IV) use reverse pipetting. To perform reverse pipetting, fully depress the pipette plunger to the

second stop before aspiration, slowly but completely release it while aspirating, and then depress it to the first stop when dispensing.

4. PEB also allows to effectively extract total proteins from pollen grains, combined with mechanical rupture of the grains by a potter or a tissue lyser. Total pollen proteins in PEB can be employed as positive controls in immunoprobing experiments.
5. When using plasticware, pollen is adsorbed on the plastic surface, causing major sample losses. For this reason, it is recommended to prefer glassware wherever possible.
6. Sucrose crystals or other aggregates that might form in the GM can pellet at $100{,}000 \times g$ and be present in the EVs fraction. These particles can have dimensions comparable to pollensomes, hence they can be counted by nanoparticle analyzers along with EVs, leading to an overestimation of pollensomes concentration. For this reason, when isolating pollensomes for nanoparticle tracking analysis (NTA), it is recommended to use 20 mL of GM as a negative control. This negative control must follow the EVs isolation protocol from **steps 2** to **13**, eventually resuspending its EVs fraction in particle-free PBS. The correct pollensomes concentration can then be calculated by subtracting the average particle concentration of the negative control from the average particle concentration of the pollen sample.
7. EVs isolation protocol can be applied to any pollen type, but GM composition and hydration and germination protocols need to be adjusted to the plant species according to the literature.
8. Some pollen grains and tubes might burst while mounting the slides, due to the pressure of the coverslip. This is normal, and burst pollen grains should not be accounted for in the germination rate. However, a high percentage of burst pollen (>5%) should be considered problematic since excessive pollen tube fragility can cause cytoplasmic contaminations in the EVs fraction. In this case, it is better to prepare another slide to exclude possible errors while mounting the first slide. If the pollen still appears damaged in the second slide, consider repeating the experiment, reducing the germination time, to work with shorter and more stable pollen tubes. To minimize pollen bursting, handle germinated pollen with maximum care, and always use wide-bore tips, which can be made by cutting the tip end of the regular pipette tips with clean scissors. The presence of cytoplasmic contamination in the isolated EVs can be checked by atomic force microscopy, electron microscopy, or by immunoblotting. In the latter case, EVs and EVF fractions can be immunoprobed for cell compartment markers, such as

H^+-ATPase for plasma membrane, UGPase for cytoplasm, ARF1 for Golgi membranes, and COXII for mitochondria.

9. Germinated pollen grains are more prone to float than non-germinated and non-hydrated ones. This implies that the germination rate will be underestimated when sampling from the bottom of the Petri dish, and overestimated when sampling from the surface of the medium. Thoroughly mixing the sample at the end of germination using a 5000 μL pipette with a wide-bore tip can reduce these differences, but it increases the risk of causing pollen rupture. Hence, it should be performed very carefully, and it is not recommended for fragile pollen grains. To bypass this problem, an alternative approach is to prepare the slides at the end of **step 4** of EV isolation. In this case, germinated pollen is sampled directly from the tube, immediately after its transfer.
10. The aim of the first centrifugation is to separate pollen grains from the germination medium in a stable pellet, without breaking the pollen tubes. If the pellet is unstable, increase the spinning velocity and duration. Different pollen types have different physical properties; for example, grains with abundant pollenkitt, like *Olea,* or with air sacs like *Pinus,* tend to float. Therefore, it might be necessary to spin the samples twice to remove them from the supernatant. Moreover, germinated grains are more hydrophilic than non-germinated ones; hence, samples with a high germination rate and pollen tube length will form a viscous, unstable pellet. In case the pollen pellet remains unstable, do not collect the supernatant in the immediate proximity of the pellet to avoid contaminations of the S fraction with pollen grains, and take note of the recovered S volume.
11. If the pollen tubes of the pelleted grains appear more damaged in **step 5** than they were in **step 3** of the EVs isolation protocol, then the centrifuge is damaging the pollen tubes, causing contamination of the EVs and EVF fractions with intracellular components. In this case, repeat the EVs isolation protocol from **step 1** to **step 4**, and modify **step 5** by reducing the spinning velocity to 4,000× *g* and extending its duration to 25 min.
12. If the sample opposes increasing resistance to filtration, up to the point where the plunger hardly moves, then the filter is probably overloaded. Filter overloading during the first filtration might occur with large samples, e.g., 40 mL, in which case it is possible to simply replace the filter with a clean one. However, overloading with smaller samples is a strong indicator of pollen rupture and severe cytoplasmic contamination of the S fraction. In this case, it is recommended to repeat the

protocol, taking greater care to avoid pollen rupture, and to check for cytoplasmic contamination at the end of isolation (*see* **Notes 8** and **11**).

13. Samples should be centrifuged at 100,000× *g* for 60 min. If the ultracentrifuge does not allow to exclude acceleration and deceleration from the centrifuge duration, then spin the samples for 70 min to account for the extra time required by the instrument to reach the final velocity and to brake.
14. EV pellet may appear as a small, yellow or transparent dot. However, in most cases it is non-visible. For this reason, when using fixed-angle rotors, it is important to mark the side of the tube on which the pellet will form, i.e., the side facing the outer radius of the rotor, to indicate the pellet position.
15. EVs must be visualized immediately after isolation, because they tend to aggregate over time. Although it is possible to store them at −80 °C, freezing and thawing could disrupt them or alter their shape [8]. After protein extraction, EV samples can be stored at −20 °C.
16. Due to adhesive forces, it is impossible to recover the same volume of buffer that was initially added. It is safe to assume that EV loss will be proportional to buffer loss, the total concentration of EVs or EV proteins per pollen sample should be calculated over the total volume of the buffer added (100 μL for particle concentration, 50 μL for protein concentration). If the volume of the buffer recovered is equal to or even higher than the one added, it means that the EV pellet was not dry enough, thus the EV fraction will have EVF contaminations. This can be verified by immunoprobing the EV sample for EVF markers, such as profilin.
17. The volume of 50% TCA required depends on the volume of EVF sample recovered at the end of the protocol, which is typically around 18 mL. In this case, 3.6 mL of 50% TCA solution should be added to the EVF sample to achieve a final concentration of 10% (w/v) TCA.
18. Precipitated proteins will form a white, fluffy pellet. This pellet may be unstable; if so, it is recommended to aspirate as much supernatant as possible without disturbing the pellet, then repeat the centrifugation and remove the remaining supernatant. In case the pellet is still unstable after the second centrifugation, do not aspirate the supernatant and proceed to the next step, adding one extra acetone washing to the protocol to eliminate TCA.

References

1. Welsh JA, Goberdhan DCI, O'Driscoll L et al (2024) Minimal information for studies of extracellular vesicles (MISEV2023): from basic to advanced approaches. J Extracell Vesicles 13: e12404
2. Qiao SA, Roth R (2025) Messenger and message: uncovering the roles, rhythm and regulation of extracellular vesicles in plant biotic interactions. Curr Opin Plant Biol 83:102672
3. Prado N, De Dios Alché J, Casado-Vela J et al (2014) Nanovesicles are secreted during pollen germination and pollen tube growth: a possible role in fertilization. Mol Plant 7:573–577
4. Prado N, De Linares C, Sanz ML et al (2015) Pollensomes as natural vehicles for pollen allergens. J Immunol 195:445–449
5. Yazdanpanah S, Romano S, Valentino A et al (2025) Plant-derived exosomes: carriers and cargo of natural bioactive compounds: emerging functions and applications in human health. Nanomaterials 15:1005
6. Ambrosone A, Barbulova A, Cappetta E et al (2023) Plant extracellular vesicles: current landscape and future directions. Plants 12:1–18
7. Suanno C, Tonoli E, Fornari E et al (2023) Small extracellular vesicles released from germinated kiwi pollen (pollensomes) present characteristics similar to mammalian exosomes and carry a plant homolog of ALIX. Front Plant Sci 14:1090026
8. Ahmadian S, Jafari N, Tamadon A et al (2024) Different storage and freezing protocols for extracellular vesicles: a systematic review. 15:1–21

Chapter 19

Preparing a Pollen Reference Collection and Standardized Protocols for Aerobiological and Melissopalynological Pollen Slides

Eleonora Clò, Paola Torri, Martina Denami, and Anna Maria Mercuri

Abstract

Palynology, the study of pollen and spores, has developed from its early foundations in the seventeenth century into a multidisciplinary science with applications in taxonomy, environmental reconstruction, aerobiology, forensic investigations, melissopalynology, and archaeology. This chapter provides a comprehensive overview of methodological approaches central to both basic and applied palynology. The preparation and organization of the reference pollen collection (Palynotheca) are described as indispensable tools for identification, teaching, and comparative research. Detailed protocols for preparing fresh, acetolysed, and permanent reference slides are outlined, emphasizing their diagnostic value and long-term stability. Aerobiological monitoring is examined with particular attention to standardized volumetric sampling method, slide preparation, and data analysis for ecological and allergenic forecasting. Melissopalynological analysis is presented through its historical development and current applications in honey authentication, environmental monitoring, and biodiversity assessment, with protocols for the analysis of fresh and acetolysed honey samples as well as pollen loads. Finally, pollen germination tests are introduced as complementary techniques for evaluating viability and reproductive biology in agronomically relevant species. The integration of classical and applied methodologies demonstrates the central role of pollen analysis in advancing both scientific research and practical applications across diverse fields.

Key words Palynotheca, Reference collection, Aerobiology, Melissopalynology, Acetolysis, Pollen slide preparation, Hirst spore trap, Pollen germination

1 Introduction

Palynology is a science that, although its origins date back at least to the seventeenth century, with significant early work by Nehemiah Grew and Marcello Malpighi [1], has only recently celebrated its first century as a formal discipline [2]. During this period of extraordinary scientific progress, the study of pollen and spores—as defined by Hyde and Williams [3], who coined the term *palynology* from the Greek roots meaning "to sprinkle" and "dust"—has undergone a remarkable transformation, rapidly expanding into

Giampiero Cai et al. (eds.), *Pollen and Pollen Tube Analysis: Methods and Protocols*, Methods in Molecular Biology, vol. 3042,
https://doi.org/10.1007/978-1-0716-5308-1_19,

multiple applications. Many of these, such as taxonomy, environmental reconstruction, and aerobiology, had already been anticipated by Erdtman [4]. According to Mercuri [5], a distinction can be made between studying pollen as an object of research (basic palynology) and using pollen as a tool to measure environmental variables (applied palynology). Nevertheless, basic and applied palynology share a common foundation in the structural complexity of the pollen wall. Pollen morphology provides the diagnostic traits essential for taxonomic identification and floristic lists and supports applied studies, including air-dispersed pollen monitoring and analyses of bee products. The reference pollen collection or Palynotheca is indispensable in the laboratory of palynology. In this chapter, we firstly describe the preparation of this fundamental tool, and then the methods in aerobiology and melissopalynology.

1.1 The Reference Pollen Collection

The preparation of a pollen reference collection or Palynotheca is an indispensable resource in any palynological laboratory (Fig. 1). Such a collection serves as a benchmark for the identification and comparison of pollen grains, providing essential support for both basic and applied research.

The reference collection should be organized into the following sections:

1. Envelopes with flowers/strobili, containing pollen from plant material collected in the field or obtained from accurately identified herbarium specimens. Precise identification of the plant is essential. Flowers or inflorescences should be stored in paper envelopes, with proper documentation. The envelopes must be labeled: information such as plant species (Latin name and authority), collection date, and locality (possibly with GPS coordinates) is crucial to ensure the scientific value of the collection. Whenever possible, a specimen (herbarium voucher) of the entire plant (if herbaceous) or a flowering branch (if woody) should also be preserved as a reference for identification and documentation.
2. Small tubes, containing acetolysed pollen preserved in glycerin jelly.
3. Slides, prepared with acetolysed pollen, fresh pollen, or also with both preparations in a slide. The slides are placed in boxes that can be organized by morphological type following the NPC system (Number, Position, Character of the apertures [4]), e.g., saccatae, monoporate, tricolpate, etc.; they can also be classified by family or other taxonomic criteria (e.g., Fabaceae, Asteraceae, or Gymnosperms), or by area or geographical context.

Fig. 1 The Palynotheca is a collection of slides. In the photo, the reference collection is organised into envelopes with plant material, small tubes containing acetolysed pollen preserved in glycerin jelly, and reference pollen slides including fresh and acetolysed pollen. (Photos: Laboratory of Palynology and Palaeobotany, University of Modena and Reggio Emilia)

1.1.1 Classification and Sections of the Collection

Each slide or sample should be carefully cataloged, including information on the preparation type (fresh or acetolysed), plant origin, and morphological notes. Slide boxes are typically stored in slide racks and classified according to the above-mentioned criteria.

A well-organized reference collection is also structured according to taxonomic hierarchy, including family, genus, and species. Sections may include: (i) the Reference Collection containing the standard reference material described above, (ii) the Research

Collection including slides related to materials and sites studied, and (iii) the Educational Collection created to support university teaching and practical training for candidate palynologists. In fact, when introducing university courses in palynology, it can be very useful to create a dedicated teaching section within the reference collection [6]. Such a teaching-oriented section facilitates the integration of theory and practice, significantly reducing the time required for students to acquire hands-on skills in pollen analysis. Key methodological aspects include (i) the provision of individual slide sets for each student, (ii) the preparation of slides containing abundant and well-preserved pollen grains with clear diagnostic features, and (iii) the inclusion of mixed preparations with species that exhibit both closely related and subtly distinct morphologies. Such an approach fosters the recognition of diagnostic traits, promotes the development of comparative analytical skills, and supports the effective integration of theoretical and practical training in palynology [7, 8].

In this chapter, the procedures for preparing reference slides of modern pollen obtained from conifers and flowering plants are described in detail. Such preparations are essential for pollen analysis, as they enable researchers to observe key morphological features—such as size, shape, apertures, and surface ornamentation—that are crucial for accurate taxonomic/evolutionary research and comparative studies. Slides may be prepared using untreated (fresh) pollen or acetolysed pollen, depending on the objectives of the study.

1.1.2 Advantages and Disadvantages

A reference pollen collection offers numerous benefits for both basic and applied palynology. It provides reliable material for comparative studies, allowing researchers to identify pollen grains with great accuracy and confidence, also reaching the species level. Such collections also serve as essential references for systematic, evolutionary and ecological investigations [9]. As said above, they can also be used to develop standardized teaching and training materials, ensuring consistency in education, dissemination, and professional practice.

However, maintaining a high level of quality in this collection implies some limitations. Reference slides may deteriorate over time due to fading or physical damage. To maintain their integrity, careful storage under controlled humidity and temperature should be considered as it is necessary along with periodic monitoring of the state of preservation of pollen. It is therefore recommended to have a curator of the Palynotheca who, on the one hand, carries out periodic checks on the quality of the slides and, on the other, periodically prepares new reference samples to replace deteriorated materials or expand the collection.

1.2 Aerobiology

Aerobiology is the scientific study of air-borne biological particles, such as pollen grains, fungal spores, and other microorganisms. This discipline is essential for understanding the dispersal, transport, and seasonal dynamics of these particles, as well as their impacts on human health. Monitoring airborne pollen and spores provides crucial information for allergy forecasting, environmental management, and ecological research.

Most aerobiological monitoring is managed by national and international monitoring networks, such as the European Aeroallergen Network (EAN; https://ean.polleninfo.eu/Ean/) and POLLnet (https://pollnet.isprambiente.it/), the Italian aerobiological monitoring network managed by the National System for Environmental Protection (SNPA). In North America, aerobiological monitoring is coordinated by the National Allergy Bureau (NAB; https://climahealth.info/resource-library/nationalallergy-bureau/), operated by the American Academy of Allergy, Asthma & Immunology (AAAAI).

Networked monitoring programs play a central role in aerobiology by enabling systematic data collection across multiple sites [10, 11]. These networks rely on standardized methodologies for sampling via artificial spore traps, sample preparation, and microscopic analysis, which ensure the comparability and reliability of data over time and across locations. By integrating data from different sites, monitoring networks allow researchers to track large-scale patterns of pollen distribution, identify trends, and assess the effects of environmental changes, such as climate variability and urbanization, on airborne particles (e.g., [12]). The establishment of coordinated aerobiological networks also supports public health initiatives. Furthermore, the continuous accumulation of long-term datasets contributes to scientific research, enabling the study of interannual variability, the impact of invasive species, and the effects of environmental policies. Recently, there has been growing interest in the development of automated tools for pollen sampling, while pollen analysis with IT-based identification is still in the experimental phase ([13], and references therein).

Studies on the airborne dispersal of pollen and spores have been facilitated by the development of sampling techniques that capture particles in the atmosphere by trapping them on surfaces suitable for direct microscopic examination. They can be volumetric or gravitational artificial spore traps. Volumetric samplers continuously draw in a known volume of air and deposit pollen grains onto adhesive-coated tapes. Gravitational traps, on the other hand, collect particles that fall naturally onto sticky surfaces. The choice of trap depends on the objective of the monitoring program, whether it is for long-term environmental studies, allergen surveillance, or seasonal pollen forecasting. Volumetric methods currently represent the state-of-the-art approach for both qualitative and quantitative investigations of pollen and fungal spores worldwide

[14]. Counts are usually expressed as grains per cubic meter of air, allowing the creation of pollen calendars or spatial-temporal distribution maps [15].

1.3 Melissopalynology

Pollen analysis of honey, or melissopalynology, is based on the principle that pollen and honeydew elements (e.g., fungal spores) in honey reflect the vegetation and environment of the foraging area. This method provides information on the botanical and geographical origin of honey through the identification of pollen grains and other structured elements in the sediment [16]. The technique of melissopalynology was first developed by Zander in 1935 and subsequently standardized by the International Commission for Bee Botany (ICBB) in 1978 [17]. Minor adaptations across laboratories later led to a harmonized protocol proposed by Von der Ohe et al. [16]. Building on these classical and harmonized protocols, de Almeida-Muradian et al. [18] provided a comprehensive reference for honey analysis, with pollen analysis serving as a key step in assessing honey authenticity. Centrifugally extracted honeys are preferred, as filtration or pressing may alter pollen spectra and prevent reliable determination of botanical origin.

Beyond honey authentication and quality control, melissopalynology has been applied in studies of atmospheric pollution [19] and plant diversity monitoring and environmental change [20]. It also supports product quality assessment by detecting extraneous particles or yeasts indicative of fermentation [17].

Because nectar pollen content varies among plant species and is influenced by multiple factors, the melissopalynologist identifies the main taxa and their relative frequencies (qualitative analysis). For accurate classification, pollen analysis is often combined with sensory and physicochemical tests, and in some cases with quantitative counts of pollen concentration.

1.4 Pollen Germination

Although pollen germination does not strictly fall within the classical methods of applied palynology, it can be a valuable technique for assessing reproductive potential and detecting anomalies in the anthers of certain plant species. This approach is particularly useful in agronomically important plants, where understanding pollen viability is crucial for crop breeding, hybridization programs, and seed production. Pollen germination tests allow researchers to evaluate the percentage of viable grains, the speed of tube growth, and the uniformity of germination, providing insights into the overall fertility of the plant.

In addition, this method can help identify environmental or physiological factors affecting pollen performance, such as temperature, humidity, or nutrient deficiencies. While not routinely applied in palynological surveys, integrating pollen germination studies with morphological and reference collections can offer a more comprehensive understanding of plant reproductive biology and support both basic botanical research and applied agricultural studies.

The germination requirements of pollen exhibit considerable variation between species. In addition to moisture, successful germination and pollen tube growth generally depend on the presence of a carbohydrate source, as well as adequate levels of boron and calcium [21]. In this chapter, a straightforward method for pollen germination applicable to a variety of species is presented.

2 Materials

2.1 Reference Slides of Modern Pollen (see Note 1)

The first five reagents are used exclusively in the acetolysis method and are not required for the preparation of fresh pollen.

1. Glacial acetic acid (CH_3COOH).
2. Acetic anhydride [$(CH_3CO)_2O$].
3. Sulfuric acid 96% (H_2SO_4).
4. Distilled water (H_2O).
5. Alcohol–glycerol solution (1:1).
6. Alcohol 96%.
7. Glycerin jelly with basic fuchsin (*see* **Note 2**).
8. Solid Paraffin block 51–53°.
9. Scissors, forceps, or scalpel.
10. Needles.
11. 10 mL centrifuge tubes.
12. Microcentrifuge tube.
13. Pipettes.
14. Microscope slides.
15. Coverslips.
16. Fume hood.
17. Water bath.
18. Centrifuge.
19. Laboratory oven.
20. Heating plate.
21. Lab vortex mixer.

2.2 Air-Borne Pollen Collection with Hirst Type Spore Trap

1. Silicone fluid (polydimethylsiloxanes).
2. Carbon tetrachloride (CCl_4).
3. Glycerin jelly with basic fuchsin.
4. Hirst volumetric sampler.

5. Transparent adhesive tape.
6. Flowmeter.
7. Flat brush (15-mm wide).
8. Cutting block.
9. Tweezers, scissors, and scalpel.
10. Microscope slides.
11. Coverslips.
12. Water bath.
13. Heating plate.

2.3 Pollen Extraction from Honeys and Pollen Loads

The first three reagents are used exclusively in the acetolysis method and are not required for the preparation of fresh pollen.

1. Glacial acetic acid (CH_3COOH).
2. Acetic anhydride [$(CH_3CO)_2O$].
3. Sulfuric acid 96% (H_2SO_4).
4. Distilled water (H_2O).
5. Alcohol–glycerol solution (1:1).
6. Glycerin jelly with basic fuchsin.
7. Solid Paraffin block 51–53°.
8. Fume hood.
9. Precision scale.
10. 50 mL centrifuge tubes.
11. Heating plate.
12. Pipettes.
13. Water bath.
14. Centrifuge.
15. Laboratory oven.
16. Microscope slides.
17. Coverslips.
18. *Lycopodium clavatum* L. spore tablets.

2.4 Pollen Tube Growth

1. Sucrose ($C_{12}H_{22}O_{11}$).
2. Calcium nitrate [$Ca(NO_3)_2$].
3. Boric acid (H_3BO_3).
4. Magnesium sulfate ($MgSO_4$).
5. Potassium nitrate (KNO_3).
6. Distilled water (H_2O).
7. Scissors, forceps, or scalpel.

8. 10 mL centrifuge tubes.
9. Pipettes.
10. Microscope slides.
11. Coverslips.
12. Laboratory oven.

3 Methods

3.1 Reference Slides of Modern Pollen: Standard Protocol for Fresh Pollen

1. Remove pollen from the anthers of well-developed flowers, possibly selecting mature buds close to anthesis. Under a stereomicroscope, carefully dissect the flower using scissors, forceps, or a scalpel. For plants with very small flowers, collect the entire flower and remove as much extraneous tissue as possible.
2. Transfer a small amount of pollen from the anthers to the center of a microscope slide.
3. Add one or more drops of alcohol and allow it to partially evaporate. The oily and resinous substances of the pollen will form a ring on the slide; this ring should be removed using cotton moistened with alcohol [22].
4. Add a drop of hot, melted glycerin jelly containing basic fuchsin as a stain to color the pollen wall. Stir the pollen with a needle to distribute it evenly, and gently heat the slide on a heating plate.

At this stage, the slide can be examined as a temporary preparation. To obtain a permanent slide suitable for long-term storage in a palynotheca, proceed with the following **steps** (**5–7**).

5. Apply a thin layer of paraffin around the edges of a coverslip using a small spatula.
6. Place the coverslip squarely over the residue drop and briefly place the slide on a heating plate to melt the paraffin, sealing the specimen.
7. Label the slide appropriately.

3.2 Reference Slides of Modern Pollen: Standard Protocol for Acetolysed Pollen

Acetolysis is a chemical treatment used to completely remove the pollen protoplast and intine [23]. This procedure enhances the visibility of morphological features, such as apertures and exine sculpturing, which are often obscured in untreated samples. Acetolysed reference pollen provides long-term stability, making it invaluable for palynological reference collections and comparative studies, as well as for the extraction of fossil pollen from sediments [9]. Following acetolysis, the exine typically changes color from nearly colorless to yellow, amber, or brown, depending on the duration of treatment and the thickness of the exine [24].

1. Remove pollen from the anthers and dissect the flower as described above (Subheading 3.1, **step 1**). For very small flowers, collect whole flowers and remove as much extraneous tissue as possible.
2. Place the pollen or anthers in a centrifuge tube.
3. Suspend the material in 5 mL of glacial acetic acid for at least 15 min (dehydratation). Gently crush against the tube wall with a glass rod to release the pollen grains.
4. If necessary, sieve the suspension using a 250 μm mesh to remove large debris (*see* **Note 3**).
5. Centrifuge for 3 min at 2500 rpm and decant the supernatant.
6. Prepare the acetolysis mixture: prepare nine parts acetic anhydride and add drop by drop 1 part of concentrated sulphuric acid. Use 4–5 mL per sample. Add the mixture carefully to the tubes (Warning: the reaction is exothermic and explosive on contact with water) and place the tubes in a water bath at 90 °C for 3–4 min (*see* **Note 4**).
7. Remove the tubes from the water bath, centrifuge (3000 rpm for 3 min), and decant the supernatant.
8. Add distilled water, centrifuge (3000 rpm for 3 min), and decant. Repeat washing and centrifugation until all traces of the acetolysis mixture are removed.
9. Add 1 mL of alcohol and transfer the pollen residue to a microcentrifuge tube (*see* **Note 5**).
10. Add 1–2 drops of alcohol–glycerol solution (1:1) and place the tube in an oven at 50 °C for 8–12 h.

The pollen residue can be mounted to prepare permanent slides following the procedure outlined above for fresh pollen (Subheading 3.1, **steps 5–7**), including application of glycerin jelly, placement and sealing of the coverslip with paraffin, and appropriate labeling.

3.3 Airborne Pollen Collection with Hirst Type Spore Trap: Sampler Installation and Preparation of the Sampling Surface

The classical spore-trap apparatus utilizing this method is described by Hirst [25], which serves as the prototype for the two most widely used artificial samplers currently available: the Burkard Spore Trap and the Lanzoni VPPS 2000. It operates by drawing a fixed volume of air through a narrow orifice. Particles suspended in the air are impacted onto an adhesive-coated surface, usually a transparent tape (Melinex) mounted on a rotating drum. The drum rotates at a constant speed, allowing for continuous sampling over 1 week, with a time-resolved record of deposition. This procedure provides a standardized method for quantifying the concentration of pollen grains and fungal spores dispersed in the atmosphere [15].

1. Choose a location where local atmospheric circulation is not influenced by nearby obstacles. Preferably, place the sampler at the centre of a flat rooftop at a height of 15–20 m above the ground. Avoid sites near public parks or areas with strong industrial emissions (*see* **Note 6**).
2. Record the location and type of vegetation immediately surrounding the building.
3. Use transparent adhesive tape, a non-hydrophilic surface such as a plastic film. Ensure the tape adheres perfectly to the drum, which is part of the spore-trap instrument (*see* **Note** 7).
4. Prepare a 3% solution of silicon fluid (viscosity 2,500,000 centistokes) in carbon tetrachloride [26] (*see* **Note 8**).
5. Apply a thin layer of silicon fluid to the tape to make the surface adhesive.
6. Brush the solution slowly and firmly over the tape only once. Avoid using too little solution, as it may compromise good sampling results. Quick evaporation of the solvent ensures homogeneous coverage. Carry out the operation under a fume hood due to the volatile and toxic nature of the carbon tetrachloride.
7. Store the drum in a container to prevent contamination until it is mounted on the sampler.
8. When changing the drum, check the airflow and the suction slot. The latter should be cleaned with a cotton swab soaked in alcohol.

The air is drawn in by a suction pump and directed, through a slit (2×14 mm), onto a specially treated sampling surface, where suspended airborne particles are deposited by impact. The pump operates at a flow rate of 10 ± 1 liters of air per min, corresponding to a total sampled volume of 14.4 m^3 over a 24-h period ($0.6\ m^3 \times 24$ h). The sampler is equipped with a rotating drum of 345 mm circumference, driven by an internal clock mechanism with seven-day autonomy, which rotates at a constant speed of 2 mm per h. The length of the strip allows the preparation of seven slides for microscopic examination, each containing 48 ± 0.48 mm of sample surface.

3.4 Airborne Pollen Collection with Hirst Type Spore Trap: Samples Mounting

1. Remove the sampled tape from the drum and place it on a cutting block marked with daily transverse divisions (*see* **Note 9**).
2. Label the slide clearly with the station name and sampling date.
3. Melt the glycerin jelly with basic fuchsin in a water bath.
4. Place a few (3–5) drops of glycerin jelly with basic fuchsin on a microscope slide.

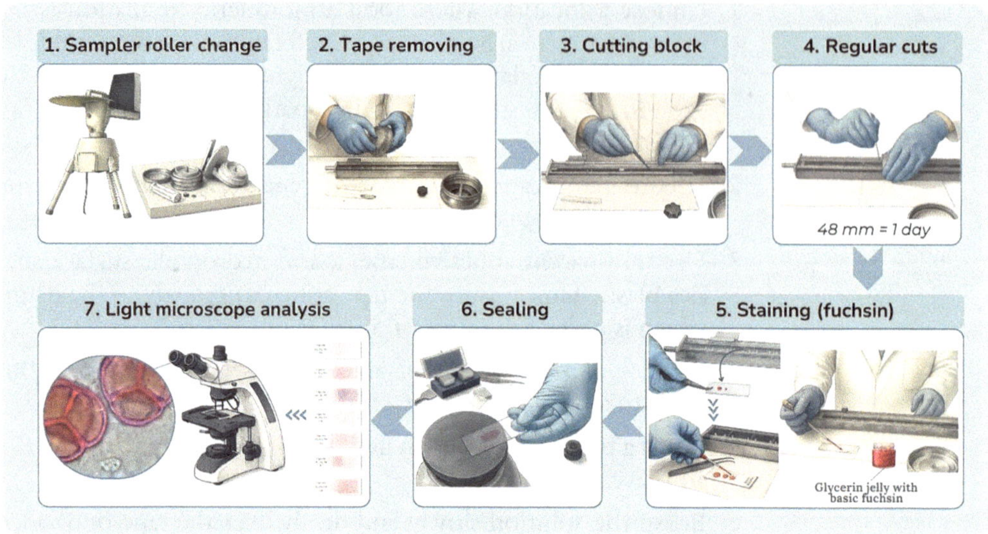

Fig. 2 Collection of airborne pollen using spore traps and preparation/staining of microscope slides for pollen analysis. (Flowchart by Cristina Ricucci)

5. Carefully lay the tape segment onto the jelly, with the silicone layer facing upwards.
6. Place a few drops of jelly on the cover glass. Quickly invert the cover glass and position it over the tape.
7. Use a heating plate to maintain the slide at a sufficiently high temperature (40–50 °C), to promote jelly fluidity, and to eliminate air bubbles.
8. Allow the slide to dry in a horizontal position for several hours before performing microscopic counts (an overview of the main steps is provided in Fig. 2).

3.5 Pollen Extraction from Honeys and Pollen Loads: Standard Protocol for Fresh Pollen (Hydrated Method)

The methodology for preparing honey sediment follows the protocol described by Louveaux et al. [17] with minor modifications. Due to the high sugar content and viscosity of honey, pollen is extracted by centrifuging a honey–water solution.

1. Weigh 10 g of honey into 50 mL conical centrifuge tubes (*see* **Note 10**).
2. When quantitative analysis is required, add an internal standard (commonly *Lycopodium* spores) to the sample to determine the absolute number of counted elements.
3. Add 20 mL of distilled water (preheated if necessary to 20–40 °C) and shake until the honey is completely dissolved. The use of an oven, in addition to hot water, can help to complete the dissolution process of the sugar crystals.

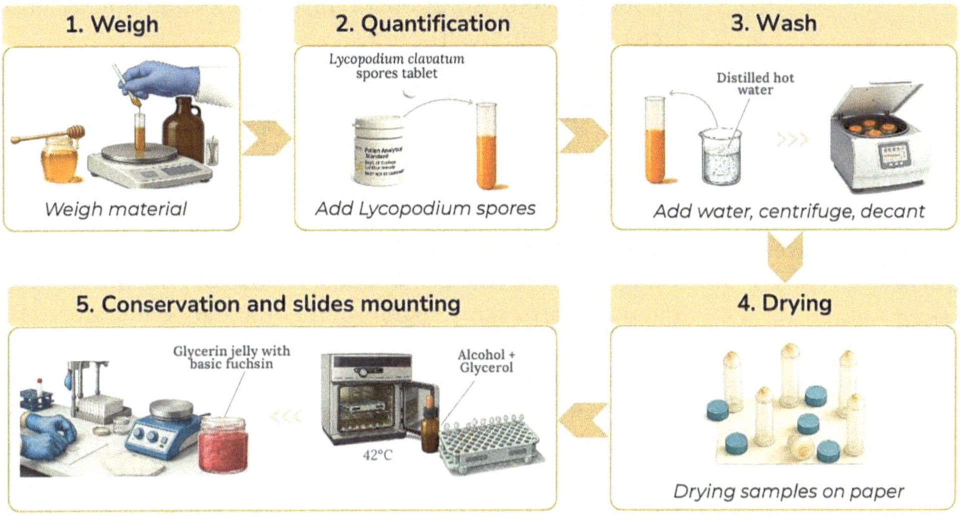

Fig. 3 Preparation of slides with fresh pollen from honey sediment. (Flowchart by Cristina Ricucci)

4. Centrifuge at 2500 rpm (revolutions per minute) for 10 min.
5. Carefully discard the supernatant using a pipette, avoiding contact with the pellet. Leave a small volume of liquid above the sediment to minimize disturbance and reduce the risk of pollen loss.
6. Resuspend the pellet in distilled water and centrifuge again under the same conditions (2500 rpm, 10 min).
7. Remove the supernatant, then invert the tubes and place them on absorbent paper to drain any remaining liquid, taking care not to disturb the pellet (an overview of the main steps is provided in Fig. 3).
8. Prepare microscopic slides and start heating a hot plate.
9. Place the slide on the heating plate and apply a small piece of glycerin jelly with added basic fuchsin (optionally, to enhance morphological identification).
10. Transfer only half of the sediment (given the high pollen concentration, only a part is used) onto the microscopic slide and leave the slide on the heating plate only for the time strictly necessary to dissolve the glycerin jelly. Stir well before collecting the sediment to ensure that a representative sample is used.
11. Sealed with paraffin to ensure long-term preservation.
12. Allow the slide to dry in a horizontal position for several hours before performing microscopic counts.

13. The unused portion of sediment can be preserved by adding an alcohol-glycerol solution (1:1) and allowing the alcohol to evaporate in an oven.
14. The slides, as well as the remaining sediment, were carefully labeled and stored in a palynotheca.

3.6 Pollen Extraction from Honeys and Pollen Loads: Standard Protocol for Acetolysed Pollen

In some cases, after the analysis of fresh pollen samples, a subsample may be subjected to acetolysis to clarify uncertain determinations. The advantages of this process, as highlighted by Koelzer et al. [27], include enhanced clarity and visibility of pollen morphology, as well as the broader availability of comparative literature. However, acetolysis also has disadvantages, such as the loss of some honeydew elements, in the case of honeydew honeys, or the removal of other non-pollen organic material that can be useful for tracing alterations or adulterations of the product. The methodology for preparing acetolysed honey samples follows the protocol described by Louveaux et al. [17] with minor modifications.

Steps (**1**) to (**7**) are carried out as described in the previous section. The procedure then continues as follows:

8. Before acetolysisAcetolysis, wash the concentrated pollen pellet with acetic acid to remove residual water.
9. Add approximately 1 mL of acetolysis mixture consisting of nine parts of acetic anhydride and 1 part of sulfuric acid; centrifuge at 2500 rpm for 3 min and carefully decant the acetic acid.
10. Heat the tubes in a water bath for 3 min until the solution darkens in color.
11. Centrifuge at 3000 rpm for 3 min. Seal the centrifuge tubes with parafilm rather than their caps.
12. Discard the supernatant containing the acetolysis mixture and add 3 mL of distilled water.
13. Centrifuge at 3000 rpm for 3 min, then carefully decant the supernatant.
14. Perform one final washing step with distilled water.

The residue can be mounted to prepare permanent slides following the procedure outlined above for fresh pollen (Subheading 3.5, **steps 8–13**).

3.7 Pollen Extraction from Honeys and Pollen Loads: Reading and Analysis

Pollen grains are counted to a total of 500 units. To ensure a representative examination, the sample is examined along a matrix of multiple random and equidistant lines, as described by Von der Ohe et al. [16].

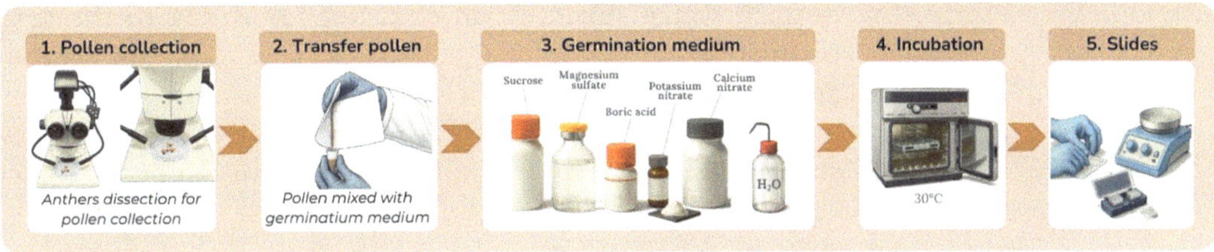

Fig. 4 Preparation of slides for studies on pollen germination. (Flowchart by Cristina Ricucci)

3.8 Pollen Extraction from Honeys and Pollen Loads: Standard Protocol for Pollen Loads

Melissopalynology can also be applied to pollen loads. The preparation methods are described by Campos et al. [28].

1. Divide the collected polyfloral pollen into monofloral groups based on visual morphological traits (e.g., color).
2. From each color group, collect approximately 1 mg of pollen and transfer it onto a clean glass slide. Prepare three separate loads for each group.
3. Add a drop of distilled water to disperse the grains, ensuring they spread in a single layer.
4. Allow the water to evaporate completely.
5. Place the slide on a heating plate, add a small piece of glycerin jelly, and keep it on the heating plate only until the gelatin has completely dissolved.

3.9 Pollen Tube Growth: Standard Protocol for Pollen Germination

1. Collect pollen from freshly opened flowers.
2. Transfer the pollen into a centrifuge tube.
3. Prepare the germination medium by dissolving 10 g sucrose, 30 mg calcium nitrate, 10 mg boric acid, 20 mg magnesium sulfate, and 20 mg potassium nitrate in 100 mL distilled water [29]. Transfer the germination medium into the centrifuge tube and mix gently.
4. Incubate the tube at 30 °C for 30–60 min.
5. Using a pipette, place a few drops of the incubated pollen onto a microscope slide and carefully cover with a coverslip (an overview of the main steps is provided in Fig. 4).

4 Notes

1. All reagents must be used in a fume hood, with the appropriate personal protective equipment, and after consulting the relevant safety data sheets.
2. The composition of fuchsin-stained glycerin gelatin is as follows: 50 ml glycerin, 7 g gelatin, 1 g phenol, and a small amount of basic Fuchsin (Acid/Basic), diluted in 42 mL

distilled water, mixed in an extractor hood due to the toxic nature of phenol. Glycerin gelatin is solid at room temperature and needs to be liquefied before use. This is most swiftly done in a bathroom or in a microwave oven, which takes only a few seconds. The glycerin jelly with basic fuchsin can be purchased already prepared (Lanzoni srl).

3. Pour the contents of the centrifuge tube into a small beaker on which a 250 μm mesh sieve is placed to retain the larger particles and allow the pollen to pass. The contents of the beaker are then poured into the centrifuge tube.
4. Acetolysis can sometimes cause swelling or damage to certain types of pollen. In this case, reduce the time spent in the water bath.
5. To recover most of the residue: add 0.5 mL of alcohol, melt with a lab vortex mixer, and pour into the microcentrifuge tube. Repeat the operation.
6. The sampler should be anchored to the ground and then connected to a permanent electric socket. The connecting cable should be well insulated.
7. The transparent adhesive tape is a commercially available product (for example, sampling pre-cut strips by Lanzoni srl). Clean the drum, place 1 cm of double-sided tape over the black line of the drum, and position the transparent adhesive tape so that the two ends of the tape are aligned with the black line. The tape must be mounted taut: run a finger over the tape to stretch it.
8. The silicon solution can be purchased already prepared (Lanzoni srl).
9. The sampled tape is removed from the drum using tweezers and a scalpel and placed on a cutting block (available from the company that manufactures the sampler). On the latter, it is possible to identify the 48 mm long tape portions corresponding to each individual day, taking care to note the sampling start and end times. If the tape was exposed for a full week, eight segments will be obtained corresponding to six full days plus two fractions of a day, the first and last. The segments relating to the six full days will each be 48 mm long, while those relating to the first and last days will have a length proportional to the number of hours of exposure.
10. For honeys with a known botanical origin that are expected to be poor in pollen (e.g., *Robinia* or *Citrus*), it is recommended to increase the sample weight by 2 grams.

Acknowledgments

The authors would like to express their sincere gratitude to the Editors Prof. Giampiero Cai, Dr. Chiara Piccini, Dr. Veronica Conti, and Dr. Sara Parri for the kind invitation to contribute to this volume.

The work of E.C. was funded by the project "National Biodiversity Future Center—NBFC," implemented under the National Recovery and Resilience Plan (NRRP), Mission 4 Component 2 Investment 1.4—Call for tender No. 3138 of December 16, 2021, rectified by Decree n.3175 of December 18, 2021 of Italian Ministry of University and Research funded by the European Union—Next Generation EU. Project code CN_00000033, Concession Decree No. 1034 of June 17, 2022, adopted by the Italian Ministry of University and Research, CUP E93C22001090001.

References

1. Manten AA (1967) Lennart Von Post and the foundation of modern palynology. Mededelingen van het Botanisch Museum en Herbarium van de Rijksuniversiteit te Utrecht. 287(1): 11–22
2. Edwards K, Fyfe R, Jackson S (2017) The first 100 years of pollen analysis. Nat Plants 3: 17001. https://doi.org/10.1038/nplants.2017.1
3. Hyde HA, Williams DA (1944) The right word Pollen Analysis Circular 8:6
4. Erdtman G (1969) Handbook of palynology: morphology, taxonomy, ecology. Munksgaard, Copenhagen
5. Mercuri AM (2015) Applied palynology as a trans-disciplinary science: the contribution of aerobiology data to forensic and palaeoenvironmental issues. Aerobiologia 31(3):323–339. https://doi.org/10.1007/s10453-015-9367-5
6. da Luz CFP, Martarello N (2022) The Palynotheca collection and three-dimensional models of pollen grains and spores as teaching resources for a meaningful learning of botany. Int J Biol Nat Sci 2(7):1–19. https://doi.org/10.22533/at.ed.813272220101
7. Torri P, Trevisan Grandi G, Accorsi CA (2005) Palinoteca e corsi universitari di palinologia. Inform Bot Ital 37:940–941. (Presented at the 100th Congress of the Italian Botanical Society, Rome, 20–23 September 2005)
8. Torri P, Torri F, Florenzano A, Montecchi MC, Rattighieri E, Rinaldi R, Bosi G, Mazzanti M, Mercuri AM (2015) The Pollen Reference Collection of LPP—UNIMORE. In: Masi A, Aranbarri J, Giardini M, Sadori L (eds) Abstract Book MedPalyno Symposium. Rome, p 113. ISBN 978-88-548-8693-3
9. Zappa J, Braga L, Gornov D., Rattighieri E, Ricucci C, Florenzano A (this book) Palynology: pollen extraction from mosses and sediments
10. Galán C, Smith M, Thibaudon M, Frenguelli G, Oteros J, Gehrig R, D'Amato G et al (2014) Minimum requirements and reproducibility of analysis in aerobiological monitoring. Aerobiologia 30(4):385–395. https://doi.org/10.1007/s10453-014-9335-5
11. Miraglia A, Donatelli G, Galán C, Bonini M, Spieksma F, Lazzarotto F (2023) Multiple monitoring stations in big cities: first example of a Mediterranean aerobiological network. Sustainability 15(5):4155. https://doi.org/10.3390/su15054155
12. Ziello C, Sparks TH, Estrella N, Belmonte J, Bergmann KC, Bucher E et al (2012) Changes to airborne pollen counts across Europe. PLoS One 7(4):e34076. https://doi.org/10.1371/journal.pone.0034076
13. Buters J, Clot B, Galán C et al (2024) Automatic detection of airborne pollen: an overview. Aerobiologia 40:13–37
14. Bastl M, Bastl K, Dirr L et al (2023) Methods and standards of pollen monitoring—significance of pollen measurements at different altitudes. Allergo J Int 32:162–166. https://doi.org/10.1007/s40629-023-00268-3

15. Mandrioli P (2000) Method for sampling and counting of airborne pollen and fungal spores. Institute of Atmospheric and Oceanic Sciences, National Research Council, Bologna
16. Von Der Ohe W, Oddo LP, Piana ML, Morlot M, Martin P (2004) Harmonized methods of melissopalynology. Apidologie 35(1):S18–S25
17. Louveaux J, Maurizio A, Vorwohl G (1978) Methods of melissopalynology. Bee World 59(4):139–157
18. Almeida-Muradian LB, Barth OM, Dietemann V, Eyer M, Freitas ADSD, Martel AC, Gasparotto Sattler JA et al (2020) Standard methods for Apis mellifera honey research. J Apic Res 59(3):1–62
19. Mercuri AM, Porrini C (1991) Melissopalynological analysis applied to air pollution studies in urban areas of Modena and Reggio Emilia (Italy). Aerobiologia 7(1):38–48
20. Di Marco G, Gismondi A, D'Agostino A, Leonardi D, Canini A (2024) Pilot study for environmental monitoring through beekeeping products of Pistoia territory. J Apic Res 63(3):533–541
21. Patel RG, Mankad AU (2014) In vitro pollen germination—A review. Int J Sci Res 3(5): 2319–7064
22. Wodehouse RP (1935) Pollen grains: their structure, identification, and significance in science and medicine. McGraw-Hill, New York
23. Erdtman G (1943) An introduction to pollen analysis. Chronica Botanica, Waltham, MA
24. Traverse A (1988) Paleopalynology. Unwin Hyman, London
25. Hirst JM (1952) An automatic volumetric spore trap. Ann Appl Biol 39:257–265
26. Mandrioli P, Puppi G (1978) Pollini allergenici in Emilia-Romagna. Collana Studi e Documentazione n.13, Dip. Ambiente e territorio R.E. R, Bologna, p 79
27. Koelzer K, Ribarits A, Weber M (2024) Comparing the acetolysed and hydrated methods for the pollen analysis of honey. Grana 63(3): 259–274
28. Campos MG, Anjos O, Chica M, Campoy P, Nozkova J, Almaraz-Abarca N, Carreck NL et al (2021) Standard methods for pollen research. J Apic Res 60(4):1–109
29. Brewbaker JL, Kwack BH (1964) The calcium ion and substances influencing pollen growth. In: Linskens HF (ed) Pollen physiology and fertilization. Elsevier North Holland, Amsterdam, pp 145–151

Chapter 20

Palynology: Extracting Pollen from Mosses and Sediments

Jessica Zappa, Lorenzo Braga, Daniil Gornov, Eleonora Rattighieri, Cristina Ricucci, and Assunta Florenzano

Abstract

Pollen grains are preserved in both moss cushions and sediments, but these archives differ in temporal resolution, preservation conditions, and extraction protocols. Moss cushions typically reflect recent pollen deposition (a few months to ~10 years), whereas sediments archive long-term pollen spectra. Each archive, differing in matrix composition, requires tailored methodological protocols for efficient and reproducible pollen extraction. In this chapter, we provide a synthesis of the standard procedures, based on what is largely in use and adopted to better concentrate pollen and other palynomorphs, detailing chemical treatments, materials, and references of existing literature for pollen extraction and analysis. The chapter describes the general guidelines for pollen extraction from mosses and sediments, also referring to the special case of sediment samples in forensic sciences.

Key words Palynology, Laboratory protocol, Pollen extraction, Moss cushions, Quaternary sediments

1 Introduction

Palynology is the study of microscopic palynomorphs—including pollen and spores—preserved in various depositional environments [1, 2]. The term "palynomorph" refers to microscopic, acid-resistant particles produced by plants, animals, and Protista. Palynology is an interdisciplinary science that combines biology and earth sciences, such as geology and hydrology, offering critical insights into paleoenvironmental reconstructions, vegetation dynamics, and the influence of climate and anthropogenic forcing on natural ecosystems. Palynology is applied across several disciplines, including geology, ecology, archaeology, forensic science, agronomy (e.g., melissopalynology), and environmental sciences (e.g., aerobiology monitoring), making it a versatile discipline that bridges natural sciences and applied research [3].

Giampiero Cai et al. (eds.), *Pollen and Pollen Tube Analysis: Methods and Protocols*, Methods in Molecular Biology, vol. 3042, https://doi.org/10.1007/978-1-0716-5308-1_20,

Palynological analysis relies on three key pollen features [4–6]:

1. Morphology: species-specific features (unit, polarity, symmetry, shape, size, apertures, ornamentation) allow identification at different taxonomic levels (e.g., family, genus, pollen type, or species).
2. Resistance: sporopollenin is made by long-chain fatty acid esters, phenolic compounds, polysaccharides, and other components, including aliphatic and aromatic compounds; therefore, it is a highly durable biopolymer characterizing the pollen wall, and confers remarkable chemical stability and preservation.
3. Abundance: high production and dispersal of pollen, especially in anemophilous plants, and depending on the species, made pollen "the most spread plant part in time and space all over the world" [1].

Both moss cushions and sediments act as traps of pollen rain [7], but differ in time resolution, depositional processes, and matrix composition, requiring archive-specific preparation protocols for accurate analysis. Moss cushions reflect more recent deposition (from a few months to max. ~10 years), while sediments can record much longer temporal sequences depending on the studied context. Furthermore, matrix composition and depositional processes can influence pollen preservation, highlighting the need for archive-specific, tailored extraction methods.

1.1 Pollen Analysis in Moss Samples

Moss cushions provide a short-term record of local airborne pollen spectra, providing a snapshot of the local vegetation over time. Moss samples are widely used to monitor contemporary vegetation patterns, calibrate regional pollen-vegetation and land-use relationships [8], and are especially valuable for creating modern analogs in paleoenvironmental research. There is a considerable debate regarding the time resolution of pollen rain captured in moss samples, which is influenced by moss growth rate and local conditions. Crowder and Cuddy [9] suggest that mosses reflect pollen deposition over a period of 5 to 15 years, whereas Bradshaw [10] argues that they represent only the last 5 years. Caseldine [11] estimates that moss samples encompass no more than 2 years of deposition, while Cundill [12] considers that the pollen present in the actively growing parts of mosses represents just 12 months, or even less (as summarized in [13]). Despite these different findings, it is generally accepted that moss samples provide valuable information on the local pollen rain and vegetation pattern, which can be used as a modern analog for near-paleoenvironmental records data interpretation [8]. The classical study contexts for this type of record are: natural environments (e.g., closed and marginal forests) and urban areas, more typically in cool, wet, shadowed environments on different surfaces such as rocks, trees, walls, and on soil where they grow.

1.2 Pollen Analyses in Quaternary Sediment Samples

Sedimentary contexts, particularly from lakes and peat bogs, provide an exceptionally long-term natural archive of past environmental conditions. In these contexts, palynology offers a unique window into past vegetation patterns, allowing researchers to trace back environmental changes and assess the influence of natural and anthropogenic forcing [14]. The high abundance, widespread dispersal, and often excellent preservation of pollen grains in these natural archives make them particularly helpful as ecological indicators, providing quantitative data useful for understanding complex environmental issues [15–17]. Its application is crucial in Quaternary research, and it can be employed in archaeological contexts to investigate ancient land-use practices and past human–landscape relationships [18, 19].

Fossil sediment matrices often include mineral particles, organic matter, and micro- and macro-remains, which can hinder the pollen grains' observation and identification. Therefore, physical and chemical treatments are necessary to extract pollen grains, removing carbonates, silicates, and humic substances, concentrating and preserving pollen. A proper extraction and treatment ensure that the structures of the pollen wall are more visible, well-preserved, facilitating a higher taxonomic-resolution identification and quantification. The typical study context for this type of analysis is lacustrine, peatlands, archaeological sites, and marine sediments.

Several methodologies and protocols are available for pollen extraction, and the choice of procedure largely depends on the type of material, the environmental context, and the chronological period under investigation. For instance, mineral-rich sediments typically require stronger chemical treatments to remove silicates and carbonates, whereas organic-rich deposits may require additional steps to eliminate the abundance of humic substances. A range of standardized protocols has been developed, balancing efficiency in isolating pollen grains with the need to minimize damage to morphological features. Most of these methods rely on chemical treatments with acids and concentration of organic matter (e.g., [20–23]). However, in some countries, restricted access to strong acids like HF, H_2SO_4, and HCl makes fossil pollen studies challenging. To address this, Santos and Ledru [24] proposed an acid-free protocol using KOH, Na-metaphosphate, Na-pyrophosphate, and $ZnCl_2$, providing a safer, cheaper, and environmentally friendly alternative to classical methods while preserving pollen and spore quality comparable to traditional acid-based methods. Moreover, a simple method based on a filtration process of the organic matter resulting from the maceration process, employing polyester filters and a vacuum inversion system, has been recently proposed [25].

In general, many protocols are reported in the literature that must be selected or adapted on a case-by-case study context to ensure optimal recovery and identification of pollen for paleoenvironmental reconstruction [26–30].

1.3 Pollen Analysis of Samples for Forensic Tasks

According to the size and abundance of pollen grains in the environment, pollen analysis can also be applied to investigate and reconstruct some events that occurred at a crime scene. The examination of forensic samples for palynological analysis can include different materials of interest, which may contain pollen and other palynomorphs [31, 32]. These materials include: soil and sediment particles from footwear, vehicle soles, carpets, mud traces, as well as soils and sediments collected from crime scenes or areas associated with criminal activities. In addition, pollen extraction can be carried out on different textiles, fabrics, vehicle interiors, rooms, bodies, and plants. Sometimes, liquid samples (water, blood stains, gastrointestinal contents, and other bodily fluids) may also be useful as a source of palynological evidence. The wide variety of possible materials that can be examined for pollen and non-pollen palynomorph analyses can be challenging for palynologists. Each sample type requires treatment strategies that maximize the recovery of palynomorphs while minimizing damage to other forms of evidence. This balance is essential, as the preservation of all evidentiary material is critical for subsequent investigative procedures and for maintaining the probative value of the biological evidence in judicial proceedings. Some of the contexts and materials that can be objects of forensic analysis are: soil, sediments, mud particles, treads on soles of footwear and tires; clothing and other fabrics, footwear, twine and rope, air filters, firearms and tools, powders, crystals and other granulated materials, plant material, human and other animal material.

2 Materials

2.1 Pollen Extraction from Mosses

2.1.1 Chemical Reagents

1. Potassium hydroxide 20% (KOH; *see* **Note 1**).
2. Distilled water.
3. Hydrochloric acid 10% (HCl; *see* **Note 2**).
4. Ethanol 96% (*see* **Note 3**).
5. Alcohol–glycerol solution (*see* **Note 4**).
6. Glycerol jelly (for the recipe, see [7]; *see* **Note 5**).
7. Solid Paraffin block 51–53°.

2.1.2 Laboratory Instruments and Equipment

1. Fume hood.
2. Precision scale.
3. 100 mL beakers.
4. Heating plate.
5. Centrifuge.
6. Fine mesh strainer.
7. Glass stirring rods.
8. 50 mL tubes.
9. 2.5 mL micro tubes.
10. Laboratory oven.

2.1.3 Other Products

Lycopodium clavatum L. spore tablets.

2.2 Pollen Extraction from Sediments [21, 22]

2.2.1 Chemical Reagents

1. Tetrasodium pyrophosphate hydrate ($Na_4P_2O_7{\cdot}xH_2O$; *see* **Note 6**).
2. Distilled water.
3. Hydrochloric acid 10% (HCl; *see* **Note 2**).
4. Glacial acetic acid (CH_3COOH; *see* **Note 7**).
5. Acetic anhydride ($(CH_3CO)_2O$; *see* **Note 8**).
6. Sulfuric acid 96% (H_2SO_4; *see* **Note 9**).
7. Sodium metatungstate hydrate ($Na_6[H_2W_{12}O_{40}]{\cdot}xH_2O$) at specific density ($\rho = 2.0\ g/cm^3$; *see* **Note 10**).
8. Hydrofluoric acid (HF 40%; *see* **Note 11**).
9. Ethanol 96% (*see* **Note 3**).
10. Alcohol–glycerol solution (*see* **Note 4**).
11. Glycerol jelly (for the recipe, see [7]; *see* **Note 5**).
12. Solid Paraffin block 51–53 °C.

2.2.2 Laboratory Instruments and Equipment

1. Fume hood.
2. Precision scale.
3. 100 mL beakers.
4. Heating plate.
5. 7 μm nylon sieve.
6. Syringes.
7. Pipettes.
8. Centrifuge.
9. Fine mesh strainer.
10. Glass stirring rods.
11. Lab vortex mixer.

12. Water bath.
13. 10 mL round-bottom tubes.
14. 10 mL conical-bottom tubes.
15. 2.5 mL micro tubes.
16. Laboratory oven.

2.2.3 Other Products

Lycopodium clavatum L. spore tablets.

2.3 Pollen Extraction from Different Materials, Surfaces, Soils, and Sediments in Forensic Cases

Chemical reagents, laboratory instruments, and equipment are the ones reported for previous methods (see Subheadings 2.1 and 2.2). Here are listed additional chemicals that are specific to forensic cases.

2.3.1 Additional Chemicals Specific to Forensic Cases

1. Citofix.
2. Hematoxylin ($C_{16}H_{14}O_6$).
3. Eosin.
4. Xylol (C_8H_{10}).

3 Methods

All the procedures described must be carried out in accordance with general safety guidelines, some of which are listed in Subheading 4 (*see* **Notes 12–16**).

3.1 Pollen Extraction from Mosses: Laboratory Protocol

The treatment of moss samples can be carried out throughout a single day. It consists of the following steps (Fig. 1):

1. Weigh: weigh ~0.5/1 g of moss, excluding soil residues, in a 50 mL beaker.
2. Deflocculation: add 5–10 mL of potassium hydroxide (KOH 10%) and boil on a heating plate under the fume hood (*see* **Note 17**).
3. Preparing material for absolute quantification: dissolve one (or more) *Lycopodium* spore tablet(s) in HCl 10%, rinse the resulting material at least two times with distilled water, centrifuge (2500 rpm for 3 min), and decant.
4. Filtering: filter the beaker solution with moss in a new 50 mL beaker using a fine mesh strainer and add the *Lycopodium* spores throughout the filter.
5. Transfer material to falcon tubes: transfer the filtered material to 50 mL falcon tubes, rinse with distilled hot water, centrifuge, and decant 2–3 times until the material appears clean.

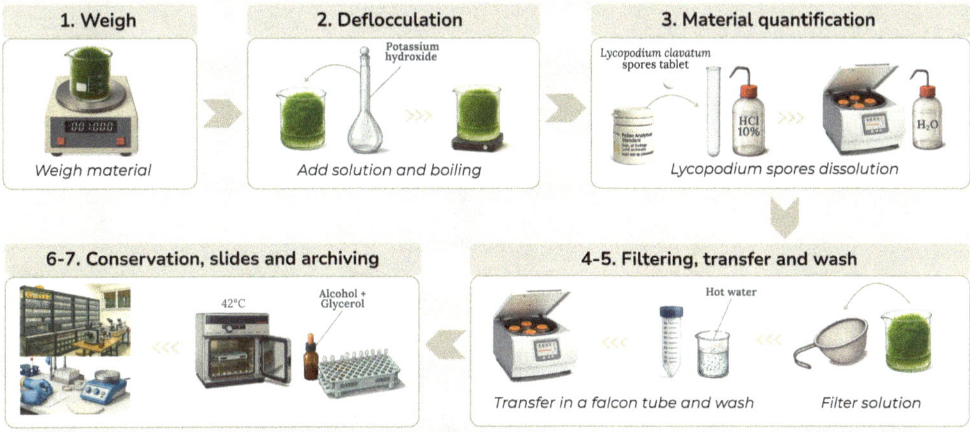

Fig. 1 Schematic workflow of pollen extraction protocol from mosses described in the text. (Schema by C.R.)

6. Transfer material to micro tubes: transfer material in the micro tubes using ethanol 96%, centrifuge, and decant.
7. Samples conservation: add 2 drops of alcohol–glycerol solution to avoid material deterioration and oven-dry.
8. Permanent slides preparation: mount residue material on permanent slides using glycerol jelly and cover slips sealed with paraffin.

If required, the moss sample can be further treated with HCl and HF to remove inorganic material.

3.2 Pollen Extraction from Sediments: Laboratory Protocol (Fig. 2)

The treatment of sediment samples is usually carried out in at least 5 days, involving the following steps (Fig. 2):

Day 1

1. Sediment sub-sampling (e.g., from sediment cores, archaeological layers) through sterile material, avoiding contamination by airborne pollen.
2. Material weighing and deflocculation:

   Weigh ~1–2 g of sediment into a 50 mL beaker. Add one/three *Lycopodium* spore tablets containing a known number of spores for concentration estimates and 20 mL of tetrasodium pyrophosphate hydrate to promote deflocculation. Leave samples to settle overnight (*see* **Notes 18** and **19**).

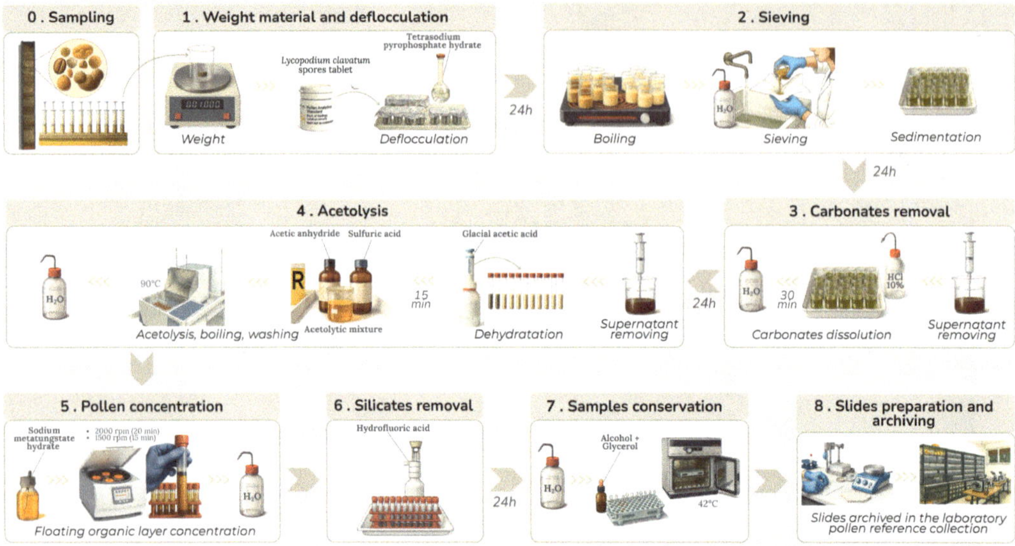

Fig. 2 Schematic workflow of pollen extraction protocol from sediments described in the text. (Schema by C.R.)

Day 2

3. Sieving:

Place the beakers on a heating plate and bring the samples to a gentle boil. Sieve the material through a 7 μm nylon mesh under running water to eliminate clays and transfer the content to the beaker. Leave samples to settle overnight.

Day 3

4. Carbonate removal:

Remove the supernatant by aspiration with a syringe. Add gradually 10% cold hydrochloric acid (HCl) to remove carbonates, wait 30 min before adding distilled water to bring to volume. Left samples to settle overnight (*see* **Note 20**).

Day 4

5. Acetolysis:

Remove the supernatant by aspiration with a syringe. Transfer the residue solution into 10 mL round-bottom tubes, centrifuge at 2500 rpm for 3 min, and decant. Dehydrate the samples using 2 mL of glacial acetic acid for at least 15 min, centrifuge (3 min at 2500 rpm), and decant. Prepare and add 3 mL of acetolysis solution per sample following the standard procedure described by Erdtman [33] (9:1 of acetic anhydride and sulfuric acid) and place samples in a water bath at 90 °C for about 10 min. Rinse the samples with distilled water, centrifuge, and decant to eliminate the acetolysis solution residues (*see* **Note 21**).

6. Pollen concentration with heavy liquid separation:

   Perform heavy liquid separation by adding 4 mL of hydrated Na-metatungstate solution ($\rho = 2.0$ g/cm^3) to each sample. Centrifuge the round-bottom tubes for 20 min at 2000 rpm, carefully transfer the floating organic layer containing the pollen into 10 mL conical-bottom tubes with 6 mL of distilled water. Centrifuge for 3 min at 2500 rpm and decant. Perform a second centrifuge of the round-bottom tubes for 15 min at 1500 rpm, transferring again the floating organic layer into the conical-bottom tubes with 6 mL of distilled water, centrifuge for 3 min at 2500 rpm, and decant. Rinse the conical-bottom tubes containing pollen and organic material with distilled water to remove heavy liquid residues (*see* **Note 22**).

7. Silicates Removal:

   Add 2 mL hydrofluoric acid (HF) to remove silicates. Leave the samples to react overnight (*see* **Note 23**).

Day 5

8. Samples Conservation:

   Wash samples with distilled water to remove HF residues. Transfer the material in 2.5 mL microtubes using 96% ethanol, centrifuge, and decant. Add 2 drops of alcohol–glycerol solution and dry in a laboratory oven at 42 °C for at least 2 h (drying time may depend on the residual water content and type of material). This step allows sample preservation.

9. Permanent slides preparation and archiving:

   Mount the material on permanent slides using glycerol jelly. Seal the coverslips with paraffin for long-term preservation. Slides can now be archived in proper boxes for future consultation.

3.3 Pollen Extraction from Different Materials, Surfaces, Soils, and Sediments in Forensic Cases

Most of the protocols relating to chemical treatments of forensic samples for pollen analysis are similar to the general methods used by palynologists [32, 34–36]. Mosses and other plant material are processed with the same protocol reported in Subheadings 3.1 and 3.2.

The main differences with the above-reported protocols lie mostly in sampling and, in some cases, the very first steps. The differing steps are reported below for some matrices relevant to forensic palynology.

The following protocols are described for each type of material. Additional notes to ensure safety and prevent contamination of both the samples and the laboratory [32] are reported in Subheading 4 (*see* **Notes 24–30**).

3.3.1 Soil, Sediments, Mud Particles, Treads on Soles of Footwear and Tires

1. Sampling and potassium hydroxide digestion:

   Weigh 1 cm^3 of material. Scrape or wash it out with a warm (ca. 70 °C) 10% solution of potassium hydroxide (KOH) in a test tube (*see* **Note 31**).

   Follow protocol in Subheading 3.2 from **step 3** to **8** (*see* **Note 32**).

3.3.2 Clothing and Other Fabrics

1. Sampling and potassium hydroxide digestion:

   Carefully brush and scratch parts containing soil or plant remains, or alternatively, cut a piece of material off; then wash out with a warm (c. 70 °C) 10% solution of potassium hydroxide, or put it into a test tube with 10% solution of potassium hydroxide. Warm it until c. 70 °C for 10–15 min, periodically stirring, and pour the resulting material into a new tube (*see* **Note 33**).
2. Filtering:

   Filter first through a 120–150 μm sieve and then through a 5 μm sieve. Most of the pollen will be retained on the upper surface of the sieve. Invert and carefully rinse with the minimum amount of distilled water necessary to transfer the material into a suitable container. Transfer the obtained suspension into test tubes and concentrate by centrifugation at 3,000 rpm for 3 minutes per run (*see* **Note 34**).

3.3.3 Powders, Crystals, Granulated Materials, Illicit Drugs

1. Sampling and potassium hydroxide digestion:

   Sample 6–7 cm^3 of the material and treat with potassium hydroxide.
2. Acetolysis:

   See **step 4** in Subheading 3.2.
3. Permanent slide preparation:

   See **steps 6–8** in Subheading 3.1.

3.3.4 Human and Other Animal Material (Hairs, Fur, Feather, Mucus, Gut, Feces, etc.)

1. Sampling and potassium hydroxide digestion:

   Put the collected material into a test tube or container filled with 10% solution of potassium hydroxide. Liquids should be collected into a test tube.
2. Follow the protocol in Subheading 3.2 from **step 3** to **6**.
3. **Steps 7–8**, additional steps: Biological fluid (nasal secretions, laryngeal, tracheal, bronchial smears, etc.) require two additional steps for conservation before being mounted on permanent slides:

   (a) Fixation and staining:

   Fix samples in Citofix in a buffered saline solution [36].

(b) Hematoxylin-eosin staining (H&E stain, Wissotzky stain):

Stain slides with hematoxylin ($C_{16}H_{14}O_6$) for 10 min and rinse with distilled water for 5 minutes. Stain slides with 1% eosin for 30 s and wash with distilled water. Dry slides in an oven (c. 50 °C for 10 min), quickly immerse in xylol (C_8H_{10}) and prepare as a permanent slide (**step 8**, Subheading 3.2) for further examination and storage.

In all washing, stirring, centrifuging, and decanting steps, distilled water (H_2O) is added if not indicated differently.

4 Notes

1. Potassium hydroxide: corrosive; can cause skin and eye irritation.
2. Hydrochloric acid: corrosive for metals, can cause skin and eye irritation, occlusion of the respiratory tracts.
3. Ethanol: inflammable, can cause eye irritation.
4. Alcohol–glycerol solution: can cause eye and skin irritation.
5. Glycerol jelly: low skin irritation.
6. Tetrasodium pyrophosphate hydrate: corrosive; can cause skin and eye irritation, occlusion of respiratory tracts.
7. Glacial acetic acid: inflammable, corrosive for metals; can cause skin and eye irritation.
8. Acetic anhydride: inflammable, can cause skin and eye irritation.
9. Sulfuric acid: corrosive for metals, can cause skin and eye irritation.
10. Sodium metatungstate hydrate: can cause skin and eye irritation.
11. Hydrofluoric acid: corrosive, penetrates the skin and tissues very rapidly, toxic. Respiration hazards.
12. Always wear a lab coat, protective gloves, and safety glasses.
13. Work on a clean, disinfected bench.
14. When handling chemical reagents, wear chemical-resistant gloves (nitrile gloves) and always use fume hoods for volatile, corrosive, or toxic chemicals.
15. Keep labels visible and reagents properly stored and with a bottle-top dispenser.
16. Be well aware of how to respond in case of an emergency.
17. KOH is bought at 20% concentration, diluted to 10% to be used in this step.

18. The amount of material depends on the sediments type: organic-rich sediments require smaller quantities, while mineral-rich and archeological sediment may require larger amounts (>2 g).
19. The number of *Lycopodium* spore tablets depends on the content of pollen and should be definitively assessed after the first test preparation of a sample sequence.
20. The volume of HCl used should correspond to the carbonate content: stronger effervescence indicates higher carbonate content and requires more acid. Pay attention: the reaction releases CO_2 gas—perform it under a fume hood and use ethanol 98% in case of excessive effervescence.
21. The acetolysis is an important step of palynological analysis that removes cytoplasm, intine, making the pollen wall more visible under the light microscopy. During the samples boiling, a change in the color of the solution generally indicates a completed acetolysis reaction.
22. This step efficiently separates floating organic material from heavier mineral particles and is particularly useful to concentrate pollen grains in organic-poor and archaeological sediments.
23. Pay attention: HF is highly toxic and corrosive. Handle with extreme care in a fume hood and use proper Personal Protective Equipment (PPE).
24. Wear protective clothing (laboratory coat, gloves, face mask, glasses, hat) and keep appropriate vaccinations up to date.
25. Maintaining positive air pressure and sealed windows.
26. Regularly clean all working surfaces with an oxidizing agent (e.g., bleach).
27. Use control slides and blank samples as monitoring traps on windowsills and at relevant places in the laboratory.
28. Use blank reagent or water "samples" in each run of processing of samples.
29. Use only single-use items.
30. Clean all reusable equipment with a hot surfactant solution and soak in an oxidizing agent (or its detergent mixture with ethanol (C_2H_5OH)).
31. For sandy soils or other suspicious low pollen content sediments, more than 2–3 cm^3 of sample must be collected.
32. Horrocks [35] advises oxidizing the sample to remove lignin compounds and brightening pollen grains that got darkened before treating the sample with hydrofluoric acid (Subheading 3.2, **step 6**).

33. Large samples must be put into a big container and immersed with 10% solution of potassium hydroxide, warming to 70 °C for 20–30 min, periodically stirring.
34. If large volumes of solution need to be passed through a 5 μm sieve, a large-volume sieve or a venturi device can be used to apply suction.

Acknowledgments

The authors would like to express their sincere gratitude to the Editors Prof. Giampiero Cai, Dr. Chiara Piccini, Dr. Veronica Conti, and Dr. Sara Parri for the kind invitation to contribute to this volume.

The work of J.Z and C.R. was funded by the project "National Biodiversity Future Center—NBFC", implemented under the National Recovery and Resilience Plan (NRRP), Mission 4 Component 2 Investment 1.4—Call for tender No. 3138 of December 16, 2021, rectified by Decree n.3175 of December 18, 2021 of Italian Ministry of University and Research funded by the European Union—Next Generation EU. Project code CN_00000033, Concession Decree No. 1034 of June 17, 2022, adopted by the Italian Ministry of University and Research, CUP E93C22001090001.

The work of D.G. was carried out in the framework of a doctoral training within the Natural Traces project, funded by the European Union's Horizon 2020 research and innovation programme under the Marie Skłodowska-Curie grant agreement No 101120165.

References

1. Erdtman G (1969) Handbook of palynology. Morphology, taxonomy, ecology. An introduction to the study of pollen grains and spores. Munksgaard, Copenhagen
2. Kneller M (2009) Palynology. In: Gornitz V (ed) Encyclopedia of paleoclimatology and ancient environments, encyclopedia of earth Sistem sciences series. Springer, Dordrecht, pp 815–823
3. Mercuri AM (2015) Applied palynology as a trans-disciplinary science: the contribution of aerobiology data to forensic and palaeoenvironmental issues. Aerobiologia 31(3):323–339
4. Erdtman G (1954) An introduction to pollen analysis. Chronica Botanica Company, Waltham
5. Forlani L (1986) La morfologia del polline. Boll Accad Gioenia Sci Nat Catania 19(329): 525–631
6. Punt W, Hoen PP, Blackmore S, Nilsson S, Le Thomas A (2007) Glossary of pollen and spore terminology. Rev Palaeobot Palynol 143(1–2): 1–81
7. Clò E, Torri P, Denami M, Mercuri AM (this book) Palynology: reference pollen collection and fresh pollen preparations
8. Gaillard MJ, Birks HJB, Emanuelsson U, Berglund BE (1992) Modern pollen/land-use relationships as an aid in the reconstruction of past land-uses and cultural landscapes: an example from South Sweden. Veget Hist Archaeobot 1:3–17

9. Crowder AA, Cuddy DG (1973) Pollen in a small river basin: Wilton Creek, Ontario. In: Quaternary Plant Ecology. 14th Symposium of the British Ecological Society, University of Cambridge, March 1972. Blackwell Scientific, Oxford, pp 61–77
10. Bradshaw RH (1981) Modern pollen-representation factors for woods in south-East England. J Ecol 69(1):45–70
11. Caseldine CJ (1981) Surface pollen studies across Bankhead moss, fife. Scotland J Biogeogr 8(1):7–25
12. Cundill PR (1991) Comparisons of moss polster and pollen trap data: a pilot study. Grana 30(2):301–308
13. Räsänen S, Hicks S, Odgaard BV (2004) Pollen deposition in mosses and in a modified 'Tauber trap' from Hailuoto, Finland: what exactly do the mosses record? Rev Palaeobot Palynol 129(1–2):103–116
14. Sadori L (2007) Pollen records, postglacial | southern Europe. In: Elias SA (ed) Encyclopedia quaternary science. Elsevier, Amsterdam, pp 2763–2773
15. Dimbleby GW (1985) The palynology of archaeological sites. Academic Press, London
16. Faegri K, Kaland PE, Krzywinski K (eds) (1989) Textbook of pollen analysis, 4th edn. Wiley, Chirchester
17. Pearsall DM (2000) Palaeoethnobotany: A handbook of procedures, 2nd edn. Academic Press, San Diego
18. Fyfe R (2023) Human impact on vegetation. In: Pollard AM, Armitage RA, Makarewicz CA (eds) Handbook of archaeological sciences, vol 1, 2nd edn. Wiley, New York, pp 715–729
19. Mercuri AM (2014) Genesis and evolution of the cultural landscape in Central Mediterranean: the 'where, when and how' through the palynological approach. Landsc Ecol 29(10): 1799–1810
20. Lowe JJ, Accorsi CA, Bandini Mazzanti M, Bishop A, Forlani L, van der Kaars S, Mercuri AM, Rivalenti C, Torri P, Watson C (1996) Pollen stratigraphy of sediment sequences from crater lakes (Lago Albano and Lago Nemi) and the C central Adriatic spanning the interval from oxygen isotope stage 2 to present day. Mem Ist Ital Idrobiol 55:71–98
21. van der Kaars S, Penny D, Tibby J, Fluin J, Dam RA, Suparan P (2001) Late quaternary palaeoecology, palynology and palaeolimnology of a tropical lowland swamp: Rawa Danau, West-Java, Indonesia. Palaeogeogr Palaeoclimatol Palaeoecol 171(3–4):185–212
22. Florenzano A, Mercuri AM, Pederzoli A, Torri P, Bosi G, Olmi L, Rinaldi R, Bandini Mazzanti M (2012) The significance of intestinal parasite remains in pollen samples from medieval pits in the piazza Garibaldi of Parma, Emilia Romagna, northern Italy. Geoarchaeology 27:34–47
23. Miola A, Mariotti M, Mercuri AM, Fiorentino G, Kustatscher E (2014) Materiali e metodi in Paleobotanica e Archeobotanica. In: Kustatscher E., Roghi G., Bertini A. & Miola ALa storia delle piante fossili in Italia—Palaeobotany of Italy. Museo di Scienze Naturali dell'Alto Adige, Bolzano, p 38–55
24. Santos RDA, Ledru MP (2022) Acid-free protocol for extracting pollen from quaternary sediments. Palynology 46(1):1–8
25. Palacios T, Martínez-Benítez B (2024) A simple and efficient method for palynological sample preparation. J Iber Geol 50(4):703–710
26. Lentfer CJ, Boyd WE (2000) Simultaneous extraction of phytoliths, pollen and spores from sediments. J Archaeol Sci 27(5):363–372
27. Halbritter H, Ulrich S, Grímsson F, Weber M, Zetter R, Hesse M, Frosch-Radivo A (2018) Methods in palynology. In: Illustrated pollen terminology. Springer, Cham, pp 97–127
28. Magri D, Di Rita F (2015) Archaeopalynological preparation techniques. In: Yeung EC, Stasolla C, Sumner MJ, Huang BQ (eds) Plant microtechniques and protocols. Springer, Cham, pp 495–506
29. Riding JB (2021) A guide to preparation protocols in palynology. Palynology 45(supl): 1–110
30. Georget M, Castéra MH, Ludovic DEVAUX, Turon JL, Stéphanie D, Goñi MFS (2025) Protocol for pollen and dinocyst analysis in marine sediments. https://doi.org/10.17504/protocols.io.x54v92qz4l3e/v1. Accessed 29 Sep 2025
31. Mildenhall DC, Wiltshire PE, Bryant VM (2006) Forensic palynology: why do it and how it works. Forensic Sci Int 163(3):163–172
32. Wiltshire PEJ (2015) Protocols for forensic palynology. Palynology 40(1):4–24
33. Erdtman G (1960) The acetolysis method. Svensk Bot Tidskr 54:561–564
34. Coyle HM (ed) (2004) Forensic botany: principles and applications to criminal casework, 1st edn. CRC Press, Boca Raton
35. Horrocks M (2004) Sub-sampling and preparing forensic samples for pollen analysis. J Forensic Sci 49(5):1024–1027
36. Trevisan Grandi G, Popoli G, Accorsi CA (2002) Criminopalinologia: guida metodologica per il medico legale. Atti Soc Nat Mat Modena 133:139–181

INDEX

Giampiero Cai et al. (eds.), *Pollen and Pollen Tube Analysis: Methods and Protocols*, Methods in Molecular Biology, vol. 3042, https://doi.org/10.1007/978-1-0716-5308-1,

D

E

T

U

V

W

X

Y

Z

Kolibri 360 GmbH
Ferdinand-Jühlke-Straße 7
99095 Erfurt, Deutschland
produktsicherheit@kolibri360.de